FAR 1989

FEDERAL AVIATION REGULATIONS

TAB/AERO STAFF

From the Editors

We are pleased that you have selected TAB/AERO's *Federal Aviation Regulations for Pilots 1989*, the most complete reprint of FARs available for general aviation pilots, aircraft owners, air taxi and commercial operators, flight instructors, ground instructors, and skydivers.

The Federal Aviation Administration publishes Federal Aviation Regulations and makes them available by subscription through the U.S. Government Printing Office. TAB/AERO annually reprints the FAR Parts of greatest interest to general aviation and provides a free mid-year update by mail. If purchased by subscription, the Parts included in this TAB/AERO edition would cost more than $262.

Major FAR changes in this edition include Mode C transponder requirements, single-class TCAs, student pilot TCA requirements, inoperative equipment rule liberalization and placarding, flight and voice recorders, ADIZ rules, and Part 135 smoking, windshear training, and drug testing rules.

This year we have added Part 99 ADIZ rules and Part 121's drug testing program appendix, which also applies to Part 135 operators. Special features this year include a listing of penalties imposed for FAR violations, a summary of new BFR guidelines, findings of the Aviation Safety Commission, a chart depicting the new Mode C requirements, a reference table for the proposed Part 91 reorganization (which is awaiting the Administrator's signature), and the popular "Best of Flight Forum" section, expanded by popular demand.

We think you'll find the TAB/AERO *FAR 1989* to be the most useful FAR reprint published, and we welcome your comments and suggestions for next year's edition.

P.S. Don't forget to mail in the Free 1989 Mid-Year Update Request Card! You'll find it at the back of the book.

FAR revised as of 12/26/88

FIRST EDITION
FIRST PRINTING

COPYRIGHT©1989 TAB BOOKS Inc.
Printed in the United States of America

Library of Congress Catalog Card Number
70-186849

FAR
ISSN 0092-6892
ISBN 0-8306-8744-0 (pbk.)

Contents

★ New features in this year's TAB/AERO edition

Key to TAB/AERO Edition Features

Vertical line indicates section changed since last TAB/AERO edition (January 1988).

[Heavy brackets indicate specific FAR words changed.]

[Heavy brackets without vertical lines indicate changes made prior to January 1988.]

All printed materials contained within double borders are advisory or informational in nature and should not be construed as having regulatory effect.

☞ F12, 46 See the numbered item(s) in ''The Best of *Flight Forum*'' section at the back of this TAB/AERO edition for an FAA interpretation, explanation, or comment regarding this section or topic.

FEDERAL AVIATION REGULATIONS

Part 1—Definitions and Abbreviations

Contents

§ 1.1 General definitions.

As used in subchapters A through K of this chapter unless the context requires otherwise:

⟦Delete "Accelerate-stop distance and its definition.⟧

"Administrator" means the Federal Aviation Administrator or any person to whom he has delegated his authority in the matter concerned.

"Aerodynamic coefficients" means nondimensional coefficients for aerodynamic forces and moments.

"Air carrier" means a person who undertakes directly by lease, or other arrangement, to engage in air transportation.

"Air commerce" means interstate, overseas, or foreign air commerce or the transportation of mail by aircraft or any operation or navigation of aircraft within the limits of any Federal airway or any operation or navigation of aircraft when directly affects, or which may endanger safety in, interstate, overseas, or foreign air commerce.

"Aircraft" means a device that is used or intended to be used for flight in the air.

"Aircraft engine" means an engine that is used or intended to be used for propelling aircraft. It includes turbosuperchargers, appurtenances, and accessories necessary for its functioning, but does not include propellers.

"Airframe" means the fuselage, booms, nacelles, cowlings, fairings, airfoil surfaces (including rotors but excluding propellers and rotating airfoils of engines), and landing gear of an aircraft and their accessories and controls.

"Airplane" means an engine-driven fixed-wing aircraft heavier than air, that is supported in flight by the dynamic reaction of the air against its wings.

"Airport" means an area of land or water that is used or intended to be used for the landing and takeoff of aircraft, and includes its buildings and facilities, if any.

"Airport traffic area" means, unless otherwise specifically designated in Part 93, that airspace within a horizontal radius of 5 statute miles from the geographic center of any airport at which a control tower is operating, extending from the surface up to, but not including, an altitude of 3,000 feet above the elevation of the airport.

"Airship" means an engine-driven lighter-than-air aircraft that can be steered.

"Air traffic" means aircraft operating in the air or on an airport surface, exclusive of loading ramps and parking areas.

"Air traffic clearance" means an authorization by air traffic control, for the purpose of preventing collision between known aircraft, for an aircraft to proceed under specified traffic conditions within controlled airspace.

"Air traffic control" means a service operated by appropriate authority to promote the safe, orderly, and expeditious flow of air traffic.

"Air transportation" means interstate, overseas, or foreign air transportation or the transportation of mail by aircraft.

"Alternate airport" means an airport at which an aircraft may land if a landing at the intended airport becomes inadvisable.

"Altitude engine" means a reciprocating aircraft engine having a rated takeoff power that is producible from sea level to an established higher altitude.

"Appliance" means any instrument, mechanism, equipment, part, apparatus, appurtenance, or accessory, including communications equipment, that is used or intended to be used in operating or controlling an aircraft in flight, is installed in or attached to the aircraft, and is not part of an airframe, engine, or propeller.

"Approved", unless used with reference to another person, means approved by the Administrator.

"Area navigation (RNAV)" means a method of navigation that permits aircraft operations on any desired course within the coverage of station-referenced navigation signals or within the limits of self-contained system capability.

"Area navigation high route" means an area navigation route within the airspace extending upward from, and including, 18,000 feet MSL to flight level 450.

"Area navigation low route" means an area navigation route within the airspace extending upward from 1,200 feet above the surface of the earth to, but not including 18,000 feet MSL.

"Armed Forces" means the Army, Navy, Air Force, Marine Corps, and Coast Guard, including their regular and reserve components and members serving without component status.

"Autorotation" means a rotorcraft flight condition in which the lifting rotor is driven entirely by action of the air when the rotorcraft is in motion.

"Auxiliary rotor" means a rotor that serves either to counteract the effect of the main rotor torque on a rotorcraft or to maneuver the rotorcraft about one or more of its three principal axes.

"Balloon" means a lighter-than-air aircraft that is not engine driven.

"Brake horsepower" means the power delivered at the propeller shaft (main drive or main output) of an aircraft engine.

"Calibrated airspeed" means indicated airspeed of an aircraft, corrected for position and instrument error. Calibrated airspeed is equal to true airspeed in standard atmosphere at sea level.

"Category"—

(1) As used with respect to the certification, ratings, privileges, and limitations of airmen, means a broad classification of aircraft. Examples include: airplane; rotorcraft; glider; and lighter-than-air; and

(2) As used with respect to the certification of aircraft, means a grouping of aircraft based upon intended use of operating limitations. Examples include: transport; normal; utility; acrobatic; limited; restricted; and provisional.

"Category II operation", with respect to the operation of aircraft, means a straight-in ILS approach to the runway of an airport under a Category II ILS instrument approach procedure issued by the Administrator or other appropriate authority.

"Category III operations", with respect to the operation of aircraft, means an ILS approach to, and landing on, the runway of an airport using a Category III ILS instrument approach procedure issued by the Administrator or other appropriate authority.

⟦"Category A," with respect to transport category rotorcraft, means multiengine rotorcraft designed with engine and system isolation features specified in Part 29 and utilizing scheduled takeoff and landing operations under a critical engine failure concept which assures adequate designated surface area and adequate performance capability for continued safe flight in the event of engine failure.

⟦"Category B," with respect to transport category rotorcraft, means single-engine or multiengine rotorcraft which do not fully meet all Category A standards. Category B rotorcraft have no guaranteed stay-up ability in the event of engine failure and unscheduled landing is assumed.⟧

"Ceiling" means the height above the earth's surface of the lowest layer of clouds or obscuring phenomena that is reported as "broken", "overcast", or "obscuration", and not classified as "thin" or "partial".

"Civil aircraft" means aircraft other than public aircraft.

"Class"—

(1) As used with respect to the certification, ratings, privileges, and limitations of airmen, means a classification of aircraft within a category having similar operating characteristics. Examples include: single engine; multiengine; land; water; gyroplane; helicopter; airship; and free balloon; and

(2) As used with respect to the certification of aircraft, means a broad grouping of aircraft having similar characteristics of propulsion, flight, or landing. Examples include: airplane; rotorcraft; glider; balloon; landplane; and seaplane.

"Clearway" means:

(1) For turbine engine powered airplanes certificated after August 29, 1959, an area beyond the runway, not less than 500 feet wide, centrally located about the extended centerline of the runway, and under the control of the airport authorities. The clearway is expressed in terms of a clearway plane, extending from the end of the runway with an upward slope not exceeding 1.25 percent, above which no object nor any terrain protrudes. However, threshold lights may protrude above the plane if their height above the runway is 26 inches or less and if they are located to each side of the runway.

(2) For turbine engine powered airplanes certificated after September 30, 1958 but before August 30, 1959, an area beyond the takeoff runway extending no less than 300 feet on either side of the extended centerline of the runway, at an elevation no higher than the elevation of the end of the runway, clear of all fixed obstacles, and under the control of the airport authorities.

【"Climbout Speed," with respect to rotorcraft, means a referenced airspeed which results in a flight path clear of the height-velocity envelope during initial climbout.】

"Commercial operator" means a person who, for compensation or hire, engages in the carriage by aircraft in air commerce of persons or property, other than as an air carrier or foreign air carrier or under the authority of Part 375 of this Title. Where it is doubtful that an operation is for "compensation or hire", the test applied is whether the carriage by air is merely incidental to the person's other business or is, in itself, a major enterprise for profit.

"Controlled airspace" means airspace designated as a continental control area, control area, control zone, terminal control area, or transition area, within which some or all aircraft may be subject to air traffic control.

"Crewmember" means a person assigned to perform duty in an aircraft during flight time.

"Critical altitude" means the maximum altitude at which, in standard atmosphere, it is possible to maintain, at a specified rotational speed, a specified power or a specified manifold pressure. Unless otherwise stated, the critical altitude is the maximum altitude at which it is possible to maintain, at the maximum continuous rotational speed, one of the following:

(1) The maximum continuous power, in the case of engines for which this power rating is the same at sea level and at the rated altitude.

(2) The maximum continuous rated manifold pressure, in the case of engines the maximum continous power of which, is governed by a constant manifold pressure.

"Critical engine" means the engine whose failure would most adversely affect the performance or handling qualities of an aircraft.

"Decision height", with respect to the operation of aircraft, means the height at which a decision must be made, during an ILS or PAR instrument approach, to either continue the approach or to execute a missed approach.

"Equivalent airspeed" means the calibrated airspeed of an aircraft corrected for adiabatic compressible flow for the particular altitude. Equivalent airspeed is equal to calibrated airspeed in standard atmosphere at sea level.

"Extended over-water operation" means—

(1) With respect to aircraft other than helicopters, and operation over water at a horizontal distance of more than 50 nautical miles from the nearest shoreline; and

(2) With respect to helicopters, an operation over water at a horizontal distance of more than 50 nautical miles from the nearest shoreline and more than 50 nautical miles from an off-shore heliport structure.

"External load" means a load that is carried, or extends, outside of the aircraft fuselage.

"External-load attaching means" means the structural components used to attach an external load to an aircraft, including external-load containers, the backup structure at the attachment points, and any quick-release device used to jettison the external load.

"Fireproof"—

(1) With respect to materials and parts used to confine fire in a designated fire zone, means the capacity to withstand at least as well as steel in dimensions appropriate for the purpose for which they are used, the heat produced when there is a severe fire of extended duration in that zone; and

(2) With respect to other materials and parts, means the capacity to withstand the heat associated with fire at least as well as steel in dimensions appropriate for the purpose for which they are used.

"Fire resistant"—

(1) With respect to sheet or structural members means the capacity to withstand the heat associated with fire at least as well as aluminum alloy in dimensions appropriate for the purpose for which they are used; and

(2) With respect to fluid-carrying lines, fluid system parts, wiring, air ducts, fittings, and powerplant controls, means the capacity to perform the intended functions under the heat and other conditions likely to occur when there is a fire at the place concerned.

"Flame resistant" means not susceptible to combustion to the point of propagating a flame, beyond safe limits, after the ignition source is removed.

"Flammable", with respect to a fluid or gas, means susceptible to igniting readily or to exploding.

"Flap extended speed" means the highest speed permissible with wing flaps in a prescribed extended position.

"Flash resistant" means not susceptible to burning violently when ignited.

"Flight crewmember" means a pilot, flight engineer, or flight navigator assigned to duty in an aircraft during flight time.

"Flight level" means a level of constant atmospheric pressure related to a reference datum of 29.92 inches of mercury. Each is stated in three digits that represent hundreds of feet. For example, flight level 250 represents a barometric altimeter indication of 25,000 feet; flight level 255, an indication of 25,500 feet.

"Flight plan" means specified information, relating to the intended flight of an aircraft, that is filed orally or in writing with air traffic control.

"Flight time" means the time from the moment the aircraft first moves under its own power for the purpose of flight until the moment it comes to rest at the next point of landing. ("Block-to-block" time.) ☞ F1,2

"Flight visibility" means the average forward horizontal distance, from the cockpit of an aircraft in flight, at which prominent unlighted objects may be seen and identified by day and prominent lighted objects may be seen and identified by night.

"Foreign air carrier" means any person other than a citizen of the United States, who undertakes directly, by lease or other arrangement, to engage in air transportation.

"Foreign air commerce" means the carriage by aircraft of persons or property for compensation or hire, or the carriage of mail by aircraft, or the operation or navigation of aircraft, in the conduct or furtherance of a business or vocation, in commerce between a place in the United States and any place outside thereof; whether such commerce moves wholly by aircraft or partly by aircraft and partly by other forms of transportation.

"Foreign air transportation" means the carriage by aircraft of persons or property as a common carrier for compensation or hire, or the carriage of mail by aircraft, in commerce between a place in the United States and any place outside of the United States, whether that commerce moves wholly by aircraft or partly by aircraft and partly by other forms of transportation.

"Glider" means a heavier-than-air aircraft, that is supported in flight by the dynamic reaction of the air against its lifting surfaces and whose free flight does not depend principally on an engine.

"Ground visibility" means prevailing horizontal visibility near the earth's surface as reported by the United States National Weather Service or an accredited observer.

"Gyrodyne" means a rotorcraft whose rotors are normally engine-driven for takeoff, hovering, and landing, and for forward flight through part of its speed range, and whose means of propulsion, consisting usually of conventional propellers, is independent of the rotor system.

"Gyroplane" means a rotorcraft whose rotors are not engine-driven except for initial starting, but are made to rotate by action of the air when the rotorcraft is moving; and whose means of propulsion, consisting usually of conventional propellers, is independent of the rotor system.

"Helicopter" means a rotorcraft that, for its horizontal motion, depends principally on its engine-driven rotors.

"Heliport" means an area of land, water, or structure used or intended to be used for the landing and takeoff of helicopters.

"Idle thrust" means the jet thrust obtained with the engine power control lever set at the stop for the least thrust position at which it can be placed.

"IFR conditions" means weather conditions below the minimum for flight under visual flight rules.

"IFR over-the-top", with respect to the operation of aircraft, means the operation of an aircraft over-the-top on an IFR flight plan when cleared by air traffic control to maintain "VFR conditions" or "VFR conditions on top".

"Indicated airspeed" means the speed of an aircraft as shown on its pitot static airspeed indicator calibrated to reflect standard atmosphere adiabatic compressible flow at sea level uncorrected for airspeed system errors.

"Instrument" means a device using an internal mechanism to show visually or aurally the attitude, altitude, or operation of an aircraft or aircraft part. It includes electronic devices

for automatically controlling an aircraft in flight.

"Interstate air commerce" means the carriage by aircraft of persons or property for compensation or hire, or the carriage of mail by aircraft, or the operation or navigation of aircraft in the conduct or furtherance of a business or vocation, in commerce between a place in any State of the United States, or the District of Columbia, and a place in any other State of the United States, or the District of Columbia; or between places in the same State of the United States through the airspace over any place outside thereof; or between places in the same territory or possession of the United States, or the District of Columbia.

"Interstate air transportation" means the carriage by aircraft of persons or property as a common carrier for compensation or hire, or the carriage of mail by aircraft, in commerce—

(1) Between a place in a State or the District of Columbia and another place in another State or the District of Columbia;

(2) Between places in the same State through the airspace of any place outside that State; or

(3) Between places in the same possession of the United States;

whether that commerce moves wholly by aircraft or partly by aircraft and partly by other forms of transportation.

["Intrastate air transportation" means the carriage of persons or property as a common carrier for compensation or hire, by turbojet-powered aircraft capable of carrying thirty or more persons, wholly within the same State of the United States.]

"Kite" means a framework, covered with paper, cloth, metal, or other material, intended to be flown at the end of a rope or cable, and having as its only support the force of the wind moving past its surfaces.

"Landing gear extended speed" means the maximum speed at which an aircraft can be safely flown with the landing gear extended.

"Landing gear operating speed" means the maximum speed at which the landing gear can be safely extended or retracted.

"Large aircraft" means aircraft of more than 12,500 pounds, maximum certificated takeoff weight.

"Lighter-than-air aircraft" means aircraft that can rise and remain suspended by using contained gas weighing less than the air that is displaced by the gas.

"Load factor" means the ratio of a specified load to the total weight of the aircraft. The specified load is expressed in terms of any of the following: aerodynamic forces, inertia forces, or ground or water reactions.

"Mach number" means the ratio of true airspeed to the speed of sound.

"Main rotor" means the rotor that supplies the principal lift to a rotorcraft.

"Maintenance" means inspection, overhaul, repair, preservation, and the replacement of parts, but excludes preventive maintenance.

"Major alteration" means an alteration not listed in the aircraft, aircraft engine, or propeller specifications—

(1) That might appreciably affect weight, balance, structural strength, performance, powerplant operation, flight characteristics, or other qualities affecting airworthiness; or

(2) That is not done according to accepted practices or cannot be done by elementary operations.

"Major repair" means a repair—

(1) That, if improperly done, might appreciably affect weight, balance, structural strength, performance, powerplant operation, flight characteristics, or other qualities affecting airworthiness; or

(2) That is not done according to accepted practices or cannot be done by elementary operations.

"Manifold pressure" means absolute pressure as measured at the appropriate point in the induction system and usually expressed in inches of mercury.

"Medical certificate" means acceptable evidence of physical fitness on a form prescribed by the Administrator.

"Minimum descent altitude" means the lowest altitude, expressed in feet above mean sea level, to which descent is authorized on final approach or during circle-to-land maneuvering in execution of a standard instrument approach procedure, where no electronic glide slope is provided.

"Minor alteration" means an alteration other than a major alteration.

"Minor repair" means a repair other than a major repair.

"Navigable airspace" means airspace at and above the minimum flight altitudes prescribed by or under this chapter, including airspace needed for safe takeoff and landing.

"Night" means the time between the end of evening civil twilight and the beginning of morning civil twilight, as published in the American Air Almanac, converted to local time.

"Non-precision approach procedure" means a standard instrument approach procedure in which no electronic glide slope is provided.

"Operate", with respect to aircraft, means use, cause to use or authorize to use aircraft, for the purpose (except as provided in § 91.10 of this chapter) of air navigation including the piloting of aircraft, with or without the right of legal control (as owner, lessee, or otherwise).

"Operational control", with respect to a flight, means the exercise of authority over initiating, conducting, or terminating a flight.

"Overseas air commerce" means the carriage by aircraft of persons or property for compensation or hire, or the carriage of mail by aircraft, or the operation or navigation of aircraft in the conduct or furtherance of a business or vocation, in commerce between a place in any State of the United States, or the

District of Columbia, and any place in a territory or possession of the United States; or between a place in a territory or possession of the United States, and a place in any other territory or possession of the United States.

"Overseas air transportation" means the carriage by aircraft of persons or property as a common carrier for compensation or hire, or the carriage of mail by aircraft, in commerce—

(1) Between a place in a State or the District of Columbia and a place in a possession of the United States; or

(2) Between a place in a possession of the United States and a place in another possession of the United States;

whether that commerce moves wholly by aircraft or partly by aircraft and partly by other forms of transportation.

"Over-the-top" means above the layer of clouds or other obscuring phenomena forming the ceiling.

"Parachute" means a device used or intended to be used to retard the fall of a body or object through the air.

"Person" means an individual, firm, partnership, corporation, company, association, joint-stock association, or governmental entity. It includes a trustee, receiver, assignee, or similar representative of any of them.

"Pilotage" means navigation by visual reference to landmarks.

"Pilot in command" means the pilot responsible for the operation and safety of an aircraft during flight time.

"Pitch setting" means the propeller blade setting as determined by the blade angle measured in a manner, and at a radius, specified by the instruction manual for the propeller.

"Positive control" means control of all air traffic, within designated airspace, by air traffic control.

"Precision approach procedure" means a standard instrument approach procedure in which an electronic glide slope is provided, such as ILS and PAR.

"Preventive maintenance" means simple or minor preservation operations and the replacement of small standard parts not involving complex assembly operations.

"Prohibited area" means designated airspace within which the flight of aircraft is prohibited.

"Propeller" means a device for propelling an aircraft that has blades on an engine-driven shaft and that, when rotated, produces a thrust approximately perpendicular to its plane of rotation. It includes control components normally supplied by its manufacturer, but does not include main and auxiliary rotors or rotating airfoils of engines.

"Public aircraft" means aircraft used only in the service of a government, or a political subdivision. It does not include any government-owned aircraft engaged in carrying persons or property for commercial purposes.

"Rated continuous OEI power," with respect to rotorcraft turbine engines, means the approved brake horsepower developed under static conditions at specified altitudes and temperatures within the operating limitations established for the engine under Part 33 of this chapter, and limited in use to the time required to complete the flight after the failure of one engine of a multiengine rotorcraft.

"Rated maximum continuous augmented thrust", with respect to turbojet engine type certification, means the approved jet thrust that is developed statically or in flight, in standard atmosphere at a specified altitude, with fluid injection or with the burning of fuel in a separate combustion chamber, within the engine operating limitations established under Part 33 of this chapter, and approved for unrestricted periods of use.

"Rated maximum continuous power", with respect to reciprocating, turbopropeller, and turboshaft engines, means the approved brake horsepower that is developed statically or in flight, in standard atmosphere at a specified altitude, within the engine operating limitations established under Part 33, and approved for unrestricted periods of use.

"Rated maximum continuous thrust", with respect to turbojet engine type certification, means the approved jet thrust that is developed statically or in flight, in standard atmosphere at a specified altitude, without fluid injection and without the burning of fuel in a separate combustion chamber, within the engine operating limitations established under Part 33 of this chapter, and approved for unrestricted periods of use.

"Rated takeoff augmented thrust", with respect to turbojet engine type certification, means the approved jet thrust that is developed statically under standard sea level conditions, with fluid injection or with the burning of fuel in a separate combustion chamber, within the engine operating limitations established under Part 33 of this chapter, and limited in use to periods of not over 5 minutes for takeoff operation.

"Rated takeoff power", with respect to reciprocating, turbopropeller, and turboshaft engine type certification, means the approved brake horsepower that is developed statically under standard sea level conditions, within the engine operating limitations established under Part 33, and limited in use to periods of not over 5 minutes for takeoff operation.

"Rated takeoff thrust", with respect to turbojet engine type certification, means the approved jet thrust that is developed statically under standard sea level conditions, without fluid injection and without the burning of fuel in a separate combustion chamber, within the engine operating limitations established under Part 33 of this chapter, and limited in use to periods of not over 5 minutes for takeoff operation.

"Rated 30-minute OEI power," with respect to rotorcraft turbine engines, means the approved brake horsepower developed under static conditions at specified altitudes and temperatures within the operating limitations established for the engine under Part 33 of this chapter, and limited in use to a period of not more than 30 minutes after the failure of one engine of a multiengine rotorcraft.

"Rated 2½-minute OEI power," with respect to rotorcraft turbine engines, means the approved brake horsepower developed under static conditions at specified altitudes and temperatures within the operating limitations established for the engine under Part 33 of this chapter, and limited in use to a period of not more than 2½ minutes after the failure of one engine of a multiengine rotorcraft.

"Rating" means a statement that, as a part of a certificate, sets forth special conditions, privileges, or limitations.

"Reporting point" means a geographical location in relation to which the position of an aircraft is reported.

"Restricted area" means airspace designated under Part 73 of this chapter within which the flight of aircraft, while not wholly prohibited, is subject to restriction.

"RNAV way point (W/P)" means a predetermined geographical position used for route or instrument approach definition or progress reporting purposes that is defined relative to a VORTAC station position.

"Rocket" means an aircraft propelled by ejected expanding gases generated in the engine from self-contained propellants and not dependent on the intake of outside substances. It includes any part which becomes separated during the operation.

"Rotorcraft" means a heavier-than-air aircraft that depends principally for its support in flight on the lift generated by one or more rotors.

["Rotorcraft-load combination" means the combination of a rotorcraft and an external-load, including the external-load attaching means. Rotorcraft-load combinations are designated as Class A, Class B, Class C, and Class D, as follows:]

(1) "Class A rotorcraft-load combination" means one in which the external load cannot move freely, cannot be jettisoned, and does not extend below the landing gear.

(2) "Class B rotorcraft-load combination" means one in which the external load is jettisonable and is lifted free of land or water during the rotorcraft operation.

(3) "Class C rotorcraft-load combination" means one in which the external load is jettisonable and remains in contact with land or water during the rotorcraft operation.

[(4) "Class D rotorcraft-load combination" means one in which the external-load is other than a Class A, B, or C and has been specifically approved by the Administrator for that operation.]

"Route segment" means a part of a route. Each end of that part is identified by—

(1) a continental or insular geographical location; or

(2) a point at which a definite radio fix can be established.

"Sea level engine" means a reciprocating aircraft engine having a rated takeoff power that is producible only at sea level.

"Second in command" means a pilot who is designated to be second in command of an aircraft during flight time.

"Show", unless the context otherwise requires, means to show to the satisfaction of the of the Administrator.

"Small aircraft" means aircraft of 12,500 pounds or less, maximum certificated takeoff weight.

"Standard atmosphere" means the atmosphere defined in *U.S. Standard Atmosphere, 1962* (Geopotential altitude tables).

"Stopway" means an area beyond the take-off runway, no less wide than the runway and centered upon the extended centerline of the runway, able to support the airplane during an aborted takeoff, without causing structural damage to the airplane, and designated by the airport authorities for use in decelerating the airplane during an aborted takeoff.

"Takeoff power"—

(1) With respect to reciprocating engines, means the brake horsepower that is developed under standard sea level conditions, and under the maximum conditions of crankshaft rotational speed and engine manifold pressure approved for the normal takeoff, and limited in continuous use to the period of time shown in the approved engine specification; and

(2) With respect to turbine engines, means the brake horsepower that is developed under static conditions at a specified altitude and atmospheric temperature, and under the maximum conditions of rotorshaft rotational speed and gas temperature approved for the normal takeoff, and limited in continuous use to the period of time shown in the approved engine specification.

["Takeoff Safety Speed" means a referenced airspeed obtained after lift-off at which the required one-engine-inoperative climb performance can be achieved.]

"Takeoff thrust", with respect to turbine engines, means the jet thrust that is developed under static conditions at a specific altitude and atmospheric temperature under the maximum conditions of rotorshaft rotational speed and gas temperature approved for the normal takeoff, and limited in continuous use to the period of time shown in the approved engine specification.

"Time in service", with respect to maintenance time records, means the time from the moment an aircraft leaves the surface of the earth until it touches it at the next point of landing.

"Traffic pattern" means the traffic flow that is prescribed for aircraft landing at, taxiing on, or taking off from, an airport.

True airspeed" means the airspeed of an aircraft relative to undisturbed air. True airspeed is equal to equivalent airspeed multiplied by $(po/p)^{1/2}$.

"Type"—

(1) As used with respect to the certification, ratings, privileges, and limitations of airmen, means a specific make and basic model of aircraft, including modifications thereto that do not change its handling or flight characteristics. Examples include: DC–7, 1049, and F–27; and

(2) As used with respect to the certification of aircraft, means those aircraft which are similar in design. Examples include: DC–7 and DC–7C; 1049G and 1049H; and F–27 and F–27F.

(3) As used with respect to the certification of aircraft engines means those engines which are similar in design. For example, JT8D and JT8D–7 are engines of the same type, and JT9D–3A and JT9D–7 are engines of the same type.

"United States", in a geographical sense, means (1) the States, the District of Columbia, Puerto Rico, and the possessions, including the territorial waters, and (2) the airspace of those areas.

"United States air carrier" means a citizen of the United States who undertakes directly by lease, or other arrangement, to engage in air transportation.

"VFR over-the-top", with respect to the operation of aircraft, means the operation of an aircraft over-the-top under VFR when it is not being operated on an IFR flight plan. ☞ **F3**

§ 1.2 Abbreviations and symbols.

In Subchapters A through K of this chapter:

"AGL" means above ground level.

"ALS" means approach light system.

"ASR" means airport surveillance radar.

"ATC" means air traffic control.

"CAS" means calibrated airspeed.

"CAT II" means Category II.

"CONSOL or CONSOLAN" means a kind of low or medium frequency long range navigational aid.

"DH" means decision height.

"DME" means distance measuring equipment compatible with TACAN.

"EAS" means equivalent airspeed.

"FAA" means Federal Aviation Administration.

"FM" means fan marker.

"GS" means glide slope.

"HIRL" means high-intensity runway light system.

"IAS" means indicated airspeed.

"ICAO" means International Civil Aviation Organization.

"IFR" means instrument flight rules.

"ILS" means instrument landing system.

"IM" means ILS inner marker.

"INT" means intersection.

"LDA" mean localizer-type directional aid.

"LFR" means low frequency radio range.

"LMM" means compass locator at middle marker.

"LOC" means ILS localizer.

"LOM" means compass locator at outer marker.

"*M*" means mach number.

"MAA" means maximum authorized IFR altitude.

"MALS" mean medium intensity approach light system.

"MALSR" means medium intensity approach light system with runway alignment indicator lights.

"MCA" means minimum crossing altitude.

"MDA" means minimum descent altitude.

"MEA" means minimum en route IFR altitude.

"MM" means ILS middle marker.

"MOCA" means minimum obstruction clearance altitude.

"MRA" means minimum reception altitude.

"MSL" means mean sea level.

"NDB(ADF)" means nondirectional beacon (automatic direction finder).

"NOPT" means no procedure turn required.

■ "OEI" means one engine inoperative.

"OM" means ILS outer marker.

"PAR" means precision approach radar.

"RAIL" means runway alignment indicator light system.

"RBN" means radio beacon.

"RCLM" means runway centerline marking.

"RCLS" means runway centerline light system.

"REIL" means runway end identification lights.

"RR" means low or medium frequency radio range station.

"RVR" means runway visual range as measured in the touchdown zone area.

"SALS" means short approach light system.

"SSALS" means simplified short approach light system.

"SSALSR" mean simplified short approach light system with runway alignment indicator lights.

"TACAN" means ultra-high frequency tactical air navigational aid.

"TAS" means true airspeed.

"TDZL" means touchdown zone lights.

"TVOR" means very high frequency terminal omnirange station.

"V_A" means design maneuvering speed.

"V_B" means design speed for maximum gust intensity.

"V_C" means design cruising speed.

"V_D" means design diving speed.

"V_{DF}/M_{DF}" means demonstrated flight diving speed.

"V_F" means design flap speed.

"V_{FC}/M_{FC}" means maximum speed for stability characteristics.

"V_{FE}" means maximum flap extended speed.

"V_H" means maximum speed in level flight with maximum continuous power.

"V_{LE}" means maximum landing gear extended speed.

"V_{LO}" means maximum landing gear operating speed.

"V_{LOF}" means lift-off speed.

"V_{MC}" means minimum control speed with the critical engine inoperative.

"V_{MO}/M_{MO}" means maximum operating limit speed.

"V_{MU}" means minimum unstick speed.

"V_{NE}" means never-exceed speed.

"V_{NO}" means maximum structural cruising speed.

"V_R" means rotation speed.

"V_S" means the stalling speed or the minimum steady flight speed at which the airplane is controllable.

"V_{SO}" means the stalling speed or the minimum steady flight speed in the landing configuration. ☞**F116**

"V_{S1}" means the stalling speed or the minimum steady flight speed obtained in a specified configuration.

[“V_{TOSS}" means takeoff safety speed for Category A rotorcraft.]

"V_X" means speed for best angle of climb.

"V_Y" means speed for best rate of climb.

V_1 means takeoff decision speed (formerly denoted as critical engine failure speed).

"V_2" means takeoff safety speed.

"$V_{2\,min}$" means minimum takeoff safety speed.

"VFR" means visual flight rules.

"VHF" means very high frequency.

"VOR" means very high frequency omnirange station.

"VORTAC" means collocated VOR and TACAN.

§ 1.3 Rules of construction.

(a) In Subchapters A through K of this chapter, unless the context requires otherwise:

(1) Words importing the singular include the plural;

(2) Words importing the plural include the singular; and

(3) Words importing the masculine gender include the feminine.

(b) In Subchapters A through K of this chapter, the word:

(1) "Shall" is used in an imperative sense;

(2) "May" is used in a permissive sense to state authority or permission to do the act prescribed, and the words "no person may . . ." or "a person may not . . ." mean that no person is required, authorized, or permitted to do the act prescribed; and

(3) "Includes" means "includes but is not limited to".

FEDERAL AVIATION REGULATIONS

Part 43—Maintenance, Preventive Maintenance, Rebuilding, and Alteration

---

§ 43.1   Applicability.

(a) Except as provided in paragraph (b), this Part prescribes rules governing the maintenance, preventive maintenance, rebuilding, and alteration of any—

[(1) Aircraft having a U.S. airworthiness certificate;

[(2) Foreign-registered civil aircraft used in common carriage or carriage of mail under the provisions of Part 121, 127, or 135 of this chapter; and

[(3) Airframe, aircraft engines, propellers, appliances, and component parts of such aircraft.]

(b) This Part does not apply to any aircraft for which an experimental airworthiness certificate has been issued, unless a different kind of airworthiness certificate had previously been issued for that aircraft.

[§ 43.2   Records of overhaul and rebuilding.

[(a) No person may describe in any required maintenance entry or form an aircraft, airframe, aircraft engine, propeller, appliance, or component part as being overhauled unless—

[(1) Using methods, techniques, and practices acceptable to the Administrator, it has been disassembled, cleaned, inspected, repaired as necessary, and reassembled; and

[(2) It has been tested in accordance with approved standards and technical data, or in accordance with current standards and technical data acceptable to the Administrator, which have been developed and documented by the holder of the type certificate, supplemental type certificate, or a material, part, process, or appliance approval under § 21.305 of this chapter.

[(b) No person may describe in any required maintenance entry or form an aircraft, airframe, aircraft engine, propeller, appliance, or component part as being rebuilt unless it has been disassembled, cleaned, inspected, repaired as necessary, reassembled, and tested to the same tolerances and limits as a new item, using either new parts or used parts that either conform to new part tolerances and limits or to approved oversized or undersized dimensions.]

§ 43.3   Persons authorized to perform maintenance, preventive maintenance, rebuilding, and alterations.

[(a) Except as provided in this section and § 43.17, no person may maintain, rebuild, alter, or perform preventive maintenance on an aircraft, airframe, aircraft engine, propeller, appliance, or component part to which this Part applies.] Those items, the performance of which is a major alteration, a major repair, or preventive maintenance, are listed in Appendix A.

(b) The holder of a mechanic certificate may perform maintenance, preventive maintenance, and alterations as provided in Part 65.

(c) The holder of a repairman certificate may perform maintenance and preventive maintenance as provided in Part 65.

(d) A person working under the supervision of a holder of a mechanic or repairman certificate may perform the maintenance, preventive maintenance, and alterations that his supervisor is authorized to perform, if the supervisor personally observes the work being done to the extent necessary to ensure that it is being done properly and if the supervisor is readily available, in person, for consultation. [However, this paragraph does not authorize the performance of any inspection required by Part 91 or Part 125 of this chapter or any inspection performed after a major repair or alteration.]

(e) The holder of a repair station certificate may perform maintenance, preventive maintenance, and alterations as provided in Part 145.

(f) The holder of an air carrier operating certificate or an operating certificate issued under Part 121, 127, or 135, may perform maintenance, preventive maintenance, and alterations as provided in Part 121, 127, or 135.

(g) The holder of a pilot certificate issued under Part 61 may perform preventive maintenance on any aircraft owned or operated by that pilot which is not used under Part 121, 127, 129, or 135.

[(h) Notwithstanding the provisions of paragraph (g) of this section, the Administrator may approve a certificate holder under Part 135 of this chapter, operating rotorcraft in a remote area, to allow a pilot to perform specific preventive maintenance items provided—

[1) The items of preventive maintenance are a result of a known or suspected mechanical difficulty or malfunction that occurred en route to or in a remote area;

[(2) The pilot has satisfactorily completed an approved training program and is authorized in writing by the certificate holder for each item of preventive maintenance that the pilot is authorized to perform;

[(3) There is no certificated mechanic available to perform preventive maintenance;

[(4) The certificate holder has procedures to evaluate the accomplishment of a preventive maintenance item that requires a decision concerning the airworthiness of the rotorcraft; and

[(5) The items of preventive maintenance authorized by this section are those listed in paragraph (c) of Appendix A of this Part.]

[(i)] A manufacturer may—

(1) Rebuild or alter any aircraft, aircraft engine, propeller, or appliance manufactured by him under a type or production certificate;

(2) Rebuild or alter any appliance or part of aircraft, aircraft engines, propellers, or appliances manufactured by him under a Technical Standard Order Authorization, an

FAA-Parts Manufacturer Approval, or Product and Process Specification issued by the Administrator; and

(3) Perform any inspection required by Part 91 or Part 125 of this chapter on aircraft it manufactures, while currently operating under a production certificate or under a currently approved production inspection system for such aircraft.

§ 43.5 Approval for return to service after maintenance, preventive maintenance, rebuilding, or alteration.

No person may approve for return to service any aircraft, airframe, aircraft engine, propeller, or appliance, that has undergone maintenance, preventive maintenance, rebuilding, or alteration unless—

(a) The maintenance record entry required by § 43.9 or § 43.11, as appropriate, has been made;

(b) The repair or alteration form authorized by or furnished by the Administrator has been executed in a manner prescribed by the Administrator; and

(c) If a repair or an alteration results in any change in the aircraft operating limitations or flight data contained in the approved aircraft flight manual, those operating limitations or flight data are appropriately revised and set forth as prescribed in § 91.31.

§ 43.7 Persons authorized to approve aircraft, airframes, aircraft engines, propellers, appliances, or component parts for return to service after maintenance, preventive maintenance, rebuilding, or alteration.

(a) Except as provided in this section and § 43.17, no person, other than the Administrator, may approve an aircraft, airframe, aircraft engine, propeller, appliance, or component part for return to service after it has undergone maintenance, preventive maintenance, rebuilding, or alteration.

(b) The holder of a mechanic certificate or an inspection authorization may approve an aircraft, airframe, aircraft engine, propeller, appliance, or component part for return to service as provided in Part 65 of this chapter.

(c) The holder of a repair station certificate may approve an aircraft, airframe, aircraft engine, propeller, appliance, or component part for return to service as provided in Part 145 of this chapter.

(d) A manufacturer may approve for return to service any aircraft, airframe, aircraft engine, propeller, appliance, or component part which that manufacturer has worked on under § 43.3(h). However, except for minor alterations, the work must have been done in accordance with technical data approved by the Administrator.

(e) The holder of an air carrier operating certificate or an operating certificate issued under Part 121, 127, or 135, may approve an aircraft,

airframe, aircraft engine, propeller, appliance, or component part for return to service as provided in Part 121, 127, or 135 of this chapter, as applicable.

(f) A person holding at least a private pilot certificate may approve an aircraft for return to service after performing preventive maintenance under the provisions of § 43.3(g).

§ 43.9 Content, form, and disposition of maintenance, preventive maintenance, rebuilding, and alteration records (except inspections performed in accordance with Part 91, Part 123, Part 125, § 135.411(a)(1), and § 135.419 of this chapter).

(a) *Maintenance record entries.* Except as provided in paragraphs (b) and (c) of this section, each person who maintains, performs preventive maintenance, rebuilds, or alters an aircraft, airframe, aircraft engine, propeller, appliance, or component part shall make an entry in the maintenance record of that equipment containing the following information:

(1) A description (or reference to data acceptable to the Administrator) of work performed.

(2) The date of completion of the work performed.

(3) The name of the person performing the work if other than the person specified in paragraph (a) (4) of this section.

(4) If the work performed on the aircraft, airframe, aircraft engine, propeller, appliance, or component part has been performed satifactorily, the signature, certificate number, and kind of certificate held by the person approving the work. The signature constitutes the approval for return to service only for the work performed.

In addition to the entry required by this paragraph, major repairs and major alterations shall be entered on a form, and the form disposed of, in the manner prescribed in Appendix B, by the person performing the work.

(b) Each holder of an air carrier operating certificate or an operating certificate issued under Part 121, 127, or 135, that is required by its approved operations specifications to provide for a continuous airworthiness maintenance program, shall make a record of the maintenance, preventive maintenance, rebuilding, and alteration, on aircraft, airframes, aircraft engines, propellers, appliances, or component parts which it operates in accordance with the applicable provisions of Part 121, 127, or 135 of this chapter, as appropriate.

(c) This section does not apply to persons performing inspections in accordance with Part 91, 123, 125, § 135.411(a) (1), or § 135.419 of this chapter.

§ 43.11 Content, form, and disposition of records for inspections conducted under Parts 91 and 125 and §§ 135.411(a)(1) and 135.419 of this chapter.

☞ **F72–75**

(a) *Maintenance record entries.* The person approving or disapproving for return to service

an aircraft, airframe, aircraft engine, propeller, appliance, or component part after any inspection performed in accordance with Part 91, 123, 125, § 135.411(a) (1), or § 135.419 shall make an entry in the maintenance record of that equipment containing the following information:

(1) The type of inspection and a brief description of the extent of the inspection.

(2) The date of the inspection and aircraft total time in service.

(3) The signature, the certificate number, and kind of certificate held by the person approving or disapproving for return to service the aircraft, airframe, aircraft engine, propeller, appliance, component part, or portions thereof.

(4) Except for progressive inspections, if the aircraft is found to be airworthy and approved for return to service, the following or a similarly worded statement—"I certify that this aircraft has been inspected in accordance with (insert type) inspection and was determined to be in airworthy condition."

(5) Except for progressive inspections, if the aircraft is not approved for return to service because of needed maintenance, noncompliance with applicable specifications, airworthiness directives, or other approved data, the following or a similarly worded statement—"I certify that this aircraft has been inspected in accordance with (insert type) inspection and a list of discrepancies and unairworthy items dated (date) has been provided for the aircraft owner or operator."

(6) For progressive inspections, the following or a similarly worded statement— "I certify that in accordance with a progressive inspection program, a routine inspection of (identify whether aircraft or components) and a detailed inspection of (identify components) were performed and the (aircraft or components) are (approved or disapproved) for return to service." If disapproved, the entry will further state "and a list of discrepancies and unairworthy items dated (date) has been provided to the aircraft owner or operator."

(7) If an inspection is conducted under an inspection program provided for in Part 91, 123, 125, or § 135.411(a) (1), the entry must identify the inspection program, that part of the inspection program accomplished, and contain a statement that the inspection was performed in accordance with the inspections and procedures for that particular program.

(b) *Listing of discrepancies and placards.* If the person performing any inspection required by Part 91 or § 135.411(a)(1) of this chapter finds that the aircraft is unairworthy or does not meet the applicable type certificate data, airworthiness directives, or other approved data upon which its airworthiness depends, that person must give the owner or lessee a signed and dated list of those discrepancies. For those items permitted to be inoperative under § 91.30(d)(2), that person shall place a placard, that meets the aircraft's airworthiness certification regulations, on each inoperative instrument and the cockpit control of each item of inoperative equipment, marking it ''Inoperative,'' and shall add the items to the signed and dated list of discrepancies given to the owner or lessee.

§ 43.12 Maintenance records: falsification, reproduction, or alteration.

(a) No person may make or cause to be made:

(1) Any fraudulent or intentionally false entry in any record or report that is required to be made, kept, or used to show compliance with any requirement under this Part;

(2) Any reproduction, for fraudulent purpose, of any record or report under this Part; or

(3) Any alteration, for fraudulent purpose, of any record or report under this Part.

(b) The commission by any person of an act prohibited under paragraph (a) of this section is a basis for suspending or revoking the applicable airman, operator, or production certificate, Technical Standard Order Authorization, FAA-Parts Manufacturer Approval, or Product and Process Specification issued by the Administrator and held by that person.

§ 43.13 Performance rules (general).

(a) Each person performing maintenance, alteration, or preventive maintenance on an aircraft, engine, propeller, or appliance shall use the methods, techniques, and practices prescribed in the current manufacturer's maintenance manual or Instructions for Continued Airworthiness prepared by its manufacturer, or other methods, techniques, and practices acceptable to the Administrator, except as noted in § 43.16. He shall use the tools, equipment, and test apparatus necessary to assure completion of the work in accordance with accepted industry practices. If special equipment or test apparatus is recommended by the manufacturer involved, he must use that equipment or apparatus or its equivalent acceptable to the Administrator.

(b) Each person maintaining or altering, or performing preventive maintenance, shall do that work in such a manner and use materials of such a quality, that the condition of the aircraft, airframe, aircraft engine, propeller, or appliance worked on will be at least equal to its original or properly altered condition (with regard to aerodynamic function, structural strength, resistance to vibration and deterioration, and other qualities affecting airworthiness).

(c) *Special provisions for holders of air carrier operating certificates and operating certificates issued under the provisions of Part 121, 127, or 135 [and Part 129 operators holding operations specifications].* Unless otherwise notified by the Administrator, the methods, techniques, and practices contained in the maintenance manual or the maintenance part of the manual of the holder of an air carrier operating certificate or an operating certificate under Part 121, 127, or 135 (that is required by its operating specifications to provide a con-

tinuous airworthiness maintenance and inspection program) [and Part 129 operators holding operations specifications] constitute acceptable means of compliance with this section.

(d) [Deleted.

§ 43.15 Additional performance rules for inspections.

(a) *General.* Each person performing an inspection required by Part 91, 123, 125, or 135 of this chapter, shall—

(1) Perform the inspection so as to determine whether the aircraft, or portion(s) thereof under inspection, meets all applicable airworthiness requirements; and

(2) If the inspection is one provided for in Part 123, 125, 135, or § 91.169(e) of this chapter, perform the inspection in accordance with the instructions and procedures set forth in the inspection program for the aircraft being inspected.

(b) *Rotorcraft.* Each person performing an inspection required by Part 91 on a rotorcraft shall inspect the following systems in accordance with the maintenance manual or Instructions for Continued Airworthiness of the manufacturer concerned:

(1) The drive shafts or similar systems.

(2) The main rotor transmission gear box for obvious defects.

(3) The main rotor and center section (or the equivalent area).

(4) The auxiliary rotor on helicopters.

(c) *Annual and 100–hour inspections.*

(1) Each person performing an annual or 100-hour inspection shall use a checklist while performing the inspection. The checklist may be of the person's own design, one provided by the manufacturer of the equipment being inspected, or one obtained from another source. This checklist must include the scope and detail of the items contained in Appendix D to this Part and paragraph (b) of this section.

(2) Each person approving a reciprocating-engine-powered aircraft for return to service after an annual or 100-hour inspection shall, before that approval, run the aircraft engine or engines to determine satisfactory performance, in accordance with the manufacturer's recommendations, of—

(i) Power output (static and idle r.p.m.);

(ii) Magnetos;

(iii) Fuel and oil pressure; and

(iv) Cylinder and oil temperature.

(3) Each person approving a turbine-engine-powered aircraft for return to service after an annual, 100-hour, progressive inspection shall, before that approval, run the aircraft engine or engines to determine satisfactory performance

in accordance with the manufacturer's recommendations.

(d) *Progressive inspection.*

(1) Each person performing a progressive inspection shall, at the start of a progressive inspection system, inspect the aircraft completely. After this initial inspection, routine and detailed inspections must be conducted as prescribed in the progressive inspection schedule. Routine inspections consist of visual examination or check of the appliances, the aircraft, and its components and systems, insofar as practicable without disassembly. Detailed inspections consist of a thorough examination of the appliances, the aircraft, and its components and systems, with such disassembly as is necessary. For the purposes of this subparagraph, the overhaul of a component or system is considered to be a detailed inspection.

(2) If the aircraft is away from the station where inspections are normally conducted, an appropriately rated mechanic, a certificated repair station, or the manufacturer of the aircraft may perform inspections in accordance with the procedures and using the forms of the person who would otherwise perform the inspection.

§ 43.16 Airworthiness Limitations.

Each person performing an inspection or other maintenance specified in an Airworthiness Limitations section of a manufacturer's maintenance manual or Instructions for Continued Airworthiness shall perform the inspection or other maintenance in accordance with that section, or in accordance with operations specifications approved by the Administrator under Parts 121, 123, 127, or 135, or an inspection program approved under § 91.169(e).

§ 43.17 Mechanical work performed on U.S. registered aircraft by certain Canadian persons.

(a) A person holding a valid mechanic certificate of competence (Aircraft Maintenance Engineer license) and appropriate ratings issued by the Canadian Government, or a person who is an authorized employee (Approved Inspector) performing work for a company whose system of quality control for the inspection and maintenance of aircraft has been approved by the Canadian Department of Transport may, in connection with aircraft of U.S. registry in Canada:

(1) Perform maintenance, preventive maintenance and alterations if those operations are done in accordance with § 43.13 and the maintenance record entries are made in accordance with § 43.9.

(2) Except for an annual inspection, perform any inspection required by § 91.169 of this chapter, if the inspection is done in accordance with § 43.15 and the maintenance record entries are made in accordance with § 43.11.

(3) Approve (certify) maintenance, preventive maintenance, and alterations performed

under this section except that a Canadian Aircraft Maintenance Engineer may not approve a major repair or major alteration.

(b) A Canadian Department of Transport Airworthiness Inspector, or an authorized employee (Approved Inspector) performing work for a company approved by the Canadian Department of Transport, may approve (certify) a major repair or major alteration performed under this section if the work was done in accordance with technical data approved by the Administrator.

(c) No person may operate in air commerce an aircraft, airframe, aircraft engine, propeller, or appliance on which maintenance, preventive maintenance, or alteration has been performed under this section unless it has been approved by a person authorized in this section.

Note.—The recordkeeping and reporting requirements contained herein have been approved by the Bureau of the Budget in accordance with the Federal Reports Act of 1942.

Appendix A
Major Alterations, Major Repairs, and Preventive Maintenance

(a) *Major alterations.*

(1) *Airframe major alterations.* Alterations of the following parts and alterations of the following types, when not listed in the aircraft specifications issued by the FAA, are airframe major alterations:

(i) Wings.
(ii) Tail surfaces.
(iii) Fuselage.
(iv) Engine mounts.
(v) Control system.
(vi) Landing gear.
(vii) Hull or floats.
(viii) Elements of an airframe including spars, ribs, fittings, shock absorbers, bracing, cowlings, fairings, and balance weights.
(ix) Hydraulic and electrical actuating system of components.
(x) Rotor blades.
(xi) Changes to the empty weight or empty balance which result in an increase in the maximum certificated weight or center of gravity limits of the aircraft.
(xii) Changes to the basic design of the fuel, oil, cooling, heating, cabin pressurization, electrical, hydraulic, de-icing, or exhaust systems.
(xiii) Changes to the wing or to fixed or movable control surfaces which affect flutter and vibration characteristics.

(2) *Powerplant major alterations.* The following alterations of a powerplant when not listed in the engine specifications issued by the FAA, are powerplant major alterations:

(i) Conversion of an aircraft engine from one approved model to another, involving any changes in compression ratio, propeller reduction gear, impeller gear ratios or the substitution of major engine

parts which requires extensive rework and testing of the engine.
(ii) Changes to the engine by replacing aircraft engine structural parts with parts not supplied by the original manufacturer or parts not specifically approved by the Administrator.
(iii) Installation of an accessory which is not approved for the engine.
(iv) Removal of accessories that are listed as required equipment on the aircraft or engine specification.
(v) Installation of structural parts other than the type of parts approved for the installation.
(vi) Conversions of any sort for the purpose of using fuel of a rating or grade other than that listed in the engine specifications.

(3) *Propeller major alterations.* The following alterations of a propeller when not authorized in the propeller specifications issued by the FAA are propeller major alterations:

(i) Changes in blade design.
(ii) Changes in hub design.
(iii) Changes in the governor or control design.
(iv) Installation of a propeller governor or feathering system.
(v) Installation of propeller de-icing system.
(vi) Installation of parts not approved for the propeller.

(4) *Appliance major alterations.* Alterations of the basic design not made in accordance with recommendations of the appliance manufacturer or in accordance with an FAA Airworthiness Directive are appliance major alterations. In addition, changes in the basic design of radio communication and navigation equipment approved under type certification or a Technical Standard Order that have an effect on frequency stability, noise level, sensitivity, selectivity, distortion, spurious radiation, AVC characteristics, or ability to meet environmental test conditions and other changes that have an effect on the performance of the equipment are also major alterations.

(b) *Major repairs.*

(1) *Airframe major repairs.* Repairs to the following parts of an airframe and repairs of the following types, involving the strengthening, reinforcing, splicing, and manufacturing of primary structural members or their replacement, when replacement is by fabrication such as riveting or welding, are airframe major repairs:

(i) Box beams.
(ii) Monocoque or semimonocoque wings or control surfaces.
(iii) Wing stringers or chord members.
(iv) Spars.
(v) Spar flanges.
(vi) Members of truss-type beams.
(vii) Thin sheet webs of beams.
(viii) Keel and chine members of boat hulls or floats.
(ix) Corrugated sheet compression

members which act as flange material of wings or tail surfaces.
(x) Wing main ribs and compression members.
(xi) Wing or tail surface brace struts.
(xii) Engine mounts.
(xiii) Fuselage longerons.
(xiv) Members of the side truss, horizontal truss, or bulkheads.
(xv) Main seat support braces and brackets.
(xvi) Landing gear brace struts.
(xvii) Axles.
(xviii) Wheels.
(xix) Skis, and ski pedestals.
(xx) Parts of the control system such as control columns, pedals, shafts, brackets, or horns.
(xxi) Repairs involving the substitution of material.
(xxii) The repair of damaged areas in metal or plywood stressed covering exceeding six inches in any direction.
(xxiii) The repair of portions of skin sheets by making additional seams.
(xxiv) The splicing of skin sheets.
(xxv) The repair of three or more adjacent wing or control surface ribs or the leading edge of wings and control surfaces, between such adjacent ribs.
(xxvi) Repair of fabric covering involving an area greater than that required to repair two adjacent ribs.
(xxvii) Replacement of fabric on fabric covered parts such as wings, fuselages, stabilizers, and control surfaces.
(xxviii) Repairing, including rebottoming, of removable or integral fuel tanks and oil tanks.

(2) *Powerplant major repairs.* Repairs of the following parts of an engine and repairs of the following types, are powerplant major repairs:

(i) Separation or disassembly of a crankcase or crankshaft of a reciprocating engine equipped with an integral supercharger.
(ii) Separation or disassembly of a crankcase or crankshaft of a reciprocating engine equipped with other than spurtype propeller reduction gearing.
(iii) Special repairs to structural engine parts by welding, plating, metalizing, or other methods.

(3) *Propeller major repairs.* Repairs of the following types to a propeller are propeller major repairs:

(i) Any repairs to, or straightening of steel blades.
(ii) Repairing or machining of steel hubs.
(iii) Shortening of blades.
(iv) Retipping of wood propellers.
(v) Replacement of outer laminations on fixed-pitch wood propellers.
(vi) Repairing elongated bolt holes in the hub of fixed-pitch wood propellers.
(vii) Inlay work on wood blades.
(viii) Repairs to composition blades.

(ix) Replacement of tip fabric.

(x) Replacement of plastic covering.

(xi) Repair of propeller governors.

(xii) Overhaul of controllable pitch propellers.

(xiii) Repairs to deep dents, cuts, scars, nicks, etc., and straightening of aluminum blades.

(xiv) The repair or replacement of internal elements of blades.

(4) *Appliance major repairs.* Repairs of the following types to appliances are appliance major repairs:

(i) Calibration and repair of instruments.

(ii) Calibration of radio equipment.

(iii) Rewinding the field coil of an electrical accessory.

(iv) Complete disassembly of complex hydraulic power valves.

(v) Overhaul of pressure type carburetors, and pressure type fuel, oil, and hydraulic pumps.

(c) *Preventive maintenance.* Preventive maintenance is limited to the following work, provided it does not involve complex assembly operations:

(1) Removal, installation, and repair of landing gear tires.

(2) Replacing elastic shock absorber cords on landing gear.

(3) Servicing landing gear shock struts by adding oil, air, or both.

(4) Servicing landing gear wheel bearings, such as cleaning and greasing.

(5) Replacing defective safety wiring or cotter keys.

(6) Lubrication not requiring disassembly other than removal of nonstructural items such as cover plates, cowlings, and fairings.

(7) Making simple fabric patches not requiring rib stitching or the removal of structural parts or control surfaces. In the case of balloons, the making of small fabric repairs to envelopes (as defined in, and in accordance with, the balloon manufacturers' instructions) not requiring load tape repair or replacement.

(8) Replenishing hydraulic fluid in the hydraulic reservoir.

(9) Refinishing decorative coating of fuselage, balloon baskets, wings, tail group surfaces (excluding balanced control surfaces), fairings, cowlings, landing gear, cabin, or cockpit interior when removal or disassembly of any primary structure or operating system is not required.

(10) Applying preservative or protective material to components where no disassembly of any primary structure or operating system is involved and where such coating is not prohibited or is not contrary to good practices.

(11) Repairing upholstery and decorative furnishings of the cabin, cockpit, or balloon interior when the repairing does not require disassembly of any primary structure or operating system or interfere with an operating system or affect primary structure of the aircraft.

(12) Making small simple repairs to fairings, nonstructural cover plates, cowlings, and small patches and reinforcements not changing the contour so as to interfere with proper airflow.

(13) Replacing side windows where that work does not interfere with the structure or any operating system such as controls, electrical equipment, etc.

(14) Replacing safety belts.

(15) Replacing seats or seat parts with replacement parts approved for the aircraft, not involving disassembly of any primary structure or operating system.

(16) Trouble shooting and repairing broken circuits in landing light wiring circuits.

(17) Replacing bulbs, reflectors, and lenses of position and landing lights.

(18) Replacing wheels and skis where no weight and balance computation is involved.

(19) Replacing any cowling not requiring removal of the propeller or disconnection of flight controls.

(20) Replacing or cleaning spark plugs and setting of spark plug gap clearance.

(21) Replacing any hose connection except hydraulic connections.

(22) Replacing prefabricated fuel lines.

(23) Cleaning or replacing fuel and oil strainers or filter elements.

(24) Replacing and servicing batteries.

[(25)] Cleaning of balloon burner pilot and main nozzles in accordance with the balloon manufacturer's instructions.

[(26)] Replacement or adjustment of nonstructural standard fasteners incidental to operations.

[(27)] The interchange of balloon baskets and burners on envelopes when the basket or burner is designated as interchangeable in the balloon type certificate data and the baskets and burners are specifically designed for quick removal and installation.]

[(28)] The installation of anti-misfueling devices to reduce the diameter of fuel tank filler openings provided the specific device has been made a part of the aircraft type certificate data by the aircraft manufacturer, the aircraft manufacturer has provided FAA-approved instructions for installation of the specific device, and installation does not involve the disassembly of the existing tank filler opening.

[(29)] Removing, checking, and replacing magnetic chip detectors.

Appendix B
Recording of Major Repairs and Major Alterations

(a) Except as provided in paragraphs (b), (c), [and (d)] of this appendix, each person performing a major repair or major alteration shall—

(1) Execute FAA Form 337 at least in duplicate;

(2) Give a signed copy of that form to the aircraft owner; and

(3) Forward a copy of that form to the local FAA District Office within 48 hours after the aircraft, airframe, aircraft engine, propeller, or appliance is approved for return to service.

(b) For major repairs made in accordance with a manual or specification acceptable to the Administrator, a certificated repair station may, in place of the requirements of paragraph (a)—

(1) Use the customer's work order upon which the repair is recorded;

(2) Give the aircraft owner a signed copy of the work order and retain a duplicate copy for a least 2 years from the date of approval for return to service of the aircraft, airframe, aircraft engine, propeller, or appliance;

(3) Give the aircraft owner a maintenance release signed by an authorized representative of the repair station and incorporating the following information:

(i) Identity of the aircraft, airframe, aircraft engine, propeller, or appliance.

(ii) If an aircraft, the make, model, serial number, nationality and registration marks, and location of the repaired area.

(iii) If an airframe, aircraft engine, propeller, or appliance, give the manufacturer's name, name of the part, model, and serial numbers (if any); and

(4) Include the following or a similarly worded statement—

"The aircraft, airframe, aircraft engine, propeller, or appliance identified above was repaired and inspected in accordance with current Regulations of the Federal Aviation Administration and is approved for return to service.

"Pertinent details of the repair are on file at this repair station under Order No. _____.   Date _____

Signed _____ for
       (signature of authorized representative)

_____   _____
(repair station name)   (certificate no.)

_____ ."
             (address)

(c) For a major repair or major alteration made by a person authorized in § 43.17, the person who performs the major repair or major alteration and the person authorized by § 43.17 to approve that work shall execute a FAA Form 337 at least in duplicate. A completed copy of that form shall be—

(1) Given to the aircraft owner; and

(2) Forwarded to the Federal Aviation Administration, Aircraft Registration Branch, P.O. Box 25082, Oklahoma City, Oklahoma 73125, within 48 hours after the work is inspected.

[(d) For extended-range fuel tanks installed within the passenger compartment or a baggage compartment, the person who performs the work and the person authorized to approve the work by § 43.7 of this part shall execute an FAA Form 337 in at least triplicate. One (1) copy of the FAA Form 337 shall be placed on board the aircraft as specified in § 91.173 of this chapter. The remaining forms shall be distributed as required by paragraph (a)(2) and (3) or (c)(1) and (2) of this paragraph as appropriate.]

Appendix C

[Revoked]

Appendix D

Scope and Detail of Items (as Applicable to the Particular Aircraft) to be Included in Annual and 100–Hour Inspections

(a) Each person performing an annual or 100-hour inspection shall, before that inspection, remove or open all necessary inspection plates, access doors, fairing, and cowling. He shall thoroughly clean the aircraft and aircraft engine.

(b) Each person performing an annual or 100-hour inspection shall inspect (where applicable) the following components of the fuselage and hull group:

(1) Fabric and skin—for deterioration, distortion, other evidence of failure, and defective or insecure attachment of fittings.

(2) Systems and components—for improper installation, apparent defects, and unsatisfactory operation.

(3) Envelope, gas bags, ballast tanks, and related parts—for poor condition.

(c) Each person performing an annual or 100-hour inspection shall inspect (where applicable) the following components of the cabin and cockpit group:

(1) Generally—for uncleanliness and loose equipment that might foul the controls.

(2) Seats and safety belts—for poor condition and apparent defects.

(3) Windows and windshields—for deterioration and breakage.

(4) Instruments—for poor condition, mounting, marking, and (where practicable) for improper operation.

(5) Flight and engine controls—for improper installation and improper operation.

(6) Batteries—for improper installation and improper charge.

(7) All systems—for improper installation, poor general condition, apparent and obvious defects, and insecurity of attachment.

(d) Each person performing an annual or 100-hour inspection shall inspect (where applicable) components of the engine and nacelle group as follows:

(1) Engine section—for visual evidence of excessive oil, fuel, or hydraulic leaks, and sources of such leaks.

(2) Studs and nuts—for improper torquing and obvious defects.

(3) Internal engine—for cylinder compression and for metal particles or foreign matter on screens and sump drain plugs. If there is weak cylinder compression, for improper internal condition and improper internal tolerances.

(4) Engine mount—for cracks, looseness of mounting, and looseness of engine to mount.

(5) Flexible vibration dampeners—for poor condition and deterioration.

(6) Engine controls—for defects, improper travel, and improper safetying.

(7) Lines, hoses, and clamps—for leaks, improper condition, and looseness.

(8) Exhaust stacks—for cracks, defects, and improper attachment.

(9) Accessories—for apparent defects in security of mounting.

(10) All systems—for improper installation, poor general condition, defects, and insecure attachment.

(11) Cowling—for cracks, and defects.

(e) Each person performing an annual or 100-hour inspection shall inspect (where applicable) the following components of the landing gear group:

(1) All units—for poor condition and insecurity of attachment.

(2) Shock absorbing devices—for improper oleo fluid level.

(3) Linkage, trusses, and members—for undue or excessive wear, fatigue, and distortion.

(4) Retracting and locking mechanism—for improper operation.

(5) Hydraulic lines—for leakage.

(6) Electrical system—for chafing and improper operation of switches.

(7) Wheels—for cracks, defects, and condition of bearings.

(8) Tires—for wear and cuts.

(9) Brakes—for improper adjustment.

(10) Floats and skis—for insecure attachment and obvious or apparent defects.

(f) Each person performing an annual or 100-hour inspection shall inspect (where applicable) all components of the wing and center section assembly for poor general condition, fabric or skin deterioration, distortion, evidence of failure, and insecurity of attachment.

(g) Each person performing an annual or 100-hour inspection shall inspect (where applicable) all components and systems that make up the complete empennage assembly for poor general condition, fabric or skin deterioration, distortion, evidence of failure, insecure attachment, improper component installation, and improper component operation.

(h) Each person performing an annual or 100-hour inspection shall inspect (where applicable) the following components of the propeller group:

(1) Propeller assembly—for cracks, nicks, binds, and oil leakage.

(2) Bolts—for improper torquing and lack of safetying.

(3) Anti-icing devices—for improper operations and obvious defects.

(4) Control mechanisms—for improper operation, insecure mounting, and restricted travel.

(i) Each person performing an annual or 100-hour inspection shall inspect (where applicable) the following components of the radio group:

(1) Radio and electronic equipment—for improper installation and insecure mounting.

(2) Wiring and conduits—for improper routing, insecure mounting, and obvious defects.

(3) Bonding and shielding—for improper installation and poor condition.

(4) Antenna including trailing antenna—for poor condition, insecure mounting, and improper operation.

(j) Each person performing an annual or 100-hour inspection shall inspect (where applicable) each installed miscellaneous item that is not otherwise covered by this listing for improper installation and improper operation.

FEDERAL AVIATION REGULATIONS

Part 61—Certification: Pilots and Flight Instructors

Contents

Subpart A—General

§ 61.1 Applicability.

(a) This Part prescribes the requirements for issuing pilot and flight instructor certificates and ratings, the conditions under which those certificates and ratings are necessary, and the privileges and limitations of those certificates and ratings.

(b) Except as provided in § 61.71 of this Part, an applicant for a certificate or rating may, until November 1, 1974, meet either the requirements of this Part, or the requirements in effect immediately before November 1, 1973. However, the applicant for a private pilot certificate with a free balloon class rating must meet the requirements of this Part.

[§ 61.2 Certification of foreign pilots and flight instructors.

[A person who is neither a United States citizen nor a resident alien is issued a certificate under this Part (other than under § 61.75 or § 61.77), outside the United States, only when the Administrator finds that the pilot certificate is needed for the operation of a U.S.-registered civil aircraft or finds that the flight instructor certificate is needed for the training of students who are citizens of the United States.]

§ 61.3 Requirements for certificates, rating, and authorizations.

(a) *Pilot certificate.* No person may act as pilot in command or in any other capacity as a required pilot flight crewmember of a civil aircraft of United States registry unless he has in his personal possession a current pilot certificate issued to him under this Part. However, when the aircraft is operated within a foreign country a current pilot license issued by the country in which the aircraft is operated may be used.

(b) *Pilot certificate: foreign aircraft.* No person may, within the United States, act as pilot in command or in any other capacity as a required pilot flight crewmember of a civil aircraft of foreign registry unless he has in his personal possession a current pilot certificate issued to him under this Part, or a pilot license issued to him or validated for him by the country in which the aircraft is registered.

(c) *Medical certificate.* Except for free balloon pilots piloting balloons and glider pilots piloting gliders, no person may act as pilot in command or in any other capacity as a required pilot flight crewmember of an aircraft under a certificate issued to him under this Part, unless he has in his personal possession an appropriate current medical certificate issued under Part 67 of this chapter. However, when the aircraft is operated within a foreign country with a current pilot license issued by that country, evidence of current medical qualification for that license, issued by that country, may be used. In the case of a pilot certificate issued on the basis of a foreign pilot license under § 61.75, evidence of current medical qualification accepted for the issue of that license is used in place of a medical certificate.

(d) *Flight instructor certificate.* Except for lighter-than-air flight instruction in lighter-than-air aircraft, and for instruction in air transportation service given by the holder of an Airline Transport Pilot Certificate under § 61.169, no person other than the holder of a flight instructor certificate issued by the Administrator with an appropriate rating on that certificate may—

(1) Give any of the flight instruction required to qualify for a solo flight, solo cross-country flight, or for the issue of a pilot or flight instructor certificate or rating;

(2) Endorse a pilot logbook to show that he has given any flight instruction; or

(3) Endorse a student pilot certificate or logbook for solo operating privileges.

(e) *Instrument rating.* No person may act as pilot in command of a civil aircraft under instrument flight rules, or in weather conditions less than the minimum prescribed for VFR flight unless—

(1) In the case of an airplane, he holds an instrument rating or an airline transport pilot certificate with an airplane category rating on it;

(2) In the case of a helicopter, he holds a helicopter instrument rating or an airline transport pilot certificate with a rotorcraft category and helicopter class rating not limited to VFR;

(3) In the case of a glider, he holds an instrument rating (airplane) or an airline transport pilot certificate with an airplane category rating; or

(4) In the case of an airship, he holds a commercial pilot certificate with lighter-than-air category and airship class ratings.

(f) *Category II pilot authorization.*

(1) No person may act as pilot in command of a civil aircraft in a Category II operation unless he holds a current Category II pilot authorization for that type aircraft or, in the case of a civil aircraft of foreign registry, he is authorized by the country of registry to act as pilot in command of that aircraft in Category II operations.

(2) No person may act as second in command of a civil aircraft in a Category II operation unless he holds a current appropriate instrument rating or an [appropriate airline transport pilot certificate] or, in the case of a civil aircraft of foreign registry, he is authorized by the country of registry to act as second in command of that aircraft in Category II operations.

This paragraph does not apply to operations conducted by the holder of a certificate issued under [Parts 121 and 135] of this chapter.

(g) *Category A aircraft pilot authorization.* The Administrator may issue a certificate of authorization to the pilot of a small [aircraft] identified as a Category A aircraft in § 97.3 (b) (1) of this chapter to use that [aircraft] in a Category II operation, if he finds that the proposed operation can be safely conducted under the terms of the certificate. Such authorization does not permit operation of the aircraft carrying persons or property for compensation or hire.

(h) *Inspection of certificate.* Each person who holds a pilot certificate, flight instructor certificate, medical certificate, authorization, or license required by this Part shall present it for inspection upon the request of the Administrator, an authorized representative of the National Transportation Safety Board, or any Federal, State, or local law enforcement officer.

§ 61.5 Certificates and ratings issued under this Part.

(a) The following certificates are issued under this Part:

(1) Pilots certificates:
  (i) Student pilot.
  (ii) Private pilot.
  (iii) Commercial pilot.
  (iv) Airline transport pilot.
(2) Flight instructor certificates.

(b) The following ratings are placed on pilot certificates (other than student pilot) where applicable:

(1) Aircraft category ratings:
  (i) Airplane.
  (ii) Rotorcraft.
  (iii) Glider.
  (iv) Lighter-than-air.
(2) Airplane class ratings:
  (i) Single-engine land.
  (ii) Multiengine land.
  (iii) Single-engine sea.
  (iv) Multiengine sea.
(3) Rotorcraft class ratings:
  (i) Helicopter.
  (ii) Gyroplane.
(4) Lighter-than-air class ratings:
  (i) Airship.
  (ii) Free balloon.
(5) Aircraft type ratings are listed in Advisory Circular 61-1 entitled "Aircraft Type Ratings." This list includes ratings for the following:
  (i) Large aircraft, other than lighter-than-air.
  (ii) Small turbojet-powered airplanes.
  (iii) Small helicopters for operations requiring an airline transport pilot certificate.
  (iv) Other aircraft type ratings specified by the Administrator through aircraft type certificate procedures.
(6) Instrument ratings (on private and commercial pilot certificates only):
  (i) Instrument—airplanes.
  (ii) Instrument—helicopter.

(c) The following ratings are placed on flight instructor certificates where applicable:

(1) Aircraft category ratings:
  (i) Airplane.
  (ii) Rotorcraft.
  (iii) Glider.
(2) Airplane class ratings:
  [(i) Single-engine.
  [(ii) Multiengine.]
(3) Rotorcraft class ratings:
  (i) Helicopter.
  (ii) Gyroplane.
(4) Instrument ratings:
  (i) Instrument—airplane.
  (ii) Instrument—helicopter.

§ 61.7 Obsolete certificates and ratings.

(a) The holder of a free balloon pilot certificate issued before November 1, 1973, may not exercise the privileges of that certificate.

(b) The holder of a pilot certificate that bears any of the following category ratings without an associated class rating, may not exercise the privileges of that category rating:

(1) Rotorcraft.
(2) Lighter-than-air.
(3) Helicopter.
(4) Autogiro.

§ 61.9 Exchange of obsolete certificates and ratings for current certificates and ratings.

(a) The holder of an unexpired free balloon pilot certificate, or an unexpired pilot certificate with an obsolete category rating listed in § 61.7 (b) of this Part may exchange that certificate for a certificate with the following applicable category and class rating, without a further showing of competency, until October 31, 1975. After that date, a free balloon pilot certificate or certificate with an obsolete rating expires.

(b) *Private or commercial pilot certificate with rotorcraft category rating.* The holder of a private or commercial pilot certificate with a rotorcraft category rating is issued that certificate with a rotorcraft category rating, and a helicopter or gyroplane class rating, depending upon whether a helicopter or a gyroplane is used to qualify for the rotorcraft category rating.

(c) *Private or commercial pilot certificate with helicopter or autogiro category rating.* The holder of a private or commercial pilot certificate with a helicopter or autogiro category rating is issued that certificate with a rotorcraft category rating and a helicopter class rating (in the case of a helicopter category rating), or a gyroplane class rating (in the case of an autogiro rating).

(d) *Airline transport pilot certificate with helicopter or autogiro category rating.* The holder of an airline transport pilot certificate with a helicopter or autogiro category rating is issued that certificate with a rotorcraft category rating (limited to VFR) and a helicopter class and type rating (in the case of a helicopter category rating), or a gyroplane class rating (in the case of an autogiro category rating).

(e) *Airline transport pilot certificate with a rotorcraft category rating (without a class rating).* The holder of an airline transport pilot certificate with a rotorcraft category rating (without a class rating) is issued that certificate with a rotorcraft category rating limited to VFR, and a helicopter and type rating or a gyroplane class rating, depending upon whether a helicopter or gyroplane is used to qualify for the rotorcraft category rating.

[(f) *Free balloon pilot certificate.* The holder of a free balloon pilot certificate is issued a commercial pilot certificate with a lighter-than-air category rating and a free balloon class rating. However, a free balloon class rating may be issued with the limitations provided in § 61.141.]

(g) *Lighter-than-air pilot certificate or pilot certificate with lighter-than-air category (without a class rating).*

(1) In the case of an application made before November 1, 1975, the holder of a lighter-than-air pilot certificate or a pilot certificate with a lighter-than-air category rating (without a class rating) is issued a private or commercial pilot certificate, as appropriate, with a lighter-than-air category rating and airship and free balloon class ratings.

(2) In the case of an application made after October 31, 1975, the holder of a lighter-than-air pilot certificate with an airship rating issued prior to November 1, 1973, may be issued a free balloon class rating upon passing the appropriate flight test in a free balloon.

§ 61.11 Expired pilot certificates and reissuance.

(a) No person who holds an expired pilot certificate or rating may exercise the privileges of that pilot certificate, or rating.

(b) Except as provided, the following certificates and ratings have expired and are not reissued:

(1) An airline transport pilot certificate issued before May 1, 1949, or containing a horsepower rating. However, an airline transport pilot certificate bearing an expiration date and issued after April 30, 1949, may be reissued without an expiration date if it does not contain a horsepower rating.

(2) A private or commercial pilot certificate, or a lighter-than-air or free balloon pilot certificate, issued before July 1, 1945. However, each of those certificates issued after June 30, 1945, and bearing an expiration date, may be reissued without an expiration date.

(c) A private or commercial pilot certificate or a special purpose pilot certificate, issued on the basis of a foreign pilot license, expires on the expiration date stated thereon. A certificate without an expiration date is issued to the holder of the expired certificate only if he meets the requirements of § 61.75 of this Part for the issue of a pilot certificate based on a foreign pilot license.

[§ 61.13 Application and qualification.

[(a) An application for a certificate and rating or for an additional rating under this Part is made on a form and in a manner prescribed by the Administrator. Each person who is neither a United States citizen nor a resident alien must show evidence that the fee prescribed by Appendix A of Part 187 of this chapter has been paid if that person—

[(1) Applies for a student pilot certificate to be issued outside the United States; or

[(2) Applies for a written or practical test to be administered outside the United States for any certificate or rating issued under this Part.]

(b) An applicant who meets the requirements of this Part is entitled to an appropriate pilot certificate with aircraft ratings. Additional aircraft category, class, type and other ratings, for which the applicant is qualified, are added to his certificate. However, the Administrator may refuse to issue certificates to persons who are not citizens of the United States and who do not reside in the United States.

(c) An applicant who cannot comply with all of the flight proficiency requirements prescribed by this Part because the aircraft used by him for his flight training or flight test is characteristically incapable of performing a required pilot operation, but who meets all other requirements for the certificate or rating sought, is issued the certificate or rating with appropriate limitations.

(d) An applicant for a pilot certificate who holds a medical certificate under § 67.19 of this chapter with special limitations on it, but who meets all other requirements for that pilot certificate, is issued a pilot certificate containing such operating limitations as the Administrator determines are necessary because of the applicant's medical deficiency.

(e) A Category II pilot authorization is issued as a part of the applicant's instrument rating or airline transport pilot certificate. Upon original issue the authorization contains a limitation for Category II operations of 1,600 feet RVR and a 150 foot decision height. This limitation is removed when the holder shows that since the beginning of the sixth preceding month he has made three Category II ILS approaches to a landing under actual or simulated instrument conditions with a 150 foot decision height.

(f) Unless authorized by the Administrator—

(1) A person whose pilot certificate is suspended may not apply for any pilot or flight instructor certificate or rating during the period of suspension; and

(2) A person whose flight instructor certificate only is suspended may not apply for any rating to be added to that certificate during the period of suspension.

(g) Unless the order of revocation provides otherwise—

(1) A person whose pilot certificate is revoked may not apply for any pilot or flight instructor certificate or rating for one year after the date of revocation; and

(2) A person whose flight instructor certificate only is revoked may not apply for any flight instructor certificate for one year after the date of revocation.

§ 61.14 Refusal to submit to a drug test.

(a) This section applies to—

(1) An employee who performs a function listed in Appendix I to Part 121 of this chapter for a Part 121 certificate holder or a Part 135 certificate holder; and

(2) An employee who performs a function listed in Appendix I to Part 121 of this chapter for an operator as defined in § 135.1(c) of this chapter. An employee of a person conducting operations of foreign civil aircraft navigated within the United States pursuant to Part 375 or emergency mail service operations pursuant to Section 405(h) of the Federal Aviation Act of 1958 is excluded from the requirements of this section.

(b) Refusal by the holder of a certificate issued under this part to take a test for a drug specified in Appendix I to Part 121 of this chapter when requested by a certificate holder, by an operator as defined in § 135.1(c) of this chapter, by a local law enforcement officer under his or her own authority, or by an FAA inspector, under the circumstances specified in that appendix, is grounds for—

(1) Denial of an application for any certificate or rating issued under this part for a period of up to 1 year after the date of that refusal; and

(2) Suspension or revocation of any certificate or rating issued under this part.

§ 61.15 Offenses involving alcohol and drugs.

(a) A conviction for the violation of any Federal or state statute relating to the growing, processing, manufacture, sale, disposition, possession, transportation, or importation of narcotic drugs, marihuana, or depressant or stimulant drugs or substances is grounds for—

(1) Denial of an application for any certificate or rating issued under this Part for a period of up to 1 year after the date of final conviction; or

(2) Suspension or revocation of any certificate or rating issued under this Part.

(b) The commission of an act prohibited by § 91.11(a) or § 91.12(a) of this chapter is grounds for—

(1) Denial of an application for a certificate or rating issued under this Part for a period of up to 1 year after the date of that act; or

(2) Suspension or revocation of any certificate or rating issued under this Part.

[§ 61.16 Refusal to submit to an alcohol test or to furnish test results.

[A refusal to submit to a test to indicate the percentage by weight of alcohol in the blood, when requested by a law enforcement officer in accordance with § 91.11(c) of this chapter, or a refusal to furnish or authorize the release of the test results requested by the Administrator in accordance wtih § 91.11(c) or (d) of this chapter, is grounds for—

[(a) Denial of an application for any certificate or rating issued under this Part for a period of up to 1 year after the date of that refusal; or

[(b) Suspension or revocation of any certificate or rating issued under this Part.]

§ 61.17 Temporary certificate.

(a) A temporary pilot or flight instructor certificate, or a rating, effective for a period of not more than 120 days, is issued to a qualified applicant pending a review of his qualifications and the issuance of a permanent certificate or rating by the Administrator. The permanent certificate or rating is issued to an applicant found qualified and a denial thereof is issued to an applicant found not qualified.

(b) A temporary certificate issued under paragraph (a) of this section expires—

(1) At the end of the expiration date stated thereon; or

(2) Upon receipt by the applicant of—

(i) The certificate or rating sought; or

(ii) Notice that the certificate or rating sought is denied.

§ 61.19 Duration of pilot and flight instructor certificates.

(a) *General.* The holder of a certificate with an expiration date may not, after that date, exercise the privileges of that certificate.

(b) *Student pilot certificate.* A student pilot certificate expires at the end of the 24th month after the month in which it is issued.

(c) *Other pilot certificates.* Any pilot certificate (other than a student pilot certificate) issued under this Part is issued without a specific expiration date. However, the holder of a pilot certificate issued on the basis of a foreign pilot license may exercise the privileges of that certificate only while the foreign pilot license on which that certificate is based is effective.

(d) *Flight instructor certificate.* A flight instructor certificate—

(1) Is effective only while the holder has a current pilot certificate and a medical certificate appropriate to the pilot privileges being exercised; and ☞ F4, 4A

(2) Expires at the end of the 24th month after the month in which it was last issued or renewed.

(e) *Surrender, suspension, or revocation.* Any pilot certificate or flight instructor certificate issued under this Part ceases to be effective if it is surrendered, suspended, or revoked.

(f) *Return of certificate.* The holder of any certificate issued under this Part that is suspended or revoked shall, upon the Administrator's request, return it to the Administrator.

§ 61.21 Duration of Category II pilot authorization.

A Category II pilot authorization expires at the end of the sixth month after it was last issued or renewed. Upon passing a practical test it is renewed for each type [aircraft] for which an authorization is held. However, an authorization for any particular type [aircraft] for which an authorization is held will not be renewed to extend beyond the end of the twelfth month after the practical test was passed in that type [aircraft]. If the holder of the authorization passes the practical test for a renewal in the month before the authorization expires, he is considered to have passed it during the month the authorization expired.

§ 61.23 Duration of medical certificates.

(a) A first-class medical certificate expires at the end of the last day of—

(1) The sixth month after the month of the date of examination shown on the certificate, for operations requiring an airline transport pilot certificate;

(2) The 12th month after the month of the date of examination shown on the certificate, for operations requiring only a commercial pilot certificate; and

(3) The 24th month after the month of the date of examination shown on the certificate,

for operations requiring only a private or student pilot certificate.

(b) A second-class medical certificate expires at the end of the last day of—

(1) The 12th month after the month of the date of examination shown on the certificate, for operations requiring a commercial pilot certificate, or an air traffic control tower operator certificate; and

(2) The 24th month after the month of the date of examination shown on the certificate, for operations requiring only a private or student pilot certificate.

(c) A third-class medical certificate expires at the end of the last day of the 24th month after the month of the date of examination shown on the certificate, for operations requiring a private or student pilot certificate.

§ 61.25 Change of name.

An application for the change of a name on a certificate issued under this Part must be accompanied by the applicant's current certificate and a copy of the marriage license, court order, or other document verifying the change. The documents are returned to the applicant after inspection. ☞ F5

§ 61.27 Voluntary surrender or exchange of certificate

The holder of a certificate issued under this Part may voluntarily surrender it for cancellation, or for the issue of a certificate of lower grade, or another certificate with specific ratings deleted. If he so requests, he must include the following signed statement or its equivalent:

"This request is made for my own reasons, with full knowledge that my [insert name of certificate or rating, as appropriate] may not be reissued to me unless I again pass the tests prescribed for its issue."

§ 61.29 Replacement of lost or destroyed certificate.

(a) An application for the replacement of a lost or destroyed airman certificate issued under this Part is made by letter to the Department of Transportation, Federal Aviation Administration, Airman Certification Branch. P.O. Box 25082, Oklahoma City, Oklahoma 73125. The letter must—

(1) State the name of the person to whom the certificate was issued, the permanent mailing address (including zip code), social security number (if any), date and place of birth of the certificate holder, and any available information regarding the grade, number, and date of issue of the certificate, and the ratings on it; and

(2) Be accompanied by a check or money order for $2.00, payable to the Federal Aviation Administration.

(b) An application for the replacement of a lost or destroyed medical certificate is made by letter to the Department of Transportation, Federal Aviation Administration, Aeromedical Certification Branch, P.O. Box 25082, Oklahoma City, Oklahoma 73125, accompanied by a check or money order for $2.00.

(c) A person who has lost a certificate issued under this Part, or a medical certificate issued under Part 67 of this chapter, or both, may obtain a telegram from the FAA confirming that it was issued. The telegram may be carried as a certificate for a period not to exceed 60 days pending his receipt of a duplicate certificate under paragraph (a) or (b) of this section, unless he has been notified that the certificate has been suspended or revoked. The request for such a telegram may be made by letter or prepaid telegram, including the date upon which a duplicate certificate was previously requested, if a request had been made, and a money order for the cost of the duplicate certificate. The request for a telegraphic certificate is sent to the office listed in paragraph (a) or (b) of this section, as appropriate. However, a request for both airman and medical certificates at the same time must be sent to the office prescribed in paragraph (a) of this section.

§ 61.31 General limitations.

(a) *Type ratings required.* A person may not act as pilot in command of any of the following aircraft unless he holds a type rating for that aircraft:

(1) A large aircraft (except lighter-than-air).

(2) A helicopter, for operations requiring an airline transport pilot certificate.

(3) A turbojet powered airplane.

(4) Other aircraft specified by the Administrator through aircraft type certificate procedures.

(b) *Authorization in lieu of a type rating.*

(1) In lieu of a type rating required under subparagraphs (a)(1), (3), and (4) of this section, an aircraft may be operated under an authorization issued by the Administrator, for a flight or series of flights within the United States, if—

(i) The particular operation for which the authorization is requested involves a ferry flight, a practice or training flight, a flight test for a pilot type rating, or a test flight of an aircraft, for a period that does not exceed 60 days;

(ii) The applicant shows that compliance with paragraph (a) of this section is impracticable for the particular operation; and

(iii) The Administrator finds that an equivalent level of safety may be achieved through operating limitations on the authorization.

(2) Aircraft operated under an authorization issued under this paragraph—

(i) May not be operated for compensation or hire; and

(ii) May carry only flight crewmembers necessary for the flight.

(3) An authorization issued under this paragraph may be reissued for an additional 60-day period for the same operation if the applicant shows that he was prevented from carrying out the purpose of the particular operation before his authorization expired.

The prohibition of subparagraph (2)(i) does not prohibit compensation for the use of an aircraft by a pilot solely to prepare for or take a flight test for a type rating.

(c) *Category and class rating: carrying another person or operating for compensation or hire.* Unless he holds a category and class rating for that aircraft, a person may not act as pilot in command of an aircraft that is carrying another person or is operated for compensation or hire. In addition, he may not act as pilot in command of that aircraft for compensation or hire.

(d) *Category and class rating: other operations.* No person may act as pilot in command of an aircraft in solo flight in operations not subject to paragraph (c) of this section, unless he meets at least one of the following:

(1) He holds a category and class rating appropriate to that aircraft.

(2) He has received flight instruction in the pilot operations required by this Part, appropriate to the category and class of aircraft for first solo, given to him by a certificated flight instructor who found him competent to solo that category and class of aircraft and has so endorsed his pilot logbook.

(3) He has soloed and logged pilot-in-command time in that category and class of aircraft before November 1, 1973.

(e) *High performance airplanes.* A person holding a private or commercial pilot certificate may not act as pilot in command of an airplane that has more than 200 horsepower, or that has a retractable landing gear, flaps, and a controllable propeller, unless he has received flight instruction from an authorized flight instructor who has certified in his logbook that he is competent to pilot an airplane that has more than 200 horsepower, or that has a retractable landing gear, flaps, and a controllable propeller, as the case may be. However, this instruction is not required if he has logged flight time as pilot in command in high performance airplanes before November 1, 1973. ☞F6,7

(f) *Exception.* This section does not require a class rating for gliders, or category and class ratings for aircraft that are not type certificated as airplanes, rotorcraft, or lighter-than-air aircraft. In addition, the rating limitations of this section do not apply to—

(1) The holder of a student pilot certificate;

(2) The holder of a pilot certificate when operating an aircraft under the authority of an experimental or provisional type certificate;

(3) An applicant when taking a flight test given by the Administrator; or

(4) The holder of a pilot certificate with a lighter-than-air category rating when operating a hot air balloon without an airborne heater.

§ 61.33 Tests: general procedure.

Tests prescribed by or under this Part are given at times and places, and by persons, designated by the Administrator.

§ 61.35 Written test: prerequisites and passing grades.

(a) An applicant for a written test must—

(1) Show that he has satisfactorily completed the ground instruction or home study course required by this Part for the certificate or rating sought;

(2) Present as personal identification an airman certificate, driver's license, or other official document; and

(3) Present a birth certificate or other official document showing that he meets the age requirement prescribed in this Part for the certificate sought not later than 2 years from the date of application for the test.

(b) The minimum passing grade is specified by the Administrator on each written test sheet or booklet furnished to the applicant.

This section does not apply to the written test for an airline transport pilot certificate or a rating associated with that certificate.

§ 61.37 Written tests: cheating or other unauthorized conduct.

(a) Except as authorized by the Administrator, no person may—

(1) Copy, or intentionally remove, a written test under this Part;

(2) Give to another, or receive from another, any part or copy of that test;

(3) Give help on that test to, or receive help on that test from, any person during the period that test is being given;

(4) Take any part of that test in behalf of another person;

(5) Use any material or aid during the period that test is being given; or ☞F8

(6) Intentionally cause, assist, or participate in any act prohibited by this paragraph.

(b) No person whom the Administrator finds to have committed an act prohibited by paragraph (a) of this section is eligible for any airman or ground instructor certificate or rating, or to take any test therefor, under this chapter for a period of one year after the date of that act. In addition, the commission of that act is a basis for suspending or revoking any airman or ground instructor certificate or rating held by that person.

§ 61.39 Prerequisites for flight tests.

(a) To be eligible for a flight test for a certificate, or an aircraft or instrument rating issued under this Part, the applicant must—

(1) Have passed any required written test since the beginning of the 24th month before the month in which he takes the flight test;

(2) Have the applicable instruction and aeronautical experience prescribed in this Part;

(3) Hold a current medical certificate appropriate to the certificate he seeks or, in the case of a rating to be added to his pilot certificate, at least a third-class medical certificate issued since the beginning of the 24th month before the month in which he takes the flight test;

(4) Except for a flight test for an airline transport pilot certificate, meet the age requirement for the issuance of the certificate or rating he seeks; and

(5) Have a written statement from an appropriately certificated flight instructor certifying that he has given the applicant flight instruction in preparation for the flight test within 60 days preceding the date of application, and finds him competent to pass the test and to have satisfactory knowledge of the subject areas in which he is shown to be deficient by his FAA airman written test report. However, an applicant need not have this written statement if he—　☞ F9

(i) Holds a foreign pilot license issued by a contracting State to the Convention on International Civil Aviation that authorizes at least the pilot privileges of the airman certificate sought by him;

(ii) Is applying for a type rating only, or a class rating with an associated type rating; or

(iii) Is applying for an airline transport pilot certificate or an additional aircraft rating on that certificate.

[(b) Notwithstanding the provisions of paragraph (a)(1) of this section, an applicant for an airline transport pilot certificate or rating may take the flight test for that certificate or rating if—

[(1) The applicant—

[(i) Within the period ending 24 calendar months after the month in which the applicant passed the first of any required written tests, was employed as a flight crewmember by a U.S. air carrier or commercial operator operating either under Part 121 or as a commuter air carrier under Part 135 (as defined in Part 298 of this title) and is employed by such a certificate holder at the time of the flight test;

[(ii) Has completed initial training, and, if appropriate, transition or upgrade training; and

[(iii) Meets the recurrent training requirements of the applicable Part; or

[(2) Within the period ending 24 calendar months after the month in which the applicant passed the first of any required written tests, the applicant participated as a pilot in a pilot training program of a U.S. scheduled military air transportation service and is currently participating in that program.]

§ 61.41 Flight instruction received from flight instructors not certificated by FAA.

Flight instruction may be credited toward the requirements for a pilot certificate or rating issued under this Part if it is received from—

(a) An Armed Force of either the United States or a foreign contracting State to the Convention on International Civil Aviation in a program for training military pilots; or

(b) A flight instructor who is authorized to give that flight instruction by the licensing authority of a foreign contracting State to the Convention on International Civil Aviation and the flight instruction is given outside the United States.

§ 61.43 Flight tests: general procedures.

(a) The ability of an applicant for a private or commercial pilot certificate, or for an aircraft or instrument rating on that certificate to perform the required pilot operations is based on the following:　☞ F10

(1) Executing procedures and maneuvers within the aircraft's performance capabilities and limitations, including use of the aircraft's systems.

(2) Executing emergency procedures and maneuvers appropriate to the aircraft.

(3) Piloting the aircraft with smoothness and accuracy.

(4) Exercising judgment.

(5) Applying his aeronautical knowledge.

(6) Showing that he is the master of the aircraft, with the successful outcome of a procedure or maneuver never seriously in doubt.

(b) If the applicant fails any of the required pilot operations in accordance with the applicable provisions of paragraph (a) of this section, the applicant fails the flight test. The applicant is not eligible for the certificate or rating sought until he passes any pilot operations he has failed.

(c) The examiner or the applicant may discontinue the test at any time when the failure of a required pilot operation makes the applicant ineligible for the certificate or rating sought. If the test is discontinued the applicant is entitled to credit for only those entire pilot operations that he has successfully performed.

§ 61.45 Flight tests: required aircraft and equipment.　☞ F11

(a) General. An applicant for a certificate or rating under this Part must furnish, for each flight test that he is required to take, an appropriate aircraft of United States registry that has a current standard or limited airworthiness certificate. However, the applicant may, at the discretion of the inspector or examiner conducting the test, furnish an aircraft of U.S. registry that has a current airworthiness certificate other than standard or limited, an aircraft of foreign registry that is properly certificated by the country of registry, or a military aircraft in an operational status if its use is allowed by an appropriate military authority.

(b) Required equipment (other than controls). Aircraft furnished for a flight test must have—

(1) The equipment for each pilot operation required for the flight test;

(2) No prescribed operating limitations that prohibit its use in any pilot operation required on the test;

(3) Pilot seats with adequate visibility for each pilot to operate the aircraft safely, except as provided in paragraph (d) of this section; and

(4) Cockpit and outside visibility adequate to evaluate the performance of the applicant, where an additional jump seat is provided for the examiner.

(c) Required controls. An aircraft (other than lighter-than-air) furnished under paragraph (a) of this section for any pilot flight test must have engine power controls and flight controls that are easily reached and operable in a normal manner by both pilots, unless after considering all the factors, the examiner determines that the flight test can be conducted safely without them. However, an aircraft having other controls such as nose-wheel steering, brakes, switches, fuel selectors, and engine air flow controls that are not easily reached and operable in a normal manner by both pilots may be used, if more than one pilot is required under its airworthiness certificate, or if the examiner determines that the flight can be conducted safely.

(d) Simulated instrument flight equipment. An applicant for any flight test involving flight maneuvers solely by reference to instruments must furnish equipment satisfactory to the examiner that excludes the visual reference of the applicant outside of the aircraft.

(e) Aircraft with single controls. At the discretion of the examiner, an aircraft furnished under paragraph (a) of this section for a flight test may, in the cases listed herein, have a single set of controls. In such case, the examiner determines the competence of the applicant by observation from the ground or from another aircraft.

(1) A flight test for addition of a class or type rating, not involving demonstration of instrument skills, to a private or commercial pilot certificate.

(2) A flight test in a single-place gyroplane for—

(i) A private pilot certificate with a rotorcraft category rating and gyroplane class rating, in which case the certificate bears the limitation "rotorcraft single-place gyroplane only;" or

(ii) Addition of a rotorcraft category rating and gyroplane class rating to a pilot certificate, in which case a certificate higher than a private pilot certificate bears the limitation "rotorcraft single-place gyroplane, private pilot privileges, only."

The limitations prescribed by this subparagraph may be removed if the holder of the certificate passes the appropriate flight test in a gyroplane with two pilot stations or otherwise passes the appropriate flight test for a rotorcraft category rating.

§ 61.47 Flight tests: status of FAA inspectors and other authorized flight examiners.

An FAA inspector or other authorized flight examiner conducts the flight test of an applicant for a pilot certificate or rating for the purpose of observing the applicant's ability to perform satisfactorily the procedures and maneuvers on the flight test. The inspector or other examiner is not pilot in command of the aircraft during the flight test unless he acts in that capacity for the flight, or portion of the flight, by prior arrangement with the applicant or other person who would otherwise act as pilot in command of the flight, or portion of the flight. Notwithstanding the type of aircraft used during a flight test, the applicant and the inspector or other examiner are not, with respect to each other (or other occupants authorized by the inspector or other examiner), subject to the requirements or limitations for the carriage of passengers specified in this chapter.

§ 61.49 Retesting after failure.

An applicant for a written or flight test who fails that test may not apply for retesting until after 30 days after the date he failed the test.

However, in the case of his first failure he may apply for retesting before the 30 days have expired upon presenting a written statement from an authorized instructor certifying that he has given flight or ground instruction as appropriate to the applicant and finds him competent to pass the test.

§ 61.51 Pilot logbooks.

(a) The aeronautical training and experience used to meet the requirements for a certificate or rating, or the recent flight experience requirements of this Part must be shown by a reliable record. The logging of other flight time is not required.

(b) *Logbook entries.* Each pilot shall enter the following information for each flight or lesson logged:

(1) *General.* ☞ **F1, 2, 12, 13, 14, 15**

(i) Date.

(ii) Total time of flight.

(iii) Place, or points of departure and arrival.

(iv) Type and identification of aircraft.

(2) *Type of pilot experience or training.*

(i) Pilot in command or solo.

(ii) Second in command.

(iii) Flight instruction received from an authorized flight instructor. ☞ **F17**

(iv) Instrument flight instruction from an authorized flight instructor.

(v) Pilot ground trainer instruction.

(vi) Participating crew (lighter-than-air).

(vii) Other pilot time.

(3) *Conditions of flight.*

(i) Day or night. ☞ **F18, 19**

(ii) Actual instrument.

(iii) Simulated instrument conditions.

(c) *Logging of pilot time.*

(1) *Solo flight time.* A pilot may log as solo flight time only that flight time when he is the sole occupant of the aircraft. However, a student pilot may also log as solo flight time that time during which he acts as the pilot in command of an airship requiring more than one flight crewmember. ☞ **F20, 21, 22, 23**

(2) *Pilot-in-command flight time.* ☞ **F6, 16, 24–34, 124, 125**

(i) A private or commercial pilot may log as pilot in command time only that flight time during which he is the sole manipulator of the controls of an aircraft for which he is rated, or when he is the sole occupant of the aircraft, or when he acts as pilot in command of an aircraft on which more than one pilot is required under the type certification of the aircraft, or the regulations under which the flight is conducted.

(ii) An airline transport pilot may log as pilot in command time all of the flight time during which he acts as pilot in command.

(iii) A certificated flight instructor may log as pilot in command time all flight time during which he acts as a flight instructor. ☞ **F35**

(3) *Second-in-command flight time.* A pilot may log as second in command time all flight time during which he acts as second in command of an aircraft on which more than one pilot is required under the type certification of the aircraft, or the regulations under which the flight is conducted. ☞ **F36**

(4) *Instrument flight time.* A pilot may log as instrument flight time only that time during which he operates the aircraft solely by reference to instruments, under actual or simulated instrument flight conditions. Each entry must include the place and type of each instrument approach completed, and the name of the safety pilot for each simulated instrument flight. An instrument flight instructor may log as instrument time that time during which he acts as instrument flight instructor in actual instrument weather conditions. ☞ **F37, 38**

(5) *Instruction time.* All time logged as flight instruction, instrument flight instruction, pilot ground trainer instruction, or ground instruction time must be certified by the appropriately rated and certificated instructor from whom it was received. ☞ **F31, 34**

(d) *Presentation of logbook.*

(1) A pilot must present his logbook (or other record required by this section) for inspection upon reasonable request by the Administrator, an authorized representative of the National Transportation Safety Board, or any State or local law enforcement officer.

(2) A student pilot must carry his logbook (or other record required by this section) with him on all solo cross-country flights, as evidence of the required instructor clearances and endorsements.

§ 61.53 Operations during medical deficiency.

No person may act as pilot in command, or in any other capacity as a required pilot flight crewmember while he has a known medical deficiency, or increase of a known medical deficiency, that would make him unable to meet the requirements for his current medical certificate. ☞ **F39, 40**

[§ 61.55 Second-in-command qualifications.

[(a) Except as provided in paragraph (d) of this section, no person may serve as second in command of a aircraft type certificated for more than one required pilot flight crewmember unless that person holds—]

(1) At least a current private pilot certificate with appropriate category and class ratings; and

(2) An appropriate instrument rating in the case of flight under IFR.

[(b) Except as provided in paragraph (d) of this section, no person may serve as second in command of an aircraft type certificated for more than one required pilot flight crewmember unless, since the beginning of the 12th calendar month before the month in which the pilot serves, the pilot has, with respect to that type of aircraft—

[(1) Become familiar with all information concerning the aircraft's powerplant, major components and systems, major appliances, performance and limitations, standard and emergency operating procedures, and the contents of the approved aircraft flight manual or approved flight manual material, placards, and markings.]

(2) Except as provided in paragraph (e), performed and logged—

[(i) Three takeoffs and three landings to a full stop in the aircraft as the sole manipulator of the flight controls; and

[(ii) Engine-out procedures and maneuvering with an engine out while executing the duties of a pilot in command. For airplanes, this requirement may be satisfied in a simulator acceptable to the Administrator.]

For the purpose of meeting the requirements of subparagraph (2) of this section, a person may act as second in command of a flight under day VFR or day IFR, if no persons or property, other than as necessary for the operation, are carried.

(c) If a pilot complies with the requirements in paragraph (b) of this section in the calendar month before, or the calendar month after, the month in which compliance with those requirements is due, he is considered to have complied with them in the month they are due.

(d) This section does not apply to a pilot who—

[(1) Meets the pilot in command proficiency check requirements of Parts 121, 125, 127, or 135 of this chapter;

[(2) Is designated as the second in command of an aircraft operated under the provisions of Parts 121, 125, 127, or 135 of this chapter; or

[(3) Is designated as the second in command of an aircraft for the purpose of receiving flight training required by this section and no passengers or cargo are carried on that aircraft.]

(e) The holder of a commercial or airline transport pilot certificate with appropriate category and class ratings need not meet the requirements of paragraph (b) (2) of this section for the conduct of ferry flights, aircraft flight tests, or airborne equipment evaluation, if no persons or property other than as necessary for the operation are carried.

§ 61.57 Recent flight experience: pilot in command.

[(a) *Flight review.* No person may act as pilot in command of an aircraft unless, within the preceding 24 calendar months, that person has—] ☞ **F30, 41–44**

(1) Accomplished a flight review given to him, in an aircraft for which he is rated, by an appropriately certificated instructor or other person designated by the Administrator; and

(2) Had his log book endorsed by the person who gave him the reveiw certifying that he has satisfactorily accomplished the flight review.

However, a person who has, within the preceding 24 [calendar] months, satisfactorily completed a pilot proficiency check conducted by the FAA, an approved pilot check airman or a U.S.

Armed Force for a pilot certificate, rating or operating privilege, need not accomplish the flight review required by this section.

(b) *Meaning of flight review.* As used in this section, a flight review consists of—

(1) A review of the current general operating and flight rules of Part 91 of this chapter; and

(2) A review of those maneuvers and procedures which in the discretion of the person giving the review are necessary for the pilot to demonstrate that he can safely exercise the privileges of his pilot certificate.

(c) *General experience.* No person may act as pilot in command of an aircraft carrying passengers, nor an aircraft certificated for more than one required pilot flight crewmember, unless within the preceding 90 days, he has made three takeoffs and three landings as the sole manipulator of the flight controls in an aircraft of the same category and class and, if a type rating is required, of the same type. If the aircraft is a tailwheel airplane, the landings must have been made to a full stop in a tailwheel airplane. For the purpose of meeting the requirements of the paragraph a person may act as pilot-in-command of a flight under day VFR or day IFR if no persons or property other than as necessary for his compliance thereunder, are carried. This paragraph does not apply to operations requiring an airline transport pilot certificate, or to operations conducted under Part 135 of this chapter.

(d) *Night experience.* No person may act as pilot in command of an aircraft carrying passengers during the period beginning 1 hour after sunset and ending 1 hour before sunrise (as published in the American Air Almanac) unless, within the preceding 90 days, he has made at least three takeoffs and three landings to a full stop during that period in the category and class of aircraft to be used. This paragraph does not apply to operations requiring an airline transport pilot certificate.

(e) *Instrument.*   ☞ **F120**

(1) *Recent IFR experience.* No pilot may act as pilot in command under IFR, nor in weather conditions less than the minimums prescribed for VFR, unless he has, within the past 6 [calender] months—

(i) In the case of an aircraft other than a glider, logged at least 6 hours of instrument time under actual or simulated IFR conditions, at least 3 of which were in flight in the category of aircraft involved, including at least 6 instrument approaches, or passed an instrument competency check in the category of aircraft involved.

    ☞ **F37, 38, 47, 48, 49, 50, 51**

(ii) In the case of a glider, logged at least 3 hours of instrument time, at least half of which were in a glider or an airplane. If a passenger is carried in the glider, at least 3 hours of instrument flight time must have been in gliders.

(2) *Instrument competency check.* A pilot who does not meet the recent instrument experience requirements of subparagraph (1) of this paragraph during the prescribed time or 6 [calender] months thereafter may not serve as pilot in command under IFR, nor in weather conditions less than the minimums prescribed for VFR, until he passes an instrument competency check in the category of aircraft involved, given by an FAA inspector, a member of an armed force in the United States authorized to conduct flight tests, an FAA-approved check pilot, or a certificated instrument flight inspector. The Administrator may authorize the conduct of part or all of this check in a pilot ground trainer equipped for instruments or an aircraft simulator.

§ 61.58 Pilot-in-command proficiency check: operation of aircraft requiring more than one required pilot.

(a) Except as provided in paragraph (e) of this section, after November 1, 1974, no person may act as pilot in command of an aircraft that is type certificated for more than one required pilot crewmember unless he has satisfactorily completed the proficiency checks or flight checks prescribed in paragraphs (b) and (c) of this section.

(b) Since the beginning of the 12th calendar month before the month in which a person acts as pilot in command of an aircraft that is type certificated for more than one required pilot crewmember he must have completed one of the following:

(1) For an airplane—a proficiency or flight check in either an airplane that is type certificated for more than one required pilot crewmember, or in an approved simulator or other training device, given to him by an FAA inspector or designated pilot examiner and consisting of those maneuvers and procedures set forth in Appendix F of Part 121 of this chapter which may be performed in a simulator or training device.

(2) For other aircraft—a proficiency or flight check in an aircraft that is type certificated for more than one required pilot crewmember given to him by an FAA inspector or designated pilot examiner which includes those maneuvers and procedures required for the original issuance of a type rating for the aircraft used in the check.

(3) A pilot in command proficiency check given to him in accordance with the provisions for that check under Parts 121, 123 or 135 of this chapter. However, in the case of a person acting as pilot in command of a helicopter he may complete a proficiency check given to him in accordance with Part 127 of this chapter.

(4) A flight test required for an aircraft type rating.

(5) An initial or periodic flight check for the purpose of the issuance of a pilot examiner or check airman designation.

(6) A military proficiency check required for pilot in command and instrument privileges in an aircraft which the military requires to be operated by more than one pilot.

(c) Except as provided in paragraph (d) of this section, since the beginning of the 24th calendar month before the month in which a person acts as pilot in command of an aircraft that is type certificated for more than one required pilot crewmember he must have completed one of the following proficiency or flight checks in the particular type aircraft in which he is to serve as pilot in command:

(1) A proficiency check or flight check given to him by an FAA inspector or a designated pilot examiner which includes the maneuvers, procedures, and standards required for the original issuance of a type rating for the aircraft used in the check.

(2) A pilot-in-command proficiency check given to him in accordance with the provisions for that check under Parts 121, 123 or 135 of this chapter. However, in the case of a person acting as pilot in command of a helicopter he may complete a proficiency check given to him in accordance with Part 127 of this chapter.

(3) A flight test required for an aircraft type rating.

(4) An initial or periodic flight check for the purpose of the issuance of a pilot examiner or check airman designation.

(5) A military proficiency check required for pilot in command and instrument privileges in an aircraft which the military requires to be operated by more than one pilot.

(d) For airplanes, the maneuvers and procedures required for the checks and test prescribed in paragraphs (c)(1), (2), (4), and (5) of this section, and paragraph (c)(3) of this section in the case of type ratings obtained in conjunction with a Part 121 of this chapter training program may be performed in a simulator or training device if—

(1) The maneuver or procedure can be performed in a simulator or training device as set forth in Appendix F to Part 121 of this chapter; and

(2) The simulator or training device is one that is approved for the particular maeuver or procedure.

(e) This section does not apply to persons conducting operations subject to Parts 121, 123, 127, 133, 135, and 137 of this chapter.

(f) For the purpose of meeting the proficiency check requirements of paragraphs (b) and (c) of this section, a person may act as pilot in command of a flight under day VFR or day IFR if no persons or property, other than as necessary for his compliance thereunder, are carried.

(g) If a pilot takes the proficiency check required by paragraph (a) of this section in the calendar month before, or the calendar month after, the month in which it is due, he is considered to have taken it in the month it is due.

§ 61.59 Falsification, reproduction or alteration of applications, certificates, logbooks, reports, or records.

(a) No person may make or cause to be made—

(1) Any fraudulent or intentionally false statement on any application for a certificate, rating, or duplicate thereof, issued under this Part;

(2) Any fraudulent or intentionally false entry in any logbook, record, or report that is required to be kept, made, or used, to show compliance with any requirement for the issuance, or exercise of the privileges, or any certificate or rating under this Part; or

☞F15

(3) Any reproduction, for fraudulent purpose, of any certificate or rating under this Part; or

(4) Any alteration of any certificate or rating under this Part.

(b) The commission by any person of an act prohibited under paragraph (a) of this section is a basis for suspending or revoking any airman or ground instructor certificate or rating held by that person.

§ 61.60 Change of address.   ☞ F51,52

The holder of a pilot or flight instructor certificate who has made a change in his permanent mailing address may not after 30 days from the date he moved, exercise the privileges of his certificate unless he has notified in writing the Department of Transportation, Federal Aviation Administration, Airman Certification Branch, Box 25082, Oklahoma City, Oklahoma 73125, of his new address.

Subpart B—Aircraft Ratings and Special Certificates

§ 61.61 Applicability.

This subpart prescribes the requirements for the issuance of additional aircraft ratings after a pilot or instructor certificate is issued, and the requirements and limitations for special pilot certificates and ratings issued by the Administrator.

§ 61.63 Additional aircraft ratings (other than airline transport pilot).

(a) *General.* To be eligible for an aircraft rating after his certificate is issued to him an applicant must meet the requirements of paragraphs (b) through (d) of this section, as appropriate to the rating sought.

(b) *Category rating.* An applicant for a category rating to be added on his pilot certificate must meet the requirements of this Part for the issue of the pilot certificate appropriate to the privileges for which the category rating is sought. However, the holder of a category rating for powered aircraft is not required to take a written test for the addition of a category rating on his pilot certificate.

(c) *Class rating.* An applicant for an aircraft class rating to be added on his pilot certificate must—

(1) Present a logbook record certified by an authorized flight instructor showing that the applicant has received flight instruction in the class of aircraft for which a rating is sought and has been found competent in the pilot operations appropriate to the pilot certificate to which his category rating applies; and

(2) Pass a flight test appropriate to his pilot certificate and applicable to the aircraft category and class rating sought.

☞ F10, 53, 54

A person who holds a lighter-than-air category rating with a free balloon class rating, who seeks an airship class rating, must meet the requirements of paragraph (b) of this section as though seeking a lighter-than-air category rating.

(d) *Type rating.* An applicant for a type rating to be added on his pilot certificate must meet the following requirements:

(1) He must hold, or concurrently obtain, an instrument rating appropriate to the aircraft for which a type rating is sought.

(2) He must pass a flight test showing competence in pilot operations appropriate to the pilot certificate he holds and to the type rating sought.

(3) He must pass a flight test showing competence in pilot operations under instrument flight rules in an aircraft of the type for which the type rating is sought or, in the case of a single pilot station airplane, meet the requirements of subdivision (i) or (ii) of this subparagraph, whichever is applicable.

(i) The applicant must have met the requirements of this subparagraph in a multiengine airplane for which the type rating is required.

(ii) If he does not meet the requirements of subdivision (i) of this subparagraph and he seeks a type rating for a single-engine airplane, he must meet the requirements of this subparagraph in either a single or multiengine airplane, and have the recent instrument experience set forth in § 61.57 (e) of this Part, when he applies for the flight test under subparagraph (2) of this paragraph.

(4) An applicant who does not meet the requirements of subparagraphs (1) and (3) of this paragraph may obtain a type rating limited to "VFR only". Upon meeting these instrument requirements or the requirements of § 61.73 (e) (2), the "VFR only" limitation may be removed for the particular type of aircraft in which competence is shown.

(5) When an instrument rating is issued to the holder of one or more type ratings, the type ratings on the amended certificate bear the limitation described in subparagraph (4) for each airplane type rating for which he has not shown his instrument competency under this paragraph.

§ 61.65 Instrument rating requirements.

(a) *General.* To be eligible for an instrument rating (airplane) or an instrument rating (helicopter), an applicant must—

[(1) Hold at least a current private pilot certificate with an aircraft rating appropriate to the instrument rating sought;]

(2) Be able to read, speak, and understand the English language; and

(3) Comply with the applicable requirements of this section.

(b) *Ground instruction.* An applicant for the written test for an instrument rating must have received ground instruction, or have logged home study in at least the following areas of aeronautical knowledge appropriate to the rating sought:

(1) The regulations of this chapter that apply to flight under IFR conditions, the Airman's Information Manual, and the IFR air traffic system and procedures;

(2) Dead reckoning appropriate to IFR navigation, IFR navigation by radio aids using the VOR, ADF, and ILS systems, and the use of IFR charts and instrument approach plates;

(3) The procurement and use of aviation weather reports and forecasts, and the elements of forecasting weather trends on the basis of that information and personal observation of weather conditions; and

(4) The safe and efficient operation of airplanes or helicopters, as appropriate, under instrument weather conditions.

(c) *Flight instruction and skill—airplanes.* An applicant for the flight test for an instrument rating (airplane) must present a logbook record certified by an authorized flight instructor showing that he has received instrument flight instruction in an airplane in the following pilot operations, and has been found competent in each of them:

(1) Control and accurate maneuvering of an airplane solely by reference to instruments.

(2) IFR navigation by the use of the VOR and ADF systems, including compliance with air traffic control instructions and procedures.

(3) Instrument approaches to published minimums using the VOR, ADF, and ILS system (instruction in the use of the ADF and ILS may be received in an instrument ground trainer and instruction in the use of the ILS glide slope may be received in an airborne ILS simulator).

(4) Cross-country flying in simulated or actual IFR conditions, on Federal airways or as routed by ATC, including one such trip of at least 250 nautical miles, including VOR, ADF, and ILS approaches at different airports.

(5) Simulated emergencies, including the recovery from unusual attitudes, equipment or instrument malfunctions, loss of communications, and engine-out emergencies if a multiengine airplane is used, and missed approach procedure.

(d) *Instrument instruction and skill—(helicopter).* An applicant for the flight test for an instrument rating (helicopter) must present a logbook record certified to by an authorized flight instructor showing that he has received instrument flight instruction in a helicopter in the following pilot operations, and has been found competent in each of them:

(1) The control and accurate maneuvering of a helicopter solely by reference to instruments.

(2) IFR navigation by the use of the VOR and ADF systems, including compliance with air traffic control instructions and procedures.

(3) Instrument approaches to published minimums using the VOR, ADF, and ILS systems (instruction in the use of the ADF and ILS may be received in an instrument ground trainer, and instruction in the use of the ILS glide slope may be received in an airborne ILS simulator).

(4) Cross-country flying under simulated or actual IFR conditions, on Federal airways or as routed by ATC, including one flight of at least 100 nautical miles, including VOR, ADF, and ILS approaches at different airports.

(5) Simulated IFR emergencies, including equipment malfunctions, missed approach procedures, and deviations to unplanned alternates.

(e) *Flight experience.* An applicant for an instrument rating must have at least the following flight time as a pilot: ☞ **F55**

(1) A total of 125 hours of pilot flight time, of which 50 hours are as pilot in command in cross-country flight in a powered aircraft with other than a student pilot certificate. Each cross-country flight must have a landing at a point more than 50 nautical miles from the original departure point. ☞ **F123**

(2) 40 hours of simulated or actual instrument time, of which not more than 20 hours may be instrument instruction by an authorized instructor in an instrument ground trainer acceptable to the Administrator.

(3) 15 hours of instrument flight instruction by an authorized flight instructor, including at least 5 hours in an airplane or a helicopter, as appropriate.

(f) *Written test.* An applicant for an instrument rating must pass a written test appropriate to the instrument rating sought on the subjects in which ground instruction is required by paragraph (b), of this section. ☞ **F56**

(g) *Practical test.* An applicant for an instrument rating must pass a flight test in an airplane or a helicopter, as appropriate. The test must include instrument flight procedures selected by the inspector or examiner conducting the test to determine the applicant's ability to perform competently the IFR operations on which instruction is required by paragraph (c) or (d) of this section.

§ 61.67 Category II pilot authorization requirements.

(a) *General.* An applicant for a Category II pilot authorization must hold—

(1) A pilot certificate with an instrument rating or an [aircraft] transport pilot certificate; and

(2) A type rating for the [aircraft] type if the authorization is requested for a large [aircraft] or a small turbojet [aircraft].

(b) *Experience requirements.* Except for

the holder of an airline transport pilot certificate, an applicant for a Category II authorization must have at least—

(1) 50 hours of night flight time under VFR conditions as pilot in command;

(2) 75 hours of instrument time under actual or simulated conditions that may include 25 hours in a synthetic trainer; and

(3) 250 hours of cross-country flight time as pilot in command.

Night flight and instrument flight time used to meet the requirements of subparagraphs (1) and (2) of this paragraph may also be used to meet the requirements of subparagraph (3) of this paragraph.

(c) *Practical test required.*

(1) The practical test must be passed by—

(i) An applicant for issue or renewal of an authorization; and

(ii) An applicant for the addition of another type [aircraft] to his authorization.

(2) To be eligible for the practical test an applicant must meet the requirements of paragraph (a) of this section and, if he has not passed a practical test since the beginning of the twelfth month before the test, he must meet the following recent experience requirements:

(i) The requirements of § 61.57(e).

(ii) At least six ILS approaches since the beginning of the sixth month before the test. These approaches must be under actual or simulated instrument flight conditions down to the minimum landing altitude for the ILS approach in the type airplane in which the flight test is to be conducted. However, the approaches need not be conducted down to the decision heights authorized for Category II operations. At least three of these approaches must have been conducted manually, without the use of an approach coupler.

The flight time acquired in meeting the requirements of subdivision (ii) of this subparagraph may be used to meet the requirements of subdivision (i) of this subparagraph.

(d) *Practical test procedures.* The practical test consists of two phases:

(1) *Phase I—oral operational test.* The applicant must demonstrate his knowledge of the following:

(i) Required landing distance.

(ii) Recognition of the decision height.

(iii) Missed approach procedures and techniques utilizing computed or fixed attitude guidance displays.

(iv) RVR, its use and limitations.

(v) Use of visual clues, their availability or limitations, and altitude at which they are normally discernible at reduced RVR readings.

(vi) Procedures and techniques related to transition from nonvisual to visual flight during a final approach under reduced RVR.

(vii) Effects of vertical and horizontal wind shear.

(viii) Characteristics and limitations of the ILS and runway lighting system.

(ix) Characteristics and limitations of the flight director system, auto approach coupler (including split axis type if equipped), auto throttle system (if equipped), and other required Category II equipment.

(x) Assigned duties of the second in command during Category II approaches.

(xi) Instrument and equipment failure warning systems.

(2) *Phase II—flight test.* The flight test must be taken in an [aircraft] that meets the requirements of Part 91 of this chapter for Category II operations. The test consists of at least two ILS approaches to 100 feet including at least one landing and one missed approach. All approaches must be made with the approved flight control guidance system. However, if an approved automatic approach coupler is installed, at least one approach must be made manually. In the case of a multiengine [aircraft] that has performance capability to execute a missed approach with an engine out, the missed approach must be executed with one engine set in idle or zero thrust position before reaching the middle marker. The required flight maneuvers must be performed solely by reference to instruments and in coordination with a second in command who holds a class rating and, in the case of a large [aircraft] or a small turbojet [aircraft], a type rating for that [aircraft].

§ 61.69 Glider towing: experience and instruction requirements.

No person may act as pilot in command of an aircraft towing a glider unless he meets the following requirements:

(a) He holds a current pilot certificate (other than a student pilot certificate) issued under this Part.

(b) He has an endorsement in his pilot logbook from a person authorized to give flight instruction in gliders, certifying that he has received ground and flight instruction in gliders and is familiar with the techniques and procedures essential to the safe towing of gliders, including airspeed limitations, emergency procedures, signals used, and maximum angles of bank.

(c) He has made and entered in his pilot logbook—

(1) At least three flights as sole manipulator of the controls of an aircraft towing a glider while accompanied by a pilot who has met the requirements of this section and made and logged at least 10 flights as pilot-in-command of an aircraft towing a glider; or

(2) At least three flights as sole manipulator of the controls of an aircraft simulating glider towing flight procedures (while accompanied by a pilot who meets the requirements of this section), and at least three flights as pilot or observer in a glider being towed by an aircraft.

However, any person who, before May 17, 1967, made, and entered in his pilot logbook, 10 or more flights as pilot in command of an aircraft towing a glider in accordance with a certificate of waiver need not comply with subparagraphs (1) and (2) of this paragraph.

(d) If he holds only a private pilot certificate he must have had, and entered in his pilot logbook at least—

(1) 100 hours of pilot flight time in powered aircraft; or

(2) 200 total hours of pilot flight time in powered or other aircraft.

(e) Within the preceding 12 months he has—

(1) Made at least 3 actual or simulated glider tows while accompanied by a qualified pilot who meets the requirements of this section; or

(2) Made at least 3 flights as pilot in command of a glider towed by an aircraft.

§ 61.71 Graduates of certificated flying schools: special rules.

(a) A graduate of a flying school that is certificated under Part 141 of this chapter is considered to meet the applicable aeronautical experience requirements of this Part if he presents an appropriate graduation certificate within 60 days after the date he is graduated. However, if he applies for a flight test for an instrument rating he must hold a commercial pilot certificate, or hold a private pilot certificate and meet the requirements of § 61.65(e) (1) and § 61.123 (except paragraphs (a) and (e) thereof). In addition, if he applies for a flight instructor certificate he must hold a commercial pilot certificate.   ☞ F9

(b) An applicant for a certificate or rating under this Part is considered to meet the aeronautical knowledge and skill requirements, or both, applicable to that certificate or rating, if he applies within 90 days after graduation from an appropriate course given by a flying school that is certificated under Part 141 of this chapter and is authorized to test applicants on aeronautical knowledge or skill, or both.

However, until January 1, 1977, a graduate of a flying school certificated and operated under the provisions of § 141.29 of this chapter, is considered to meet the aeronautical experience requirements of this Part, and may be tested under the requirements of Part 61 that were in effect prior to November 1, 1973.

§ 61.73 Military pilots or former military pilots: special rules.   ☞ F57

(a) *General.* A rated military pilot or former rated military pilot who applies for a private or commercial pilot certificate, or an aircraft or instrument rating, is entitled to that certificate with appropriate ratings or to the addition of a rating on the pilot certificate he holds, if he meets the applicable requirements of this section. This section does not apply to a military pilot or former military pilot who has been removed from flying status for lack of proficiency or because of disciplinary action involving aircraft operation.

(b) *Military pilots on active flying status within 12 months.* A rated military pilot or former rated military pilot who has been active

flying status within the 12 months before he applies must pass a written test on the parts of this chapter relating to pilot privileges and limitations, air traffic and general operating rules, and accident reporting rules. In addition, he must present documents showing that he meets the requirements of paragraph (d) of this section for at least one aircraft rating, and that he is, or was at any time since the beginning of the twelfth month before the month in which he applies—

(1) A rated military pilot on active flying status in an armed force of the United States; or

(2) A rated military pilot of an armed force of a foreign contracting State to the Convention on International Civil Aviation, assigned to pilot duties (other than flight training) with an armed force of the United States who holds, at the time he applies, a current civil pilot license issued by that foreign State authorizing at least the privileges of the pilot certificate he seeks.

(c) *Military pilots not an active flying status within previous 12 months.* A rated military pilot or former military pilot who has not been on active flying status within the 12 months before he applies must pass the appropriate written and flight tests prescribed in this Part for the certificate or rating he seeks. In addition, he must show that he holds an FAA medical certificate appropriate to the pilot certificate he seeks and present documents showing that he was, before the beginning of the twelfth month before the month in which he applies, a rated military pilot as prescribed by either subparagraph (1) or (2) of paragraph (b) of this section.

(d) *Aircraft ratings: other than airplane category and type.* An applicant for a category, class, or type rating (other than airplane category and type rating) to be added on the pilot certificate he holds, or for which he has applied, is issued that rating if he presents documentary evidence showing one of the following:

(1) That he has passed an official United States military checkout as pilot in command of aircraft of the category, class, or type for which he seeks a rating since the beginning of the twelfth month before the month in which he applies.

(2) That he has had at least 10 hours of flight time serving as pilot in command of aircraft of the category, class, or type for which he seeks a rating since the beginning of the twelfth month before the month in which he applies and previously has had an official United States military checkout as pilot in command of that aircraft.

(3) That he has met the requirements of subparagraph (1) or (2) of paragraph (b) of this section, has had an official United States military checkout in the category of aircraft for which he seeks a rating, and that he passes an FAA flight test appropriate to that category and the class or type rating

he seeks. To be eligible for that flight test, he must have a written statement from an authorized flight instructor, made not more than 60 days before he applies for the flight test, certifying that he is competent to pass the test.

A type rating is issued only for aircraft types that the Administrator has certificated for civil operations. Any rating placed on an airline transport pilot certificate is limited to commercial pilot privileges.

(e) *Airplane category and type ratings.*

(1) An applicant for a commercial pilot certificate with an airplane category rating, or an applicant for the addition of an airplane category rating on his commercial pilot certificate, must hold an airplane instrument rating, or his certificate is endorsed with the following limitation: "not valid for the carriage of passengers or property for hire in airplanes on cross-country flights of more than 50 nautical miles, or at night."

(2) An applicant for a private or commercial pilot certificate with an airplane type rating, or for the addition of an airplane type rating on his private or commercial pilot certificate who holds an instrument rating (airplane), must present documentary evidence showing that he has demonstrated instrument competency in the type of airplane for which the type rating is sought, or his certificate is endorsed with the following limitation: "VFR only."

(f) *Instrument rating.* An applicant for an airplane instrument rating or a helicopter instrument rating to be added on the pilot certificate he holds, or for which he has applied, is entitled to that rating if he has, within the 12 months preceding the month in which he applies, satisfactorily accomplished an instrument flight check of a U.S. Armed Force in an aircraft of the category for which he seeks the instrument rating and is authorized to conduct IFR flights on Federal airways. A helicopter instrument rating added on an airline transport pilot certificate is limited to commercial pilot privileges.

(g) *Evidentiary documents.* The following documents are satisfactory evidence for the purposes indicated:

(1) To show that the applicant is a member of the armed forces, an official identification card issued to the applicant by an armed force may be used.

(2) To show the applicant's discharge or release from an armed force, or his former membership therein, an original or a copy of a certificate of discharge or release may be used.

(3) To show current or previous status as a rated military pilot on flying status with a U.S. Armed Force, one of the following may be used:

(i) An official U.S. Armed Force order to flight duty as a military pilot.

(ii) An official U.S. Armed Force form or logbook showing military pilot status.

(iii) An official order showing that the applicant graduated from a United States military pilot school and is rated as a military pilot.

(4) To show flight time in military aircraft as a member of a U.S. Armed Force, an appropriate U.S. Armed Force form or summary of it, or a certified United States military logbook may be used.

(5) To show pilot-in-command status, an official U.S. Armed Force record of a military checkout as pilot in command, may be used.

(6) To show instrument pilot qualification, a current instrument card issued by a U.S. Armed Force, or an official record of the satisfactory completion of an instrument flight check within the 12 months preceding the month of the application may be used. However, a Tactical (Pink) instrument card issued by the U.S. Army is not acceptable.

§ 61.75 Pilot certificate issued on basis of a foreign pilot license. ☞ F58, 59

(a) *Purpose.* The holder of a current private, commercial, senior commercial, or airline transport pilot license issued by a foreign contracting State to the Convention on International Civil Aviation may apply for a pilot certificate under this section authorizing him to act as a pilot of a civil aircraft of U.S. registry.

(b) *Certificate issued.* A pilot certificate is issued to an applicant under this section, specifying the number and State of issuance of the foreign pilot license on which it is based. An applicant who holds a foreign private pilot license is issued a private pilot certificate, and an applicant who holds a foreign commercial, senior commercial, or airline transport pilot license is issued a commercial pilot certificate, if—

(1) He meets the requirements of this section;

(2) His foreign pilot license does not contain an endorsement that he has not met all of the standards of ICAO for that license; and

(3) He does not hold a U.S. pilot certificate of private pilot grade or higher.

(c) *Limitation on licenses used as basis for U.S. certificate.* Only one foreign pilot license may be used as a basis for issuing a pilot certificate under this section.

(d) *Aircraft ratings issued.* Aircraft ratings listed on the applicant's foreign pilot license, in addition to any issued after testing under the provisions of this Part, are placed on the applicant's pilot certificate.

(e) *Instrument rating issued.* An instrument rating is issued to an applicant if—

(1) His foreign pilot license authorizes instrument privileges; and

(2) Within 24 months preceding the month in which he makes application for a certificate, he passed a test on the instrument flight rules

in Subpart B of Part 91 of this chapter, including the related procedures for the operation of the aircraft under instrument flight rules.

(f) *Medical standards and certification.* An applicant must submit evidence that he currently meets the medical standards for the foreign pilot license on which the application for a certificate under this section is based. A current medical certificate issued under Part 67 of this chapter is accepted as evidence that the applicant meets those standards. However, a medical certificate issued under Part 67 of this chapter is not evidence that the applicant meets those standards outside the United States, unless the State that issued the applicant's foreign pilot license also accepts that medical certificate as evidence of meeting the medical standards for his foreign pilot license.

(g) *Limitations placed on pilot certificate.*

(1) If the applicant cannot read, speak, and understand the English language, the Administrator places any limitation on the certificate that he considers necessary for safety.

(2) A certificate issued under this section is not valid for agricultural aircraft operations, or the operation of an aircraft in which persons or property are carried for compensation or hire. This limitation is also placed on the certificate.

(h) *Operating privileges and limitations.* The holder of a pilot certificate issued under this section may act as a pilot of a civil aircraft of U.S. registry in accordance with the pilot privileges authorized by the foreign pilot license on which that certificate is based, subject to the limitations of this Part and any additional limitations placed on his certificate by the Administrator. He is subject to these limitations while he is acting as a pilot of the aircraft within or outside the United States. However, he may not act as pilot in command, or in any other capacity as a required pilot flight crewmember, of a civil aircraft of U.S. registry that is carrying persons or property for compensation or hire.

(i) *Flight instructor certificate.* A pilot certificate issued under this section does not satisfy any of the requirements of this Part for the issuance of a flight instructor certificate.

[§ 61.77 Special purpose pilot certificate: operation of U.S.-registered civil airplanes leased by a person not a U.S. citizen.

[(a) *General.* The holder of a current foreign pilot certificate or license issued by a foreign contracting State to the Convention on International Civil Aviation, who meets the requirements of this section, may hold a special purpose pilot certificate authorizing the holder to perform pilot duties on a civil airplane of U.S. registry, leased to a person not a citizen of the United States, carrying persons of property for compensation or hire. Special purpose pilot certificates are issued under this section only for airplane types that can have a maximum passenger seating configuration, excluding any flight crewmember seat, of more than 30 seats or a maximum payload capacity (as defined in

§ 135.2(e) of this chapter) of more than 7,500 pounds.

[(b) *Eligibility.* To be eligible for the issuance or renewal of a certificate under this section, an applicant or a representative of the applicant must present the following to the Administrator:

[(1) A current foreign pilot certificate or license, issued by the aeronautical authority of a foreign contracting State to the Convention on International Civil Aviation, or a facsimile acceptable to the Administrator. The certificate or license must authorize the applicant to perform the pilot duties to be authorized by a certificate issued under this section on the same airplane type as the leased airplane.

[(2) A current certification by the lessee of the airplane—

[(i) Stating that the applicant is employed by the lessee;

[(ii) Specifying the airplane type on which the applicant will perform pilot duties; and

[(iii) Stating that the applicant has received ground and flight instruction which qualifies the applicant to perform the duties to be assigned on the airplane.

[(3) Documentation showing that the applicant has not reached the age of 60 and that the applicant currently meets the medical standards for the foreign pilot certificate or license required by paragraph (b)(1) of this section, except that a U.S. medical certificate issued under Part 67 of this chapter is not evidence that the applicant meets those standards unless the State which issued the applicant's foreign pilot certificate or license accepts a U.S. medical certificate as evidence of medical fitness for a pilot certificate or license.

[(c) *Privileges.* The holder of a special purpose pilot certificate issued under this section may exercise the same privileges as those shown on the certificate or license specified in paragraph (b)(1), subject to the limitations specified in this section. The certificate holder is not subject to the requirements of §§ 61.55, 61.57, and 61.58 of this Part.

[(d) *Limitations.* Each certificate issued under this section is subject to the following limitations:

[(1) It is valid only—

[(i) For flights between foreign countries or for flights in foreign air commerce;

[(ii) While it and the foreign pilot certificate or license required by paragraph (b)(1) of this section are in the certificate holder's personal possession and are current;

[(iii) While the certificate holder is employed by the person to whom the airplane described in the certification required by paragraph (b)(2) of this section is leased;

[(iv) While the certificate holder is performing pilot duties on the U.S.-registered civil airplane described in the certification required by paragraph (b)(2) of this section;

[(v) While the medical documentation required by paragraph (b)(3) of this section is in the certificate holder's personal possession and is currently valid; and

[(vi) While the certificate holder is under 60 years of age.

[(2) Each certificate issued under this section contains the following:

[(i) The name of the person to whom the U.S.-registered civil aircraft is leased.

[(ii) The type of aircraft.

[(iii) The limitation: "Issued under, and subject to, § 61.77 of the Federal Aviation Regulations."

[(iv) The limitation: "Subject to the privileges and limitations shown on the holder's foreign pilot certificate or license."

[(3) Any additional limitations placed on the certificate which the Administrator considers necessary.

[(e) *Termination.* Each special purpose pilot certificate issued under this section terminates—

[(1) When the lease agreement for the airplane described in the certification required by paragraph (b) (2) of this section terminates;

[(2) When the foreign pilot certificate or license, or the medical documentation, required by paragraph (b) of this section is suspended, revoked, or no longer valid;

[(3) When the certificate holder reaches the age of 60; or

[(4) After 24 months after the month in which the special purpose pilot certificate was issued.

[(f) *Surrender of certificate.* The certificate holder shall surrender the special purpose pilot certificate to the Administrator within 7 days after the date it terminates.

[(g) *Renewal.* The certificate holder may have the certificate renewed by complying with the requirements of paragraph (b) of this section at the time of application for renewal.]

Subpart C—Student Pilots

§ 61.81 Applicability.

This subpart prescribes the requirements for the issuance of student pilot certificates, the conditions under which those certificates are necessary, and the general operating rules for the holders of those certificates.

§ 61.83 Eligibility requirements: general.

To be eligible for a student pilot certificate, a person must—

(a) Be at least 16 years of age, or at least 14 years of age for a student pilot certificate limited to the operation of a glider or free balloon;

(b) Be able to read, speak, and understand English language, or have such operating limitations placed on his pilot certificate as are necessary for the safe operation of aircraft, to be removed when he shows that he can read,

speak, and understand the English language; and

(c) Hold at least a current third-class medical certificate issued under Part 67 of this chapter, or, in the case of glider or free balloon operations, certify that he has no known medical defect that makes him unable to pilot a glider or a free balloon.

§ 61.85 Application.

An application for a student pilot certificate is made on a form and in a manner provided by the Administrator and is submitted to—

(a) A designated aviation medical examiner when applying for an FAA medical certificate in the United States; or ☞ **F60**

(b) An FAA operations inspector or designated pilot examiner, accompanied by a current FAA medical certificate, or in the case of an application for a glider or free balloon pilot certificate it may be accompanied by a certification by applicant that he has no known medical defect that makes him unable to pilot a glider or free balloon.

§ 61.87 Requirements for solo flight.

(a) *General.* A student pilot may not operate an aircraft in solo flight until he has complied with the requirements of this section. As used in this subpart the term solo flight means that flight time during which a student pilot is the sole occupant of the aircraft, or that flight time during which he acts as pilot in command of an airship requiring more than one flight crewmember.

(b) *Aeronautical knowledge.* He must have demonstrated to an authorized instructor that he is familiar with the flight rules of Part 91 of this chapter which are pertinent to student solo flights.

(c) *Flight proficiency training.* He must have received ground and flight instruction in at least the following procedures and operations:

(1) *In airplanes:*

(i) Flight preparation procedures, including preflight inspection and powerplant operation;

(ii) Ground maneuvering and runups;

(iii) Straight and level flight, climbs, turns, and descents;

(iv) Flight at minimum controllable airspeeds, and stall recognition and recovery;

(v) Normal takeoffs and landings;

(vi) Airport traffic patterns, including collision avoidance precautions and wake turbulence; and

(vii) Emergencies, including elementary emergency landings.

Instruction must be given by a flight instructor who is authorized to give instruction in airplanes.

[(2) *In rotorcraft other than single-place gyroplane:*]

(i) Flight preparation procedures, including preflight inspections and powerplant operation;

[(ii) Ground maneuvering and runups;]

(iii) Hovering turns and air taxi (helicopter only);

(iv) Straight and level flight, turns, climbs, and descents;

[(v) Rapid decelerations (helicopters only);]

[(vi)] Maneuvering by ground references, airport traffic patterns including collision avoidance precautions;

[(vii)] Normal takeoffs and landings; and

[(viii)] Simulated emergency procedures, including autorotational descents with a power recovery or running landing in gyroplanes, a power recovery to a hover in single-engine helicopters, or approaches to a hover or landing with one engine inoperative in multiengine helicopters.]

Instruction must be given by a flight instructor who is authorized to give instruction in helicopters or gyroplanes, as appropriate.

[(3) *In single-place gyroplanes:*

[(i) Flight preparation procedures, including preflight inspection and powerplant operation;

[(ii) Ground maneuvering and runups;

[(iii) Straight and level flight, turns, climbs, and decents;

[(iv) Navigation by ground references, airport traffic patterns, and collision avoidance procedures;

[(v) Normal takeoffs and landings;

[(vi) Simulated emergency procedures, including autorotational descents with a power recovery or a running landing; and

[(vii) At least three successful flights in a gyroplane under the observation of a qualified instructor. Items in paragraphs (c)(3)(iii) and (iv) of this section may be accomplished in a dual-control helicopter or gyroplane. Instruction must be given by a flight instructor who is authorized to give instruction in helicopters or gyroplanes, as appropriate.]

Instruction must be given by a flight instructor who is authorized to give instruction in airplanes or rotorcraft.

(4) *In gliders:*

(i) Flight preparation procedures, including preflight inspections, towline rigging, signals, and release procedures;

(ii) Aero tows or ground tows;

(iii) Straight glides, turns, and spirals;

(iv) Flight at minimum controllable airspeeds, and stall recognition and recoveries;

(v) Traffic patterns, including collision avoidance precautions; and

(vi) Normal landings.

Instructions must be given by a flight instructor who is authorized to give instruction in gliders.

(5) *In airships:*

(i) Flight preparation, procedures, including preflight inspection and powerplant operation;

(ii) Rigging, ballasting, controlling pressure in the ballonets, and superheating;

(iii) Takeoffs and ascents;

(iv) Straight and level flight, climbs, turns, and descents; and

(v) Landings with positive and with negative static balance.

Instruction must be given by an authorized flight instructor or the holder of a commercial pilot certificate with a lighter-than-air category and airship class rating.

(6) *In free balloons:*

(i) Flight preparation procedures, including preflight operations;

(ii) Operation of hot air or gas source, ballast, valves, and rip panels, as appropriate;

(iii) Liftoffs and climbs; and

(iv) Descents, landings, and emergency use of rip panel (may be simulated).

Instruction must be given by an authorized flight instructor or the holder of a commercial pilot certificate with a lighter-than-air category and free balloon class rating.

(d) *Flight instructor endorsements.* A student pilot may not operate an aircraft in solo flight unless his student pilot certificate is endorsed, and unless within the preceding 90 days his pilot logbook has been endorsed, by an authorized flight instructor who— ☞ **F62**

(1) Has given him instruction in the make and model of aircraft in which the solo flight is made;

(2) Finds that he has met the requirements of this section; and

(3) Finds that he is competent to make a safe solo flight in that aircraft.

§ 61.89 General limitations.

(a) A student pilot may not act as pilot in command of an aircraft—

(1) That is carrying a passenger;

☞ **F61, 114, 115**

(2) That is carrying property for compensation or hire;

(3) For compensation or hire;

(4) In furtherance of a business; or

(5) On an international flight, except that a student pilot may make solo training flights from Haines, Gustavus, or Juneau, Alaska, to White Horse, Yukon, Canada, and return, over the province of British Columbia.

(b) A student pilot may not act as a required flight crewmember on any aircraft for which more than one pilot is required, except when receiving flight instruction from an authorized flight instructor on board an airship and no person other than a required flight crewmember is carried on the aircraft.

§ 61.91 Aircraft limitations: pilot in command.

A student pilot may not serve as pilot in command of any airship requiring more than one flight crewmember unless he has met the pertinent requirements prescribed in § 61.87.

§ 61.93 Cross-country flight requirements.

(a) *General.* A student pilot may not operate an aircraft in a solo cross-country flight, nor may he, except in emergency, make a solo flight landing at any point other than the airport of takeoff, until he meets the requirements prescribed in this section. However, an authorized flight instructor may allow a student

pilot to practice solo landings and takeoffs at another airport within 25 nautical miles from the airport at which the student pilot receives instruction if he finds that the student pilot is competent to make those landings and takeoffs. As used in this section the term cross-country flight means a flight beyond a radius of 25 nautical miles from the point of takeoff. ☞ **F14**

(b) *Flight training.* A student pilot must receive instruction from an authorized instructor in at least the following pilot operations pertinent to the aircraft to be operated in a solo cross-country flight:

(1) For solo cross-country in airplanes—

(i) The use of aeronautical charts, pilotage, and elementary dead reckoning using the magnetic compass;

(ii) The use of radio for VFR navigation, and for two-way communication;

(iii) Control of an airplane by reference to flight instruments;

(iv) Short field and soft field procedure, and cross wind takeoffs and landings;

(v) Recognition of critical weather situations, estimating visibility while in flight, and the procurement and use of aeronautical weather reports and forecasts; and

(vi) Cross-country emergency procedures.

(2) For solo cross-country in rotorcraft—

(i) The use of aeronautical charts and the magnetic compass for pilotage and for elementary dead reckoning;

(ii) The recognition of critical weather situations, estimating visibility while in flight, and the procurement and use of aeronautical weather reports and forecasts;

(iii) Radio communications; and

(iv) Cross-country emergencies.

(3) For solo cross-country in gliders—

(i) The recognition of critical weather situations and conditions favorable for soaring flight, and the procurement and use of aeronautical weather reports and forecasts;

(ii) The use of aeronautical charts and the magnetic compass for pilotage; and

(iii) Cross-country emergency procedures.

(4) For student cross-country in airships—

(i) The use of aeronautical charts and the magnetic compass for pilotage and dead reckoning, and the use of radio for navigation and two-way communications;

(ii) The control of an airship solely by reference to flight instruments;

(iii) The control of gas pressures, with regard to superheating and altitude changes;

(iv) The recognition of critical weather situations, and the procurement and use of aeronautical weather reports and forecasts; and

(v) Cross-country emergency procedures.

(5) For solo cross-country in free balloons—

(i) The use of aeronautical charts and the magnetic compass for pilotage;

(ii) The recognition of critical weather situations, and the procurement and use of aeronautical weather reports and forecasts; and

(iii) Cross-country emergency procedures.

(c) *Flight instructor endorsements.* A student pilot must have the following endorsements from an authorized flight instructor:

(1) An endorsement on his student pilot certificate stating that he has received instruction in solo cross-country flying and the applicable training requirements of this section, and is competent to make cross-country solo flights in the category of aircraft involved.

(2) An endorsement in his pilot logbook that the instructor has reviewed the preflight planning and preparation for each solo cross-country flight, and he is prepared to make the flight safely under the known circumstances and the conditions listed by the instructor in the logbook. The instructor may also endorse the logbook for repeated solo cross-country flights under stipulated conditions over a course not more than 50 nautical miles from the point of departure if he has given the student flight instruction in both directions over the route, including takeoffs and landings at the airports to be used.

☞ **F63**

§ 61.95 Operations in a terminal control area and at airports located within a terminal control area.

(a) A student pilot may not operate an aircraft on a solo flight in the airspace of a terminal control area unless—

(1) The student pilot has received both ground and flight instruction from an authorized instructor on that terminal control area and the flight instruction was received in the specific terminal control area for which solo flight is authorized;

(2) The logbook of that student pilot has been endorsed within the preceding 90 days for conducting solo flight in that specific terminal control area by the instructor who gave the flight training; and

(3) The logbook endorsement specifies that the student pilot has received the required ground and flight instruction and has been found competent to conduct solo flight in that specific terminal control area.

(b) Pursuant to § 91.90(b), a student pilot may not operate an aircraft on a solo flight to, from, or at an airport located within a terminal control area unless—

(1) That student pilot has received both ground and flight instruction from an authorized instructor to operate at that airport and the flight and ground instruction has been received at the specific airport for which the solo flight is authorized;

(2) The logbook of that student pilot has been endorsed within the preceding 90 days for conducting solo flight at that specific airport by the instructor who gave the flight training; and

(3) The logbook endorsement specifies that the student pilot has received the required ground and flight instruction and has been found competent to conduct solo flight operations at that specific airport.

Subpart D—Private Pilots

§ 61.101 Applicability.

This subpart prescribes the requirements for the issuance of private pilot certificates and ratings, the conditions under which those certificates and ratings are necessary, and the general operating rules for the holders of those certificates and ratings.

§ 61.103 Eligibility requirements: general.

To be eligible for a private pilot certificate, a person must—

(a) Be at least 17 years of age, except that a private pilot certificate with a free balloon or a glider rating only may be issued to a qualified applicant who is at least 16 years of age;

(b) Be able to read, speak, and understand the English language, or have such operating limitations placed on his pilot certificate as are necessary for the safe operation of aircraft, to be removed when he shows that he can read, speak, and understand the English language;

(c) Hold at least a current third-class medical certificate issued under Part 67 of this chapter, or, in the case of a glider or free balloon rating, certify that he has no known medical defect that makes him unable to pilot a glider or free balloon, as appropriate;

(d) Pass a written test on the subject areas on which instruction or home study is required by § 61.105;

(e) Pass an oral and flight test on procedures and maneuvers selected by an FAA inspector or examiner to determine the applicant's competency in the flight operations on which instruction is required by the flight proficiency provisions of § 61.107; and

(f) Comply with the sections of this Part that apply to the rating he seeks.

§ 61.105 Aeronautical knowledge.

An applicant for a private pilot certificate must have logged ground instruction from an authorized instructor, or must present evidence showing that he has satisfactorily completed a course of instruction or home study in at least the following areas of aeronautical knowledge appropriate to the category of aircraft for which a rating is sought.

[(a) *Airplanes and rotorcraft.*

[(1) The accident reporting requirements of the National Transprotation Safety Board and the Federal Aviation Regulations applicable to private pilot privileges, limitations, and flight operations for airplanes or rotorcraft, as appropriate, the use of the "Airman's Information Manual," and FAA Advisory Circulars;

[(2) VFR navigation, using pilotage, dead reckoning, and radio aids;

[(3) The recognition of critical weather situations from the ground and in flight, the procurement and use of aeronautical weather reports and forecasts;

[(4) The safe and efficient operation of airplanes or rotorcraft, as appropriate, including high density airport operations, collision avoidance precautions, and radio communication procedures; and

[(5) Basic aerodynamics and the principles of flight which apply to airplanes or rotorcraft, as appropriate.]

[(b)] *Gliders.*

(1) The accident reporting requirements of the National Transportation Safety Board and the Federal Aviation Regulations applicable to glider pilot privileges, limitations, and flight operations;

(2) Glider navigation, including the use of aeronautical charts and the magnetic compass;

(3) Recognition of weather situations of concern to the glider pilot, and the procurement and use of aeronautical weather reports and forecasts; and

(4) The safe and efficient operation of gliders, including ground and aero tow procedures, signals, and safety precautions.

[(c)] *Airships.*

(1) The Federal Aviation Regulations applicable to private lighter-than-air pilot privileges, limitations, and airship flight operations;

(2) Airship navigation, including pilotage, dead reckoning, and the use of radio aids;

(3) The recognition weather conditions of concern to the airship pilot, and the procurement and use of aeronautical weather reports and forecasts; and

(4) Airship operations, including free ballooning, the effects of superheating, and positive and negative lift.

[(d)] *Free balloons.*

(1) The Federal Aviation Regulations applicable to private free balloon pilot privileges, limitations, and flight operations;

(2) The use of aeronautical charts and the magnetic compass for free balloon navigation;

(3) The recognition of weather conditions of concern to the free balloon pilot, and the procurement and use of aeronautical weather reports and forecasts appropriate to free balloon operations; and

(4) Operating principles and procedures of free balloons, including gas and hot air inflation systems.

§ 61.107 Flight proficiency.

The applicant for a private pilot certificate must have logged instruction from an authorized flight instructor in at least the following pilot operations. In addition, his logbook must contain an endorsement by an authorized flight instructor who has found him competent to perform each of those operations safely as a private pilot.

(a) *In airplanes.*

(1) Preflight operations, including weight and balance determination, line inspection, and airplane servicing;

(2) Airport and traffic pattern operations, including operating at controlled airports, radio communications, and collision avoidance precautions;

(3) Flight maneuvering by reference to ground objects;

(4) Flight at critically slow airspeeds, and the recognition of and recovery from imminent and full stalls entered from straight flight and from turns;

(5) Normal and crosswind takeoffs and landings;

(6) Control and maneuvering an airplane solely by reference to instruments, including descents and climbs using radio aids or radar directives;

(7) Cross-country-flying, using pilotage, dead reckoning, and radio aids, including one 2-hour flight;

(8) Maximum performance takeoffs and landings;

(9) Night flying, including takeoffs, landings, and VFR navigation; and

(10) Emergency operations, including simulated aircraft and equipment malfunctions.

(b) *In helicopters.*
(1) Preflight operations, including the line inspection and servicing of helicopters;

(2) Hovering, air taxiing, and maneuvering by ground references;

(3) Airport and traffic pattern operations, including collision avoidance precautions;

[(4) Cross-country fling, using pilotage, dead reckoning, and radio aids, including one 1-hour flight;

[(5) Operations in confined areas and on pinnacles, rapid decelerations, landings on slopes, high-altitude takeoffs, and run-on landings;

[(6) Night flying, including takeoffs, landings, and VFR navigation; and

[(7) Simulated emergency procedures, including aircraft and equipment malfunctions, approaches to a hover or landing with an engine inoperative in a multiengine helicopter, or autorotational descents with a power recovery to a hover in single-engine helicopters.]

(c) *In gyroplanes.*
(1) Preflight operations, including the line inspection and servicing of gyroplanes;

(2) Flight maneuvering by ground references;

(3) Maneuvering at critically slow airspeeds, and the recognition of and recovery from high rates of descent at low airspeeds;

(4) Airport and traffic pattern operations, including collision avoidance precautions and radio communication procedures;

(5) Cross-country flying by pilotage, dead reckoning, and the use of radio aids; and

(6) Emergency procedures, including maximum performance takeoffs and landings.

(d) *In gliders.*

[(1) Preflight operations including the installation of wings and tail surfaces specifically designed for quick removal and installation by pilots, and line inspection;]

(2) Ground (auto or winch) tow or aero tow (the applicant's certificate is limited to the kind of tow selected);

(3) Precision maneuvering, including steep turns and spirals in both directions;

(4) The correct use of critical sailplane performance speeds;

(5) Flight at critically slow airspeeds, and the recognition of and recovery from imminent and full stalls entered from straight and from turning flight; and

(6) Accuracy approaches and landings with the nose of the glider stopping short of and within 200 feet of a line or mark.

(e) *In airships.*

(1) Ground handling, mooring, rigging, and preflight operations;

(2) Takeoffs and landing with static lift, and with negative and positive lift, and the use of two-way radio;

(3) Straight and level flight, climbs, turns, and descents;

(4) Precision flight maneuvering;

(5) Navigation, using pilotage, dead reckoning, and radio aids; and

(6) Simulated emergencies, including equipment malfunction, the valving of gas, and the loss of power on one engine.

(f) *In free balloons.*

[(1) Rigging and tethering, including the installation of baskets and burners specifically designed for quick removal or installation by a pilot; and the interchange of baskets or burners, when provided for in the type certificate data, classified as preventive maintenance, and subject to the recording requirements of § 43.9 of this chapter;]

(2) Operation of burner, if airborne heater used;

(3) Ascents and descents;

(4) Landing; and

(5) Emergencies, including the use of the ripcord (may be simulated).

§ 61.109 Airplane rating: aeronautical experience. ⌖ F64

An applicant for a private pilot certificate with an airplane rating must have had at least a total of 40 hours of flight instruction and solo flight time which must include the following:

(a) Twenty hours of flight instruction from an authorized flight instructor, including at least—

(1) Three hours of cross-country;

(2) Three hours at night, including 10 takeoffs and landings for applicants seeking night flying privileges; and

(3) Three hours in airplanes in preparation for the private pilot flight test within 60 days prior to that test.

An applicant who does not meet the night flying requirement in paragraph (a) (2) is issued a private pilot certificate bearing the limitation "Night flying prohibited." This limitation may be removed if the holder of the certificate shows that he has met the requirements of paragraph (a) (2),

[(b) Twenty hours of solo flight time, including at least:

[(1) Ten hours in airplanes.

[(2) Ten hours of cross-country flights, each flight with a landing at a point more than 50 nautical miles from the original departure point. One flight must be of at least 300 nautical miles with landings at a minimum of three points, one of which is at least 100 nautical miles from the original departure point.] ⌖ F14

(3) Three solo takeoffs and landings to a full stop at an airport with an operating control tower.

§ 61.111 Cross-country flights: pilots based on small islands.

(a) An applicant who shows that he is located on an island from which the required flights cannot be accomplished without flying over water more than 10 nautical miles from the nearest shoreline need not comply with paragraph (b) (2) of § 61.109. However, if other airports that permit civil operations are available to which a flight may be made without flying over water more than 10 nautical miles from the nearest shoreline, he must show that he has completed two round trip solo flights between these two airports that are farthest apart, including a landing at each airport on both flights.

(b) The pilot certificate issued to a person under paragraph (a) of this section contains an endorsement with the following limitation which may be subsequently amended to include another island if the applicant complies with paragraph (a) of this section with respect to that island:

"Passenger carrying prohibited on flights more than 10 nautical miles from (appropriate island)."

(c) If an applicant for a private pilot certificate under paragraph (1) of this section does not have at least 3 hours of solo cross-country flight time, including a round trip flight to an airport at least 50 nautical miles from the place of departure with at least two full stop landings at different points along the route, his pilot certificate is also endorsed as follows:

"Holder does not meet the cross-country flight requirements of ICAO."

(d) The holder of a private pilot certificate with an endorsement described in paragraph (b) or (c) of this section, is entitled to a removal of the endorsement, if he presents satisfactory evidence to an FAA inspector or designated pilot examiner that he has complied with the applicable solo cross-country flight requirements and has passed a practical test on cross-country flying.

§ 61.113 Rotorcraft rating: aeronautical experience.

An applicant for a private pilot certificate with a rotorcraft category rating must have at least the following aeronautical experience:

(a) [After April 30, 1987,] *For a helicopter class rating,* 40 hours of flight instruction and solo flight time in aircraft, including at least—

(1) 20 hours of flight instruction from an authorized flight instructor, 15 hours of which must be in a helicopter including—

(i) 3 hours of cross-country flying helicopters;

(ii) 3 hours of night flying in helicopters including 10 takeoffs and landings, each of which must be separate by an enroute phase of flight.

(iii) 3 hours in helicopters in preparation for private pilot flight test within 60 days before that test; and

(iv) A flight in a helicopter with a landing at point other than an airport; and

(2) 20 hours of solo flight time, 15 hours of which must be in a helicopter, including at least—

(i) 3 hours of cross-country flying in helicopters, including one flight with a landing at three or more points, each of which must be more than 25 nautical miles from each of the other two points; and

(ii) Three takeoffs and landings in helicopters at an airport with an operating control tower, each of which must be separated by an en route phase of flight.

(b) [After April 30, 1987,] *For a gyroplane class rating,* 40 hours of flight instruction and solo flight time in aircraft, including at least—

(1) 20 hours of flight instruction from an authorized flight instructor, 15 hours of which must be in a gyroplane, including—

(i) 3 hours of cross-country flying in gyroplanes;

(ii) 3 hours of night flying in gyroplanes, including 10 takeoffs and landings; and

(iii) 3 hours in gyroplanes in preparation for the private pilot flight within 60 days before that test; and

(2) 20 hours of solo flight time, 10 hours of which must be in a gyroplane, including—

(i) 3 hours of cross-country flying in gyroplanes, inlcuding one flight with a landing at three or more points, each of which must be more than 25 nautical miles from each of the other two points; and

(ii) Three takeoffs and landings in gyroplanes at an airport with an operating control tower.

(c) An applicant who does not meet the night flying requirement in paragraph (a)(1)(ii) or paragraph (b)(1)(ii) of this section is issued a private pilot certificate bearing the limitation "night flying prohibited." This limitation may be removed if the holder of the certificate demonstrates compliance with the requirements of paragraph (a)(1)(ii) or paragraph (b)(1)(ii) of this section, as appropriate. [This limitation is not placed on the certificate of an applicant who qualifies under paragraph (d) or (e) of this section.]

[(d)] Until May 1, 1987, for a helicopter rating an applicant must have at least a total of 40 hours of flight instruction and solo flight time in aircraft with at least 15 hours of solo flight time in helicopters, which must include—

[(1) A takeoff and landing at an airport that serves both airplanes and helicopters;

[(2) A flight with a landing at a point other than an airport; and

[(3) Three hours of cross-country flying, including one flight with landings at three or more points, each of which must be more than 25 nautical miles from each of the other two points.

[(e)] Until May 1, 1987, for a gyroplane rating an applicant must have at least a total of 40 hours of flight instruction and solo flight time in aircraft with at least 10 hours of solo flight time in gyroplane, which must include—

[(1) Flights with takeoffs and landings at paved and unpaved airports; and

[(2) Three hours of cross-country flying, including a flight with landings at three or more points, each of which must be more than 25 nautical miles from each of the other two points.]

§ 61.115 Glider rating: aeronautical experience.

An applicant for a private pilot certificate with a glider rating must have logged at least one of the following:

(a) Seventy solo glider flights, including 20 flights during which 360° turns were made.

(b) Seven hours of solo flight in gliders, including 35 glider flights launched by ground tow, or 20 glider flights launched by aero tows.

(c) Forty hours of flight time in gliders and single-engine airplanes, including 10 solo glider flights during which 360° turns were made.

§ 61.117 Lighter-than-air rating: aeronautical experience.

An applicant for a private pilot certificate with a lighter-than-air category rating must have at least the aeronautical experience prescribed in paragraph (a) or (b) of this section, appropriate to the rating sought.

(a) *Airships.*

A total of 50 hours of flight time as pilot with at least 25 hours in airships, which must include 5 hours of solo flight time in airships, or time performing the functions of pilot in command of an airship for which more than one pilot is required.

(b) *Free balloons.*

(1) If a gas balloon or a hot air balloon with an airborne heater is used, a total of 10 hours in free balloons with at least 6 flights under the supervision of a person holding a

commercial pilot certificate with a free balloon rating. These flights must include:

(i) Two flights, each of at least one hour's duration, if a gas balloon is used, or of 30 minutes' duration, if a hot air balloon with an airborne heater is used;

(ii) One ascent under control to 5,000 feet above the point of takeoff, if a gas balloon is used, or 3,000 feet above the point of takeoff, if a hot air balloon with an airborne heater is used; and

(iii) One solo flight in a free balloon.

(2) If a hot air balloon without an airborne heater is used, six flights in a free balloon under the supervision of a commercial balloon pilot, including at least one solo flight.

§ 61.118 Private pilot privileges and limitations: pilot in command.

Except as provided in paragraphs (a) through (d) of this section, a private pilot may not act as pilot in command of an aircraft that is carrying passengers or property for compensation or hire; nor may he, for compensation or hire, act as pilot in command of an aircraft.   ☞ F118

(a) A private pilot may, for compensation or hire, act as pilot in command of an aircraft in connection with any business or employment if the flight is only incidental to that business or employment and the aircraft does not carry passengers or property for compensation or hire.   ☞ F65

(b) A private pilot may share the operating expenses of a flight with his passengers.   ☞ F66

(c) A private pilot who is an aircraft salesman and who has at least 200 hours of logged flight time may demonstrate an aircraft in flight to a prospective buyer.

(d) A private pilot may act as pilot in command of an aircraft used in a passenger-carrying airlift sponsored by a charitable organization, and for which the passengers make a donation to the organization, if—

(1) The sponsor of the airlift notifies the FAA General Aviation District Office having jurisdiction over the area concerned, at least 7 days before the flight, and furnishes any essential information that the office request;

(2) The flight is conducted from a public airport adequate for the aircraft used, or from another airport that has been approved for the operation by an FAA inspector;

(3) He has logged at least 200 hours of flight time;

(4) No acrobatic or formation flights are conducted;

(5) Each aircraft used is certificated in the standard category and complies with the 100-hour inspection requirement of § 91.169 of this chapter; and

(6) The flight is made under VFR during the day.

For the purpose of paragraph (d) of this section, a "charitable organization" means an organization listed in Publication No. 78 of the Department of the Treasury called the "Cumulative List of Organization described in section 170(c) of the Internal Revenue Code of 1954," as amended from time to time by published supplemental lists.

§ 61.119 Free balloon rating: limitations.

(a) If the applicant for a free balloon rating takes his flight test in a hot air balloon with an airborne heater, his pilot certificate contains an endorsement restricting the exercise of the privilege of that rating to hot air balloons with airborne heaters. The restrictions may be deleted when the holder of the certificate obtains the pilot experience required for a rating on a gas balloon.

(b) If the applicant for a free balloon rating takes his flight test in a hot air balloon without an airborne heater, his pilot certificate contains an endorsement restricting the exercise of the privileges of that rating to hot air balloons without airborne heaters. The restriction may be deleted when the holder of the certificate obtains the pilot experience and passes the tests required for a rating on a free balloon with an airborne heater or a gas balloon.

§ 61.120 Private pilot privileges and limitations: second in command of aircraft requiring more than one required pilot.

Except as provided in paragraphs (a) through (d) of § 61.118 a private pilot may not, for compensation or hire, act as second in command of an aircraft that is type certificated for more than one required pilot, nor may he act as second in command of such an aircraft that is carrying passengers or property for compensation or hire.

Subpart E—Commercial Pilots

§ 61.121 Applicability.

This subpart prescribes the requirements for the issuance of commercial pilot certificates and ratings, the conditions under which those certificates and ratings are necessary, and the limitations upon these certificates and ratings.

§ 61.123 Eligibility requirements: general

To be eligible for a commercial pilot certificate, a person must—

(a) Be at least 18 years of age;

(b) Be able to read, speak, and understand the English language, or have such operating limitations placed on his pilot certificate as are necessary for safety, to be removed when he shows that he can read, speak, and understand the English language;

(c) Hold at least a valid second-class medical certificate issued under Part 67 of this chapter, or, in the case of a glider or free balloon rating, certify that he has no known medical deficiency that makes him unable to pilot a glider or a free balloon, as appropriate;

(d) Pass a written examination appropriate to the aircraft rating sought on the subjects in which ground instruction is required by § 61.125;

(e) Pass an oral and flight test appropriate to the rating he seeks, covering items selected by the inspector or examiner from those on which training is required by § 61.127; and

(f) Comply with the provision of this subpart which apply to the rating he seeks.

§ 61.125 Aeronautical knowledge.

An applicant for a commercial pilot certificate must have logged ground instruction from an authorized instructor, or must present evidence showing that he has satisfactorily completed a course of instruction or home study, in at least the following areas of aeronautical knowledge appropriate to the category of aircraft for which a rating is sought.

(a) *Airplanes.*

(1) The regulations of this chapter governing the operations, privileges, and limitations of a commercial pilot, and the accident reporting requirements of the National Transportation Safety Board.

(2) Basic aerodynamics and the principles of flight which apply to airplanes; and

(3) Airplane operations, including the use of flaps, retractable landing gears, controllable propellers, high altitude operation with and without pressurization, loading and balance computations, and the significance and use of airplane performance speeds.

(b) *Rotorcraft.*

(1) The regulations of this chapter which apply to the operations, privileges, and limitations of a commercial rotorcraft pilot, and the accident reporting requirements of the National Transportation Safety Board;

(2) Meteorology, including the characteristics of air masses and fronts, elements of weather forecasting, and the procurement and use of aeronautical weather reports and forecasts;

[(3) The use of aeronautical charts and the magnetic compass for pilotage and dead reckoning, and the use of radio aids for VFR navigation;]

(4) The safe and efficient operation of helicopters or gyroplanes, as appropriate to the rating sought; [and]

[(5) Basic aerodynamics and principles of flight which apply to rotorcraft and the significance and use of performance charts.]

(c) *Gliders.*

(1) The regulations of this chapter pertinent to commercial glider pilot operations, privilege, and limitations, and the accident reporting requirements of the National Transportation Safety Board;

(2) Glider navigation, including the use of aeronautical charts and the magnetic compass, and radio orientation;

(3) The recognition of weather situations of concern to the glider pilot from the ground and in flight, and the procurement and use of aeronautical weather reports and forecasts; and

(4) The safe and efficient operation of gliders, including ground and aero tow procedures, signals, critical sailplane performance speeds, and safety precautions.

(d) *Airships.*

(1) The regulations of this chapter pertinent to airship operations, VFR and IFR, including the privileges and limitations of a commercial airship pilot;

(2) Airship navigation, including pilotage, dead reckoning, and the use of radio aids for VFR and IFR navigation, and IFR approaches;

(3) The use and limitations of the required flight instruments;

(4) ATC procedures for VFR and IFR operations, and the use of IFR charts and approach plates;

(5) Meteorology, including the characteristics of air masses and fronts, and the procurement and use of aeronautical weather reports and forecasts;

(6) Airship ground and flight instruction procedures; and

(7) Airship operating procedures and emergency operations, including free ballooning procedures.

(e) *Free balloons.*

(1) The regulations of this chapter pertinent to commercial free balloon piloting privileges, limitations, and flight operations;

(2) The use of aeronautical charts and the magnetic compass for free balloon navigation;

(3) The recognition of weather conditions significant to free balloon flight operations, and the procurement and use of aeronautical weather reports and forecasts appropriate to free ballooning;

(4) Free balloon flight and ground instruction procedures; and

(5) Operating principles and procedures for free balloons, including emergency procedures such as crowd control and protection, high wind and water landings, and operations in proximity to buildings and power lines.

§ 61.127 Flight proficiency.

The applicant for a commercial pilot certificate must have logged instruction from an authorized flight instructor in at least the following pilot operations. In addition, his logbook must contain an endorsement by an authorized flight instructor who has given him the instruction certifying that he has found the applicant prepared to perform each of those operations competently as a commercial pilot.

(a) *Airplanes.*

(1) Preflight duties, including load and balance determination, line inspection, and aircraft servicing;

(2) Flight at critically slow airspeeds, recognition of imminent stalls, and recovery from stalls with and without power;

(3) Normal and crosswind takeoffs and landings, using precision approaches, flaps, power as appropriate, and specified approach speeds;

(4) Maximum performance takeoffs and landings, climbs, and descents;

(5) Operation of an airplane equipped with a retractable landing gear, flaps, and controllable propeller(s), including normal and emergency operations; and

(6) Emergency procedures, such as coping with power loss or equipment malfunction, fire in flight, collision avoidance precautions, and engine-out procedures if a multiengine airplane is used.

(b) *Helicopters.*

(1) Preflight duties, including line inspection and helicopter servicing;

(2) Straight and level flight, climbs, turns, and descents;

(3) Air taxiing, hovering, and maneuvering by ground references;

(4) Normal and crosswind takeoffs and landings;

[(5) Recognition of and recovery from imminent flight at critical/rapid descent with power (settling with power);]

(6) Airport and traffic pattern operations, including collision avoidance precautions and radio communications;

[(7) Cross-country flight operations;

[(8) Operations in confined areas and on pinnacles, rapid decelerations, landing on slopes, high-altitude takeoffs, and run-on landings; and

[(9) Simulated emergency procedures, including failure of an engine or other component or system, and approaches to a hover or landing with one engine inoperative in multiengine helicopters, or autorotational descents with a power recovery to a hover in single-engine helicopters.]

(c) *Gyroplanes.*

(1) Preflight operations, including line inspection and gyroplane servicing;

(2) Straight and level flight, turns, climbs, and descents;

(3) Flight maneuvering by ground references;

(4) Maneuvering at critically slow airspeeds, and the recognition of and recovery from high rates of descent at slow airspeeds;

(5) Normal and crosswind takeoffs and landings;

(6) Airport and traffic pattern operations, including collision avoidance precautions and radio communications;

(7) Cross-country flight operations; and

(8) Emergency procedures, such as power failures, equipment malfunctions, maximum performance takeoffs and landings and simulated liftoffs at low airspeed and high angles of attack.

(d) *Gliders.*

(1) Preflight duties, including glider assembly and preflight inspection;

(2) Glider launches by ground (auto or winch) or by aero tows (the applicant's certificate is limited to the kind of tow selected);

(3) Precision maneuvering, including straight glides, turns to headings, steep turns and spirals in both directions;

(4) The correct use of sailplane performance speeds, flight at critically slow airspeeds, and the recognition of and recovery from stalls entered from straight flight and from turns; and

(5) Accuracy approaches and landings, with the nose of the glider coming to rest short of and within 100 feet of a line or mark.

(e) *Airships.*

(1) Ground handling, mooring, and preflight operations;

(2) Straight and level flight, turns, climbs, and descents, under VFR and simulated IFR conditions;

(3) Takeoffs and landings with positive and with negative static lift;

(4) Turns and figure eights;

(5) Precision turns to headings under simulated IFR conditions;

(6) Preparing and filing IFR flight plans, and complying with IFR clearances;

(7) IFR radio navigation and instrument approach procedures;

(8) Cross-country flight operations, using pilotage, dead reckoning, and radio aids; and

(9) Emergency operations, including engine-out operations, free ballooning an airship, and ripcord procedures (may be simulated).

(f) *Free balloons.*

[(1) Assembly of basket and burner to the envelope, and rigging, inflating, and tethering of a free balloon;]

(2) Ground and flight crew briefing;

(3) Ascents;

(4) Descents;

(5) Landings;

(6) Operation of airborne heater, if balloon is so equipped; and

(7) Emergency operations, including the use of the ripcord (may be simulated), and recovery from a terminal velocity descent if a balloon with an airborne heater is used.

§ 61.129 Airplane rating: aeronautical experience. ☞ F14

(a) *General.* An applicant for a commercial pilot certificate with an airplane rating must hold a private pilot certificate with an airplane rating. If he does not hold that certificate and rating he must meet the flight experience requirements for a private pilot certificate and airplane rating and pass the applicable written and practical test prescribed in Subpart D of this Part. In addition, the applicant must hold an instrument rating (airplane), or the commercial pilot certificate that is issued is endorsed with a limitation prohibiting the carriage of passengers for hire in airplanes on cross-country flights of more than 50 nautical miles, or at night.

(b) *Flight time as pilot.* An applicant for a commercial pilot certificate with an airplane rating must have a total of at least 250 hours of flight time as pilot, which may include not more than 50 hours of instruction from an authorized instructor in a ground trainer acceptable to the Administrator. The total flight time as pilot must include— ☞ F48

(1) 100 hours in powered aircraft, including at least—

(i) 50 hours in airplanes, and

(ii) 10 hours of flight instruction and practice given by an authorized flight instructor in an airplane having a retractable landing gear, flaps and a controllable pitch propeller; and ☞ F46

(2) 50 hours of flight instruction given by an authorized flight instructor, including—

(i) 10 hours of instrument instruction, of which at least 5 hours must be in flight in airplanes, and

(ii) 10 hours of instruction in preparation for the commercial pilot flight test; and

(3) 100 hours of pilot in command time, including at least:

(i) 50 hours in airplanes.

(ii) 50 hours of cross-country flights, each flight with a landing at a point more than 50 nautical miles from the original departure point. One flight must have landings at a minimum of three points, one of which is at least 150 nautical miles from the original departure point if the flight is conducted in Hawaii, or at least 250 nautical miles from the original departure point if it is conducted elsewhere.

(iii) 5 hours of night flying including at least 10 takeoffs and landings as sole manipulator of the controls.

§ 61.131 Rotorcraft rating: aeronautical experience.

An applicant for a commercial pilot certificate with a rotorcraft category rating must have at least the following aeronautical experience as a pilot:

(a) [After April 30, 1987,] *For a helicopter class rating,* 150 hours of flight time, including at least 100 hours in powered aircraft, 50 hours of which must be in a helicopter, including at least—

(1) 40 hours of flight instruction from an authorized flight instructor, including 15 hours of which must be in a helicopters, including—

(i) 3 hours of cross-country flying in helicopters;

(ii) 3 hours of night flying in helicopters, including 10 takeoffs and landings, each of which must be separated by an en route phase of flight;

(iii) 3 hours in helicopters preparing for the commercial pilot flight test within 60 days before that test; and

(iv) Takeoffs and landings at three points other than airports; and

(2) 100 hours of pilot-in-command flight time, 35 hours of which must be in a helicopter, including at least—

(i) 10 hours of cross-country flying in helicopters, including one flight with a landing at three or more points, each of which must be more than 50 nautical miles from each of the other two points; and

(ii) Three takeoffs and landings in helicopters, each of which must be separated by an en route phase of flight, at an airport with an operating control tower.

(b) [After April 30, 1987,] *For a gyroplane class rating,* 150 hours of flight time in aircraft, including at least 100 hours in powered aircraft, 25 hours of which must be in a gyroplane, including at least—

(1) 40 hours of flight instruction from a authorized flight instructor, 10 hours of which must be in a gyroplane, including at least—

(i) 3 hours of cross country flying in gyroplanes;

(ii) 3 hours of night flying in gyroplanes, including 10 takeoffs and landings; and

(iii) 3 hours in gyroplanes preparing for the commercial pilot flight test within 60 days before that test; and

(2) 100 hours of pilot-in-command flight time, 15 hours of which must be in a gyroplane, including at least—

(i) 10 hours of cross-country flying in gyroplanes, including one flight with a landing at three or more points, each of which is more than 50 nautical miles from each of the other two points; and

(ii) Three takeoffs and landings in gyroplanes at an airport with an operating control tower.

[(c)[Until May 1, 1987, for a helicopter rating at least 150 hours of flight time as pilot, including at least—

[(1) 100 hours in powered aircraft and at least 50 hours in helicopters;

[(2) 100 hours of pilot in command time, including a cross-country flight with landings at three points, each of which is more than 50 nautical miles from each of the other points;

[(3) 40 hours of flight instruction from an authorized flight instructor, including 15 hours in helicopters; and

[(4) 10 hours as pilot in command in helicopters, including—

[(i) Five takeoffs and landings at night;

[(ii) Takeoffs and landings at three different airports that serve both airplanes and helicopters; and

[(iii) Takeoffs and landings at three points other than airports.

[(d)] Until May 1, 1987, for a gyroplane rating at least 200 hours of flight time as pilot, including—

[(1) 100 hours in powered aircraft;

[(2) 100 hours as pilot in command, including a cross-country flight with landings at three points, each of which is more than 50 nautical miles from each of the other two points;

[(3) 75 hours as pilot in command in gyroplanes, including—

[(i) Flights with takeoffs and landings at three different paved airports and three unpaved airports; and

[(ii) Three flights with takeoffs and landings at an airport with an operating control tower; and

[(4) Twenty hours of flight instruction in gyroplanes, including 5 hours in preparation for the commercial pilot flight test.]

§ 61.133 Glider rating: aeronautical experience.

An applicant for a commercial pilot certificate with a glider rating must meet either of the following aeronautical experience requirements:

(a) A total of at least 25 hours of pilot time in aircraft, including 20 hours in gliders, and a total of 100 glider flights as pilot in command, including 25 flights during which 360° turns were made; or

(b) A total of 200 hours of pilot time in heavier-than-air aircraft, including 20 glider flights as pilot in command during which 360° turns were made.

§ 61.135 Airship rating: aeronautical experience.

An applicant for a commercial pilot certificate with an airship rating must have a total of at least 200 hours of flight time as pilot, including—

(a) 50 hours of flight time as pilot in airships;

(b) 30 hours of flight time, performing the duties of pilot in command in airships, including—

(1) 10 hours of cross-country flight; and

(2) 10 hours of night flight; and

(c) 40 hours of instrument time, of which at least 20 hours must be in flight with 10 hours of that flight time in airships.

§ 61.137 Free balloon rating: aeronautical experience.

An applicant for a commercial pilot certificate with a free balloon rating must have the following flight time as pilot:

(a) If a gas balloon or a hot air balloon with an airborne heater is used, a total of at least 35 hours of flight time as pilot, including—

(1) 20 hours in free balloons; and

(2) 10 flights in free balloons, including—

(i) Six flights under the supervision of a commercial free balloon pilot;

(ii) Two solo flights;

(iii) Two flights of at least 2 hours duration if a gas balloon is used, or at least 1 hour duration if a hot air balloon with an airborne heater is used; and

(iv) One ascent under control to more than 10,000 feet above the takeoff point if a gas balloon is used or 5,000 feet above the takeoff point if a hot air balloon with an airborne heater is used.

(b) If a hot air balloon without an airborne heater is used, ten flights in free balloons, including—

(1) Six flights under the supervision of a commercial free balloon pilot; and

(2) Two solo flights.

§ 61.139 Commercial pilot privileges and limitations: general.

The holder of a commercial pilot certificate may: ☞ **F118**

(a) Act as pilot in command of an aircraft carrying persons or property for compensation or hire;

(b) Act as pilot in command of an aircraft for compensation or hire; and

(c) Give flight instruction in an airship if he holds a lighter-than-air category and an airship class rating, or in a free balloon if he holds a free ballon class rating.

§ 61.141 Airship and free balloon ratings: limitations.

(a) If the applicant for a free balloon class rating takes his flight test in a hot air balloon without an airborne heater, his pilot certificate contains an endorsement restricting the exercise of the privileges of that rating to hot air balloons without airborne heaters. The restriction may be deleted when the holder of the certificate obtains the pilot experience and passes the test required for a rating on a free balloon with an airborne heater or a gas balloon.

(b) If the applicant for a free balloon class rating takes his flight test in a hot air balloon with an airborne heater, his pilot certificate

contains an endorsement restricting the exercise of the privileges of that rating to hot air balloons with airborne heaters. The restriction may be deleted when the holder of the certificate obtains the pilot experience required for a rating on a gas balloon.

Subpart F—Airline Transport Pilots

§ 61.151 Eligibility requirements: general.

To be eligible for an airline transport pilot certificate, a person must—

(a) Be at least 23 years of age;

(b) Be of good moral character;

(c) Be able to read, write, and understand the English language and speak it without accent or impediment of speech that would interfere with two-way radio conversation;

(d) Be a high school graduate, or its equivalent in the Administrator's opinion, based on the applicant's general experience and aeronautical experience, knowledge, and skill;

(e) Have a first-class medical certificate issued under Part 67 of this chapter within the 6 months before the date he applies; and

(f) Comply with the sections of this Part that apply to the rating he seeks.

§ 61.153 Airplane rating: aeronautical knowledge.

An applicant for an airline transport pilot certificate with an airplane rating must, after meeting the requirements of §§ [61.151] (except paragraph (a) thereof) and [61.155], pass a written test on—

(a) The sections of this Part relating to airline transport pilots and Part 121, subpart C of Part 65, and §§ 91.1 through 91.9 and subpart B of Part 91 of this chapter, and so much of Parts 21 and 25 of this chapter as relate to the operations of air carrier aircraft;

(b) The fundamentals of air navigation and use of formulas, instruments, and other navigational aids, both in aircraft and on the ground, that are necessary for navigating aircraft by instruments;

(c) The general system of weather collection and dissemination;

(d) Weather maps, weather forecasting, and weather sequence abbreviations, symbols, and nomenclature;

(e) Elementary meteorology, including knowledge of cyclones as associated with fronts;

(f) Cloud forms;

(g) National Weather Service Federal Meteorological Handbook No. 1, as amended;

(h) Weather conditions, including icing conditions and upper-air winds, that affect aeronautical activities;

(i) Air navigation facilities used on Federal airways, including rotating beacons, course lights, radio ranges, and radio marker beacons;

(j) Information from airplane weather observations and meteorological data reported from observations made by pilots on air carrier flights;

(k) The influence of terrain on meteorological conditions and developments, and their relation to air carrier flight operations;

(l) Radio communication procedure in aircraft operations; and

(m) Basic principles of loading and weight distribution and their effect on flight characteristics.

§ 61.155 Airplane rating: aeronautical experience.

(a) An applicant for an airline transport pilot certificate with an airplane rating must hold a commercial pilot certificate or a foreign airline transport pilot or commercial pilot license without limitations, issued by a member state of ICAO, or he must be a pilot in an Armed Force of the United States whose military experience qualifies him for a commercial pilot certificate under [§ 61.73] of this Part.

(b) An applicant must have had—

(1) At least 250 hours of flight time as pilot in command of an airplane, or as copilot of an airplane performing the duties and functions of a pilot in command under the supervision of a pilot in command, or any combination thereof, at least 100 hours of which were cross-country time and 25 hours of which were night flight time; and

(2) At least 1500 hours of flight time as a pilot, including at least—

(i) 500 hours of cross-country flight time;   ☞ F122

(ii) 100 hours of night flight time; and

(iii) 75 hours of actual or simulated instrument time, at least 50 hours of which were in actual flight.   ☞ F48

Flight time used to meet the requirements of subparagraph (1) of this paragraph may also be used to meet the requirements of subparagraph (2) of this paragraph. Also, an applicant who has made at least 20 night takeoffs and landings to a full stop may substitute one additional night takeoff and landing to a full stop for each hour of night flight time required by subparagraph (2)(ii) of this paragraph. However, not more than 25 hours of night flight time may be credited in this manner.

(c) If an applicant with less than 150 hours of pilot-in-command time otherwise meets the requirements of paragraph (b)(1) of this section, his certificate will be endorsed "Holder does not meet the pilot-in-command flight experience requirements of ICAO", as prescribed by Article 39 of the "Convention on International Civil Aviation". Whenever he presents satisfactory written evidence that he has accumulated the 150 hours of pilot-in-command time, he is entitled to a new certificate without the endorsement.

[(d) A commercial pilot may credit the following flight time toward the 1,500 hours total flight time requirement of paragraph (b)(2) of this section:]

(1) All second-in-command time acquired in airplanes required to have more than one pilot by their approved Aircraft Flight Manuals or airworthiness certificates; and

(2) Flight engineer time acquired in airplanes required to have a flight engineer by their approved Aircraft Flight Manuals, while participating at the same time in an approved pilot training program approved under Part 121 of this chapter.

However, the applicant may not credit under subparagraph (2) of this paragraph more than 1 hour for each 3 hours of flight engineer flight time so acquired, nor more than a total of 500 hours.

(e) If an applicant who credits second-in-command or flight engineer time under paragraph (d) of this section toward the 1500 hours total flight time requirements of subparagraph (b)(2) of this section—

(1) Does not have at least 1200 hours of flight time as a pilot including no more than 50 percent of his second-in-command time and none of his flight engineer time; but

(2) Otherwise meets the requirements of subparagraph (b)(2) of this section,

his certificate will be endorsed "Holder does not meet the pilot flight experience requirements of ICAO", as prescribed by Article 39 of the "Convention on International Civil Aviation." Whenever he presents satisfactory evidence that he has accumulated 1200 hours of flight time as a pilot including no more than 50 percent of his second-in-command time and none of his flight engineer time, he is entitled to a new certificate without the endorsement.

(f) [Reserved]

§ 61.157 Airplane rating: aeronautical skill.

(a) An applicant for an airline transport pilot certificate with a single-engine or multi-engine class rating or an additional type rating must pass a practical test that includes the items set forth in Appendix A of this part. The FAA inspector or designated examiner may modify any required maneuver where necessary for the reasonable and safe operation of the airplane being used and, unless specifically prohibited in Appendix A, may combine any required maneuvers and may permit their performance in any convenient sequence.

(b) Whenever an applicant for an airline transport pilot certificate does not already have an instrument rating he shall, as part of the oral part of the practical test, comply with § 61.65(g) and, as part of the flight part, perform each additional maneuver required by § 61.65(g) that is appropriate to the airplane type and not required in Appendix A of this Part.

(c) Unless the Administrator requires certain or all maneuvers to be performed, the person giving a flight test for an airline transport pilot certificate or additional airplane class

or type rating may, in his discretion, waive any of the maneuvers for which a specific waiver authority is contained in Appendix A of this Part if a pilot being checked—

(1) Is employed as a pilot by a Part 121 certificate holder; and

(2) Within the preceding six calendar months, has successfully completed that certificate holder's approved training program for the airplane type involved.

(d) The items specified in paragraph (a) of this section may be performed in the airplane simulator or other training device specified in Appendix A to this Part for the particular item if—

(1) The airplane simulator or other training device meets the requirements of § 121.407 of this chapter; and

(2) In the case of the items preceded by an asterisk (*) in Appendix A, the applicant has successfully completed the training set forth in § 121.424(d) of this chapter.

However, the FAA inspector or designated examiner may require Items II(d), V(f), or V(g) of Appendix A to this Part to be performed in the airplane if he determines that action is necessary to determine the applicant's competence with respect to that maneuver.

(e) An approved simulator may be used instead of the airplane to satisfy the in-flight requirements of Appendix A of this Part, if the simulator—

(1) Is approved under § 121.407 of this chapter and meets the appropriate simulator requirements of Appendix H of Part 121; and

(2) Is used as part of an approved program that meets the training requirements of § 121.424(a) and (c) and Appendix H of Part 121 of this chapter.

[§ 61.159 Rotorcraft rating: aeronautical knowledge.

[An applicant for an airline transport pilot certificate with a rotorcraft category and a helicopter class rating must pass a written test on—

[(a) So much of this chapter as relates to air carrier rotorcraft operations;

[(b) Rotorcraft design, components, systems and performance limitations;

[(c) Basic principles of loading and weight distribution and their effect on rotorcraft flight characteristics;

[(d) Air traffic control systems and procedures relating to rotorcraft;

[(e) Procedures for operating rotorcraft in potentially hazardous meteorological conditions;

[(f) Flight theory as applicable to rotorcraft; and

[(g) The items listed under paragraphs (b) through (m) of §61.153.

[§ 61.161 Rotorcraft rating: aeronautical experience.

[(a) An applicant for an airline transport pilot certificate with a rotorcraft category and helicopter class rating must hold a commercial pilot certificate, or a foreign airline transport pilot or commercial pilot certificate with a rotorcraft category and helicopter class rating issued by a member of ICAO, or be a pilot in an armed force of the United States whose military experience qualifies that pilot for the issuance of a commercial pilot certificate under § 61.73.

[(b) An applicant must have had at least 1,200 hours of flight time as a pilot, including at least—

[(1) 500 hours of cross-country flight time;

[(2) 100 hours at night, including at least 15 hours in rotorcraft; and

[(3) 200 hours in helicopters, including at least 75 hours as pilot in command, or as second in command performing the duties and functions of a pilot in command under the supervision of a pilot in command, or any combination thereof; and

[(4) 75 hours of instrument time under actual or simulated instrument conditions of which at least 50 hours were completed in flight with at least 25 hours in helicopters as pilot in command, or as second in command performing the duties of a pilot in command under the supervision of a pilot in command, or any combination thereof.

[§ 61.163   Rotorcraft rating: aeronautical skill.

[(a) An applicant for an airline transport pilot certificate with a rotorcraft category and helicopter class rating, or additional aircraft rating, must pass a practical test on those maneuvers set forth in Appendix B of this Part in a helicopter. The FAA inspector or designated examiner may modify or waive any maneuver where necessary for the reasonable and safe operation of the rotorcraft being used and may combine any maneuvers and permit their performance in any convenient sequence to determine the applicant's competency.

[(b) Whenever an applicant for an airline transport pilot certificate with a rotorcraft category and helicopter class rating does not already have an instrument rating, the applicant shall, as part of the practical test, comply with §61.65(g).]

[§ 61.165   Additional category ratings.

[(a) *Rotorcraft category with a helicopter class rating.* The holder of an airline transport pilot certificate (airplane category) who applies for a rotorcraft category with a helicopter class rating must meet the applicable requirements of §§ 61.159 and 61.161, and 61.163 and—]

(1) Have at least 100 hours, including at least 15 hours at night, of rotorcraft flight time as pilot in command or as second in command performing the duties and functions of a pilot in command under the supervision of a pilot in command who holds an airline transport pilot certificate with an appropriate rotorcraft rating, or any combination thereof; or

(2) Complete a training program conducted by a certificated air carrier or other

approved agency requiring at least 75 hours of rotorcraft flight time as pilot in command, second in command, or as flight instruction from an appropriately rated FAA certificated flight instructor or an airline transport pilot, or any combination thereof, including at least 15 hours of night flight time.

[(b) *Airplane rating.* The holder of an airline transport pilot certificate (rotorcraft category) who applies for an airplane category must comply with §§ 61.153, 61.155 (except § 61.155(b)(1)), and 61.157 and—]

(1) Have at least 100 hours, including at least 15 hours at night, of airplane flight time as pilot in command or as second in command performing the duties and functions of a pilot in command under the supervision of a pilot in command who holds an airline transport pilot certificate with an appropriate airplane rating, or any combination thereof; or

(2) Complete a training program conducted by a certificated air carrier or other approved agency required at least 75 hours of airplane flight time as pilot in command, second in command, or as flight instruction from an appropriately rated FAA certificated flight instructor or an airline transport pilot, or any combination thereof, including at least 15 hours of night flight time.

§ 61.167   Tests.

(a) Each applicant for an airline transport pilot certificate must pass each practical and theoretical test to the satisfaction of the Administrator. The minimum passing grade in each subject is 70 percent. Each flight maneuver is graded separately. Other tests are graded as a whole.

(b) Information collected incidentally to such a test shall be treated as a confidential matter by the persons giving the test and by employees of the FAA.

§ 61.169   Instruction in air transportation service.

An airline transport pilot may instruct other pilots in air transportation service in aircraft of the category, class, and type for which he is rated. However, he may not instruct for more than 8 hours in one day nor more than 36 hours in any 7-day period. He may instruct under this section only in aircraft with functioning dual controls. Unless he has a flight instructor certificate, an airline transport pilot may instruct only as provided in this section.

§ 61.171   General privileges and limitations.

An airline transport pilot has the privileges of a commercial pilot with an instrument rating. The holder of a commercial pilot certificate who qualifies for an airline transport pilot certificate retains the ratings on his commercial pilot certificate, but he may

exercise only the privileges of a commercial pilot with respect to them.

Subpart G—Flight Instructors

§ 61.181   Applicability.

This subpart prescribes the requirements for the issuance of flight instructor certificates and ratings, the conditions under which those certificates and ratings are necessary, and the limitations upon these certificates and ratings.

§ 61.183   Eligibility requirements: general.

To be eligible for a flight instructor certificate a person must—

(a) Be at least 18 years of age;

(b) Read, write, and converse fluently in English;

(c) Hold—

(1) A commercial or airline transport pilot certificate with an aircraft rating appropriate to the flight instructor rating sought, and

(2) An instrument rating, if the person is applying for an airplane or an instrument instructor rating;

(d) Pass a written test on the subjects in which ground instruction is required by § 61.185; and

(e) Pass an oral and flight test on those items in which instruction is required by § 61.187.

§ 61.185   Aeronautical knowledge.

(a) Present evidence showing that he has satisfactorily completed a course of instruction in at least the following subjects:

(1) The learning process

(2) Elements of effective teaching.

(3) Student evaluation, quizzing, and testing.

(4) Course development.

(5) Lesson planning.

(6) Classroom instructing techniques.

(b) Have logged ground instruction from an authorized ground or flight instructor in all of the subjects in which ground instruction is required for a private and commercial pilot certificate, and for an instrument rating, if an airplane or instrument instructor rating is sought.

§ 61.187   Flight proficiency.

(a) An applicant for a flight instructor certificate must have received flight instruction, appropriate to the instructor rating sought in the subjects listed in this paragraph by a person authorized in paragraph (b) of this section. In addition, his logbook must contain an endorsement by the person who has given him the instruction certifying that he has found the applicant competent to pass a practical test on the following subjects:

(1) Preparation and conduct of lesson plans for students with varying backgrounds and levels of experience and ability.

(2) The evaluation of student flight performance.

(3) Effective preflight and postflight instruction.

(4) Flight instructor responsibilities and certifying procedures.

(5) Effective analysis and correction of common student pilot flight errors.
☞ **F67, 68**

(6) Performance and analysis of standard flight training procedures and maneuvers appropriate to the flight instructor rating sought.

(b) The flight instruction required by paragraph (a) of this section must be given by a person who has held a flight instructor certificate during the 24 months immediately preceding the date the instruction is given, who meets the general requirements for a flight instructor certificate prescribed in § 61.183, and who has given at least 200 hours of flight instruction, or 80 hours in the case of glider instruction, as a certificate flight instructor.

§ 61.189 Flight Instructor records.

(a) Each certificated flight instructor shall sign the logbook of each person to whom he has given flight or ground instruction and specify in that book the amount of the time and the date on which it was given. In addition, he shall maintain a record in his flight instuctor logbook, or in a separate document containint the following:
☞ **F69**

(1) The name of each person whose logbook or student pilot certificate he has endorsed for solo fight privileges. The record must include the type and date of each endorsement.

(2) The name of each person for whom he has signed a certification for a written, flight, or practical test, including the kind of test, date of his certification, and the result of the test.

(b) The record required by this section shall be retained by flight instructor separately or in his logbook for at least 3 years.

§ 61.191 Additional flight instructor ratings.

The holder of a flight instructor certificate who applies for an additional rating on that certificate must—

(a) Hold an effective pilot certificate with ratings appropriate to the flight instructor rating sought.

(b) Have had at least 15 hours as pilot in command in the category and class of aircraft appropriate to the rating sought; and

(c) Pass the written and practical test prescribed in this subpart for the issuance of a flight instructor certificate with the rating sought.

§ 61.193 Flight instructor authorizations.

(a) The holder of a flight instructor certificate is authorized, within the limitations of his

instructor certificate and ratings, to give—

(1) In accordance with his pilot ratings, the flight instruction required by this Part for a pilot certificate or rating;

(2) Ground instruction or a home study course required by this Part for a pilot certificate and rating;
☞ **F70**

(3) Ground and flight instruction required by this subpart for a flight instructor certificate and rating, if he meets the requirements prescribed in § 61.187 for the issuance of a flight instructor certificate;

(4) The flight instruction required for an initial solo or cross-country flight; and

(5) The flight review required in § 61.57 (a).

(b) The holder of a flight instructor certificate is authorized within the limitations of his instructor certificate to endorse—

(1) In accordance with §§ 61.87(d)(1) and 61.93(c)(1), the pilot certificate of a student pilot he has instructed authorizing the student to conduct solo or solo cross-country flights, or act as pilot-in-command of an airship requiring more than one flight crewmember;

(2) In accordance with § 61.87(d)(1), the logbook of a student pilot he has instructed authorizing single or repeated solo flights;

(3) In accordance with § 61.93(c)(2), the logbook of a student pilot whose preparation and preflight planning for a solo cross-country flight he has reviewed and found adequate for a safe fight under the conditions he has listed in the logbook;

(4) In accordance with § 61.95, the logbook of a student pilot he has instructed authorizing solo flights in a terminal control area or at an airport within a terminal control area.

(5) The logbook of a pilot or flight instructor he has examined certifying that the pilot or flight instructor is prepared for a written or flight test required by this part; and

(6) In accordance with § 61.187, the logbook of an applicant for a flight instructor certificate certifying that he has examined the applicant and found him competent to pass the practical test required by this Part.

(c) A flight instructor with a rotorcraft and helicopter rating or an airplane single-engine rating may also endorse the pilot certificate and logbook of a student pilot he has instructed authorizing the student to conduct solo and cross-country flights in a single-place gyroplane.

§ 61.195 Flight instructor limitations.

The holder of a flight instructor certificate is subject to the following limitations:

(a) *Hours of instruction.* He may not conduct more than eight hours of flight instruction in any period of 24 consecutive hours.

(b) *Ratings.* He may not conduct flight instruction in any aircraft for which he does not hold a category, class, and type rating, if appropriate, on his pilot and flight instructor certificate. However, the holder of a flight instructor certificate effective on November 1, 1973, may continue to exercise the privileges of that certificate until it expires, but not later than November 1, 1975.
☞ **F17, 71**

(c) *Endorsement of student pilot certificate.* He may not endorse a student pilot certificate for initial solo or solo cross-country flight privileges, unless he has given that student pilot flight instruction required by this Part for the endorsement, and considers that the student is prepared to conduct the flight safely with the aircraft involved.

(d) *Logbook endorsement.* He may not endorse a student pilot's logbook—

(1) For solo flight unless he has given that student flight instruction and found that student pilot prepared for solo flight in the type of aircraft involved;

(2) For a cross-country flight, unless he has reviewed the student's flight preparation, planning, equipment, and proposed procedures and found them to be adequate for the flight proposed under existing circumstances; or

(3) For solo flights in a terminal control area or at an airport within the surface area of a terminal control area unless he has given that student ground and flight instruction and has found that student prepared and competent to conduct the operations authorized.

(e) *Solo flights.* He may not authorize any student pilot to make a solo flight unless he possesses a valid student pilot certificate endorsed for solo in the make and model aircraft to be flown. In addition, he may not authorize any student pilot to make a solo cross-country flight unless he possesses a valid student pilot certificate endorsed for solo cross-country flight in the category of aircraft to be flown.

(f) *Instruction in multiengine airplane or helicopter.* He may not give flight instruction required for the issuance of a certificate or a category, or class rating, in a multiengine airplane or a helicopter, unless he has a least five hours of experience as pilot in command in the make and model of that airplane or helicopter, as the case may be.

§ 61.197 Renewal of flight instructor certificates.

The holder of a flight instructor certificate may have his certificate renewed for an additional period of 24 months if he passes the practical test for a flight instructor certificate and the rating involved, or those portions of that test that the Administrator considers necessary to determine his competency as a flight instructor. His certificate may be renewed without taking the practical test if—

(a) His record of instruction shows that he is a competent flight instructor;

(b) He has a satisfactory record as a company check pilot, chief flight instructor, pilot-in-command of an aircraft operated under Part 121 of this chapter, or other activity involving the regular evaluation of pilots, and passes any oral test that may be necessary to determine that instructor's knowledge of current pilot training and certification requirements and standards; or

(c) He has successfully completed, within 90 days before the application for the renewal of his certificate, an approved flight instructor refresher course consisting of not less than 24 hours of ground of flight instruction, or both.

§ 61.199 Expired flight instructor certificates and ratings.

(a) *Flight instructor certificates.* The holder of an expired flight instructor certificate may exchange that certificate for a new certificate by passing the practical test prescribed in § 61.187.

(b) *Flight instructor ratings.* A flight instructor rating or a limited flight instructor rating on a pilot certificate is no longer valid and may not be exchanged for a similar rating or a flight instructor certificate. The holder of either of those ratings is issued a flight instructor certificate only if he passes the written and practical test prescribed in this subpart for the issue of that certificate.

§ 61.201 Conversion to new system of instructor ratings.

(a) *General.* The holder of a flight instructor certificate that does not bear any of the new class or instrument ratings listed in § 61.5(c) (2), (3), or (4) of this Part for an instructor certificate, may not exercise the privileges of that certificate after November 1, 1975. Before that date he may exchange a certificate which has not expired for a flight instructor certificate with the appropriate new ratings in accordance with the provisions of this section. The holder of a flight instructor certificate with a glider rating need not convert that rating to a new class rating to exercise the privileges of that certificate and rating.

(b) *Airplane—single-engine.* An airplane—single-engine rating may be issued to the holder of an effective flight instructor certificate with an airplane rating who has passed the flight instructor practical test in a single-engine airplane, or who has given at least 20 hours of flight instruction in single-engine airplanes as a certificated flight instructor.

(c) *Airplane—multiengine.* An airplane—multiengine class rating may be issued to the holder of an effective flight instructor certificate with an airplane rating who has passed the flight instructor practical test in a multiengine airplane, or who has given at least 20 hours of flight instruction in multiengine airplane as a certificated flight instructor.

(d) *Rotorcraft—helicopter.* A rotorcraft—helicopter class rating may be issued to the holder of an effective flight instructor certificate with a rotorcraft rating who has passed the flight instructor practical test in a helicopter, or who has given at least 20 hours of flight instruction in helicopters as a certificated flight instructor.

(e) *Rotorcraft—gyroplane.* A rotorcraft—gyroplane class rating may be issued to the holder of an effective flight instructor certificate with a rotorcraft rating who has passed the flight instructor practical test in a gyroplane, or who has given at least 20 hours of flight instruction in gyroplanes as a certificated flight instructor.

(f) *Instrument—airplane.* An instrument—airplane instructor rating may be issued to the holder of an effective flight instructor certificate with an instrument rating who has passed the instrument instructor practical test in an airplane, or who has given at least 20 hours of instrument instruction in an airplane as a certificated flight instructor.

(g) *Instrument—helicopter.* An instrument—helicopter rating may be issued to the holder of an effective flight instructor certificate with an instrument rating who has passed the instrument instructor practical test in a helicopter, or who has given at least 20 hours of instrument flight instruction in helicopters as a certificated flight instructor.

NOTE: The reporting and recordkeeping requirements contained herein have been approved by the Office of Management and Budget in accordance with the Federal Reports Act of 1942.

Appendix A

[Practical Test Requirements for Airplane Transport Pilot Certificates and Associated Class and Type Ratings]

Throughout the maneuvers prescribed in this appendix, good judgment commensurate with a high level of safety must be demonstrated. In determining whether such judgment has been shown, the FAA inspector or designated examiner who conducts the check considers adherence to approved procedures, actions based on analysis of situations for which there is no prescribed procedure or recommended practice, and qualities of prudence and care in selecting a course of action.

Each maneuver or procedure must be performed inflight except to the extent that certain maneuvers or procedures may be performed in an airplane simulator with a visual system (visual simulator) or an airplane simulator without a visual system (non-visual simulator) or may be waived as indicated by an X in the appropriate columns. A maneuver authorized to be performed in a non-visual simulator may be performed in a visual simulator, and a maneuver authorized to be performed in a training device may be performed in a non-viusal or a visual simulator.

An asterisk (*) preceding a maneuver or procedure indicates that the maneuver or procedure may be performed in an airplane simulator or other training device as indicated, provided the applicant has successfully completed the training set forth in § 121.424(d) of this chapter.

When a maneuver or procedure is preceded by this symbol (#), it indicates that the FAA inspector or designated examiner may require the maneuver or procedure to be performed in the airplane if he determines such action is necessary to determine the applicant's competence with respect to that maneuver.

An X and asterisk (X*) indicates that a particular condition is specified in connection with the maneuver, procedure, or waiver provisions.

The procedures and maneuvers set forth in this appendix must be performed in a manner that satisfactorily demonstrates knowledge and skill with respect to—

(1) The airplane, its systems and components;

(2) Proper control of airspeed, configuration, direction, altitude, and attitude in accordance with procedures and limitations contained in the approved Airplane Flight Manual, check lists, or other approved material appropriate to the airplane type; and

(3) Compliance with approved en route, instrument approach, missed approach, ATC, or other applicable procedures.

I. Preflight.

(a) Equipment examination (oral). As part of the practical test the equipment examination must be closely coordinated with, and related to, the flight maneuvers portion but may not be given during the flight maneuvers portion. Notwithstanding § 61.21 the equipment examination may be given to an applicant who has completed a ground school that is part of an approved training program under Federal Aviation Regulations Part 121 for the airplane type involved and who is recommended by his instructor. The equipment examination must be repeated if the flight maneuvers portion is not satisfactorily completed within 60 days. The equipment examination must cover—

(1) Subjects requiring a practical knowledge of the airplane, its powerplants, systems, components, operational, and performance factors;

(2) Normal, abnormal, and emergency procedures, and the operations and limitations relating thereto; and

(3) The appropriate provisions of the approved Airplane Flight Manual.

(b) Preflight inspection. The pilot must—

(1) Conduct an actual visual inspection of the exterior and interior of the airplane, locating each item and explaining briefly the purpose of inspecting it; and

(2) Demonstrate the use of the prestart check list, appropriate control system checks, starting procedures, radio and electronic control system checks, and the selection of proper navigation and communications radio facilities and frequencies prior to flight.

If a flight is a required crewmember for the particular type airplane, the actual visual inspection may either be waived or it may be replaced by using an approved pictorial means that realistically portrays the location and detail of inspection items.

(c) Taxiing. This maneuver includes taxiing, sailing, or docking procedures in compliance with instructions issued by the appropriate traffic control authority or by the FAA inspector or designated examiner.

(d) Powerplant checks. As appropriate to the airplane type.

II. Takeoffs.

(a) Normal. One normal takeoff which, for the purpose of this maneuver, begins when the airplane is taxied into position on the runway to be used.

*(b) Instrument. One takeoff with instrument conditions simulated at or before reaching an altitude of 100 feet above the airport elevation.

(c) Cross wind. One cross wind takeoff, if practicable under the existing meteorological, airport, and traffic conditions.

*▲(d) Powerplant failure. One takeoff with a simulated failure of the most critical powerplant—

(1) At a point after V_1 and before V_2 that in the judgment of the person conducting the check is appropriate to the airplane type under the prevailing conditions; or

(2) At a point as close as possible after V_1 when V_1, and V_2 or V_1 and V_2 are identical; or

(3) At the appropriate speed for non-transport category airplanes.

For additional type rating in an airplane group with engines mounted in similar positions or from wing-mounted engines to aft fuselage-mounted engines this maneuver may be performed in a non-visual simulator.

(e) Rejected. A rejected takeoff performed in an airplane during a normal takeoff run after reaching a reasonable speed determined by giving due consideration to aircraft characteristics, runway length, surface conditions, wind direction and velocity, brake heat energy, and any other pertinent factors that may adversely affect safety or the airplane.

III. Instrument Procedures.

*(a) Area departure and area arrival. During each of these maneuvers the applicant must—

(1) Adhere to actual or simulated ATC clearances (including assigned radials); and

(2) Properly use available navigation facilities.

Either area arrival or area departure, but not both, may be waived under § 61.157(c).

*(b) Holding. This maneuver includes entering, maintaining, and leaving holding patterns. It may be performed under either area departure or area arrival.

(c) ILS and other instrument approaches. There must be the following:

*(1) At least one-normal ILS approach.

#(2) At least one manually controlled ILS approach with a simulated failure of one powerplant. The simulated

Maneuvers/Procedures	Required		Permitted			Waiver Provisions of §61.157(c)
	Simulated Instrument Conditions	Inflight	Visual Simulator	Non-Visual Simulator	Training Device	
I.(a) Equipment examination (oral)					X	
(b) Preflight inspection		X				
(b)(2) Demonstrate use of prestart check list				X		X*
II.(a) Normal		X	X			
*(b) Instrument	X			X		
*▲(d) Powerplant failure (1)		X*				
For additional type rating			X	X		X*
III.*(a) Area departure and area arrival	X		X	X		X*
*(b) Holding	X			X		X*
(c)*(1) At least one-normal ILS approach	X		X			
#(2) At least one manually controlled ILS approach	X		X			

Maneuvers/Procedures	Required		Permitted			Waiver Provisions of §61.157(c)
	Simulated Instrument Conditions	Inflight	Visual Simulator	Non-Visual Simulator	Training Device	

failure should occur before initiating the final approach course and must continue to touchdown or through the missed approach procedure.

However, either the normal ILS approach or the manually controlled ILS approach must be performed in flight.

(3) At least one nonprecision approach procedure that is representative of the nonprecision approach procedures that the applicant is likely to use. *(Simulated Instrument Conditions: X)*

(4) Demonstration of at least one nonprecision approach procedure on a letdown aid other than the approach procedure performed under subparagraph (3) of this paragraph that the applicant is likely to use. If performed in a synthetic instrument trainer, the procedures must be observed by the FAA inspector or designated examiner, or if the applicant has completed an approved training course under Part 121 of this chapter for the airplane type involved, the procedures may be observed by a person qualified to act as an instructor or check airman under that approved training program. *(Simulated Instrument Conditions: X; Visual Simulator: X; Training Device: X)*

Each instrument approach must be performed according to any procedures and limitations approved for the approach facility used. The instrument approach begins when the airplane is over the initial approach fix for the approach procedure being used (or turned over to the final approach controller in the case of GCA approach) and ends when the airplane touches down on the runway or when transition to a missed approach configuration is completed. Instrument conditions need not be simulated below 100 feet above touchdown zone elevation.

(d) Circling approaches. At least one circling approach must be made under the following conditions: *(Simulated Instrument Conditions: X)*

(1) The portion of the circling approach to the authorized minimum circling approach altitude must be made under simulated instrument conditions.

(2) The approach must be made to the authorized minimum circling approach altitude followed by a change in heading and the necessary maneuvering (by visual reference) to maintain a flight path that permits a normal landing on a runway at least 90 degrees from the final approach course of the simulated instrument portion of the approach.

(3) The circling approach must be performed without excessive maneuvering, and without exceeding the normal operating limits of the airplane. The angle of bank must not exceed 30 degrees.

When the maneuver is performed in an airplane, it may be waived as provided in § 61.157(c) if local conditions beyond the control of the pilot prohibit the maneuver or prevent it from being performed as required.

The circling approach maneuver is not required for a pilot employed by a certificate holder subject to the operating *(Waiver Provisions of §61.157(c): X*)*

rules of Part 121 of this chapter, if the certificate holder's manual prohibits a circling approach in weather conditions below 1000-3 (ceiling and visibility).

*(e) Missed approaches. Each applicant must perform at least two missed approaches, with at least one missed approach from an ILS approach. A complete approved missed approach procedure must be accomplished at least once and, at the discretion of the FAA inspector or designated examiner, a simulated powerplant failure may be required during any of the missed approaches. These maneuvers may be performed either independently or in conjunction with maneuvers required under Sections III or V of this appendix. At least one must be performed inflight. *(Simulated Instrument Conditions: X; Inflight: X*; Visual Simulator: X*; Waiver Provisions of §61.157(c): X)*

IV. Inflight Maneuvers.

*(a) Steep turns. At least one steep turn in each direction must be performed. Each steep turn must involve a bank angle of 45 degrees with a heading change of at least 180 degrees but not more than 360 degrees. *(Simulated Instrument Conditions: X; Non-Visual Simulator: X; Waiver Provisions of §61.157(c): X)*

*(b) Approaches to stalls. For the purpose of this maneuver the required approach to a stall is reached when there is a perceptible buffet or other response to the initial stall entry. Except as provided below, there must be at least three approaches to stalls as follows: *(Simulated Instrument Conditions: X; Non-Visual Simulator: X; Waiver Provisions of §61.157(c): X*)*

(1) One must be in the takeoff configuration (except where the airplane uses only a zero-flap takeoff configuration).

(2) One in a clean configuration.

(3) One in a landing configuration.

At the discretion of the FAA inspector or designated examiner, one approach to a stall must be performed in one of the above configurations while in a turn with a bank angle between 15 and 30 degrees. Two out of the three approaches required by this paragraph may be waived as provided in § 61.157(c).

*(c) Specific flight characteristics. Recovery from specific flight characteristics that are peculiar to the airplane type.

*(d) Powerplant failures. In addition to the specific requirements for maneuvers with simulated powerplant failures, the FAA inspector or designated examiner may require a simulated powerplant failure at any time during the check. *(Inflight: X; Non-Visual Simulator: X; Waiver Provisions of §61.157(c): X)*

V. Landings and Approaches to Landings.

Notwithstanding the authorizations for combining of maneuvers and for waiver of maneuvers, at least three actual landings (one to a full stop) must be made. These landings must include the types listed below but more than one type can be combined where appropriate:

(a) Normal landing.

*(b) Landing in sequence from an ILS instrument approach except that if circumstances beyond the control of the pilot prevent an actual landing, the person conducting the check *(Inflight: X*; Visual Simulator: X*)*

(Left portion of table)

Maneuvers/Procedures	Required		Permitted			
	Simulated Instrument Conditions	Inflight	Visual Simulator	Non-Visual Simulator	Training Device	Waiver Provisions of §61.157(c)
may accept an approach to a point where in his judgement a landing to a full stop could have been made. In addition, where a simulator approved for the landing maneuver out of an ILS approach is used, the approach may be continued through the landing and credit given for one of the three landings required by this section.						
(c) Cross wind landing, if practical under existing meteorological, airport, and traffic conditions.		X*				
*(d) Maneuvering to a landing with simulated powerplant failure, as follows: (1) In the case of 3-engine airplanes, maneuvering to a landing with an approved procedure that approximates the loss of two powerplants (center and one outboard-engine); or (2) In the case of other multiengine airplanes, maneuvering to a landing with a simulated failure of 50 percent of available powerplants, with the simulated loss of power on one side of the airplane. However, before January 1, 1975, in the case of a four engine turbojet powered airplane, maneuvering to a landing with a simulated failure of the most critical powerplant may be substituted therefor, if a flight instructor in an approved training program under Part 121 of this chapter certifies to the Administrator that he has observed the applicant satisfactorily perform a landing in that type airplane with a simulated failure of 50 percent of the available powerplants. The substitute maneuver may not be used if the Administrator determines that training in the two-engine out landing maneuver provided in the training program is unsatisfactory. If an applicant performs this maneuver in a visual simulator, he must, in addition, maneuver in flight to a landing with a simulated failure of the most critical powerplant.			X*			
*(e) Except as provided in paragraph (f), landing under simulated circling approach conditions, except that if circumstances beyond the control of the pilot prevent a landing, the person conducting the check may accept an approach to a point where in his judgment a landing to a full stop could have been made. The circling approach maneuver is not required for a pilot employed by a certificate holder subject to the operating rules of Part 121 of this chapter, if the certificate holder's manual prohibits a circling approach in weather conditions below 1000-3 (ceiling and visibility).			X*			
*(f) A rejected landing including a normal missed approach procedure, that is rejected approximately 50 feet over the runway and approximately over the runway threshold. This maneuver may be combined with instrument, circling, or missed approach procedures, but instrument conditions need not be simulated below 100 feet above the runway.	X*					

(Right portion of table)

Maneuvers/Procedures	Required		Permitted			
	Simulated Instrument Conditions	Inflight	Visual Simulator	Non-Visual Simulator	Training Device	Waiver Provisions of §61.157(c)
*(2) A zero-flap visual approach to a point where, in the judgment of the person conducting the check, a landing to a full stop on the appropriate runway could be made. This maneuver is not required for a particular airplane type if the Administrator has determined that the probability of flap extension failure on that type is extremely remote due to system design. In making its determination, the Administrator determines whether checking on slats only and partial flap approaches is necessary.			X*			
(b) For a single powerplant rating only, unless the applicant holds a commercial pilot certificate, he must accomplish accuracy approaches and spot landings that include a series of three landings from an altitude of 1000 feet or less, with the engine throttled and 180 degrees change in direction. The airplane must touch the ground in a normal landing attitude beyond and within 200 feet from a designated line. At least one landing must be from a forward slip. One hundred eighty degree approaches using two 90 degree turns with a straight base leg are preferred although circular approaches are acceptable.		X				
VI. Normal and Abnormal Procedures. Each applicant must demonstrate the proper use of as many of the systems and devices listed below as the FAA inspector or designated examiner finds are necessary to determine that the person being checked has a practical knowledge of the use of the systems and devices appropriate to the aircraft type:						
(a) Anti-icing and de-icing systems.				X		
(b) Auto-pilot systems.				X		
(c) Automatic or other approach aid systems.				X		
(d) Stall warning devices, stall avoidance devices, and stability augmentation devices.						
(e) Airborne radar devices.				X		
(f) Any other systems, devices, or aids available.				X		
(g) Hydraulic and electrical system failures and malfunctions.					X	
(h) Landing gear and flap systems failures or malfunctions.					X	
(i) Failure of navigation or communications equipment.				X		
VII. Emergency Procedures. Each applicant must demonstrate the proper emergency procedures for as many of the emergency situations listed below as the FAA inspector or designated examiner finds are necessary to determine that the person being checked has an adequate knowledge of, and ability to perform, such procedures:						
(a) Fire inflight.				X		
(b) Smoke control.				X		
(c) Rapid decompression.				X		
(d) Emergency descent.				X		
(e) Any other emergency procedures outlined in the appropriate approved airplane flight manual.						

Appendix B

[Practical Test Requirements for Rotorcraft Airline Transport Pilot Certificates with a Helicopter Class Rating and Associated Type Ratings]

[Throughout the maneuvers prescribed in this Appendix, good judgment commensurate with a high level of safety must be demonstrated. In determining whether such judgment has been shown, the FAA inspector or designated pilot examiner who conducts the check considers adherence to approved procedures, actions based on analysis of situations for which there is no prescribed procedure or recommended practice, and qualities of prudence and care in selecting a course of action. The successful outcome of a procedure or maneuver will never be in doubt.

Maneuvers/Procedures

The maneuvers and procedures in this appendix must be performed in a manner that satisfactorily demonstrates knowledge and skill with respect to—

(1) The helicopter, its systems and components;

(2) Proper control of airspeed, direction, altitude, and attitude in accordance with procedures and limitations contained in the approved Rotorcraft Flight Manual, checklists, or other approved material appropriate to the rotorcraft type; and

(3) Compliance with approved en route, instrument approach, missed approach, ATC, and other applicable procedures.

I. Preflight

(a) *Equipment examination (oral).* The equipment examination must be repeated if the flight maneuvers portion is not satisfactorily completed within 60 days. The equipment examination must cover—

(1) Subjects requiring a practical knowledge of the helicopter, its powerplants, systems, components, and operational and performance factors;

(2) Normal, abnormal, and emergency procedures and related operations and limitations; and

(3) The appropriate provisions of the approved helicopter Flight Manual or manual material.

(b) *Preflight inspection.* The pilot must—

(1) Conduct an actual visual inspection of the exterior and interior of the helicopter, locating each item and explaining briefly the purpose of inspecting it; and

(2) Demonstrate the use of the prestart checklist, appropriate control system checks, starting procedures, radio and electronic equipment checks, and the selection of proper navigation and communications radio facilities and frequencies before flight.

(c) *Taxiing.* The maneuver includes ground taxiing, hover taxiing (including performance checks), and docking procedures, as appropriate, in compliance with instructions

issued by ATC, the FAA inspector, or the designated pilot examiner.

(d) *Powerplant checks.* As appropriate to the helicopter type in accordance with the Rotorcraft Flight Manual procedures.

II. Takeoffs

(a) *Normal.* One normal takeoff from a stabilized hover which begins when the helicopter is taxied into position for takeoff.

(b) *Instrument.* One takeoff with instrument conditions simulated at or before reaching 100 feet above airport elevation.

(c) *Crosswind.* One crosswind takeoff from a stabilized hover, if practical under the existing meteorological, airport, and traffic conditions.

(d) *Powerplant failure.*

(1) For single-engine rotorcraft, one normal takeoff with simulated powerplant failure.

(2) For multiengine rotorcraft, one normal takeoff with simulated failure of one engine—

(i) At an appropriate airspeed that would allow continued climb performance in forward flight; or]

[(ii) At an appropriate airspeed that is 50 percent of normal cruise speed, if there is no published single-engine climb airspeed for that type of helicopter.

(e) *Rejected.* One normal takeoff that is rejected after simulated engine failure at a reasonable airspeed, determined by giving due consideration to the helicopter's characteristics, length of landing area, surface conditions, wind direction and velocity, and any other pertinent factors that may adversely affect safety.

III. Instrument Procedures

(a) *Area departure and arrival.* During each of these maneuvers, the applicant must—

(1) Adhere to actual or simulated ATC clearance (including assigned bearings or radials); and

(2) Properly use available navigation facilities.

(b) *Holding.* This maneuver includes entering, maintaining, and leaving holding patterns.

(c) *ILS and other instrument approaches.* The instrument approach begins when the helicopter is over the initial approach fix for the approach procedure being used (or turned over to the final controller in case of a surveillance or precision radar approach) and ends when the helicopter terminates at a hover or touches down or where transition to a missed approach is completed. The following approaches must be performed:

(1) At least on normal ILS approach.

(2) For multiengine rotorcraft, at least one manually controlled ILS approach with a simulated failure of one powerplant. The simulated engine failure should occur before initiating the final approach course and continue to a hover to touchdown or through the missed approach procedure.

(3) At least one nonprecision approach procedure that is representative of the nonprecision approach procedure that the applicant is likely to use.

(4) At least one nonprecision approach procedure on a letdown aid other than the approach procedure performed under subparagraph (3) of this paragraph that the applicant is likely to use.

(d) *Circling approaches.* At least one circling approach must be made under the following conditions:

(1) The portion of the circling approach to the authorized minimum circling approach altitude must be made under simulated instrument conditions.

(2) The approach must be made to the authorized minimum circling approach altitude followed by a change in heading and the necessary maneuvering (by visual reference) to maintain a flight path that permits a normal landing on a runway at least 90 degrees from the final approach course of the simulated instrument portion of the approach.

(3) The circling approach must be performed without excessive maneuvering and without exceeding the normal operating limits of the rotorcraft. The angle of bank should not exceed 30 degrees.

(e) *Missed approaches.* Each applicant must perform at least two missed approaches with at least one missed approach from an ILS approach. At the discretion of the FAA inspector or designated examiner, a simulated powerplant failure may be required during any of the missed approaches. The maneuvers may be performed either independently or in conjunction with maneuvers required under section III or V of the Appendix. At least one must be performed in flight.

IV. In-flight Maneuvers

(a) *Steep turns.* At least one steep turn in each direction must be performed. Each steep turn must involve a bank angle of 30 degrees with a heading change of at least 180 degrees but not more than 360 degrees.

(b) *Settling with power.* Demonstrate recognition of and recovery from imminent flight at critical/rapid descent with power. For the purpose of this maneuver, settling with power is reached when a perceptive buffet or other indications of imminent settling with power have been induced.

(c) *Powerplant failure.* In addition to the specific requirements for maneuvers with simulated powerplant failures, the FAA inspector or designated examiner may require a simulated powerplant failure at any time during the check.

(d) *Recovery from unusual attitudes.*]

V. Approaches and Landings

[(a) *Normal.* One normal approach to a stabilized hover or to the ground must be performed.

(b) *Instrument.* One approach to a hover or

to a landing in sequence from an ILS instrument approach.

(c) *Crosswind.* One crosswind approach to a hover or to the ground, if practical under the existing meteorological, airport, or traffic conditions.

(d) *Powerplant failure.* For a multiengine rotorcraft, maneuvering to a landing with simulated powerplant failure of one engine.

(e) *Rejected.* Rejected landing, include a normal missed approach procedure at approximately 50 feet above the runway. This maneuver may be combined with instrument or missed approach procedures, but instrument conditions need not be simulated below 100 feet above the runway or landing area.

(f) *Autorotative landings.* Autorotative landings in a single-engine helicopter. The applicant may be required to accomplish at least one autorotative approach and landing from any phase of flight as specified by the FAA inspector or designated examiner.

VI. Normal and Abnormal Procedures

Each applicant must demonstrate the proper use of as many systems and devices listed below as the FAA inspector or designated examiner finds are neccessary to determine that the applicant has a practical knowledge of the use of the systems and devices appropriate to helicopter type:

(a) Anti-icing or deicing system.

(b) Autopilot or other stability augmentation devices.

(c) Airborne radar devices.

(d) Hydraulic and electrical systems failures or malfunctions.

(e) Landing gear failures or malfunctins.

(f) Failure of navigation or communications equipment.

(g) Any other system appropriate to the helicopter as outlined in the approved Rotorcraft Flight Manual.

VII. Emergency Procedrues

Each applicant must demonstrate the proper emergency procedures for as many of the emergency situations listed below as the FAA inspector or designated examiner finds are necessary to determine that the applicant has adequate knowledge of, and ability to perform, such procedures:

(a) Fire or smoke control in flight.

(b) Ditching.

(c) Evacuation.

(d) Operations of emergency equipment.

(e) Emergency descent.

(f) Any other emergency procedure outline in the approved Rotorcrat Flight Manual.]

FEDERAL AVIATION REGULATIONS

Part 67—Medical Standards and Certification

Contents

Subpart A—General

§ 67.1 Applicability.

This subpart prescribes the medical standards for issuing medical certificates for airmen.

[§ 67.11 Issue.

[Except as provided in § 67.12, an applicant who meets the medical standards prescribed in this Part, based on medical examination and evaluation of his history and condition, is entitled to an appropriate medical certificate.

[§ 67.12 Certification of foreign airmen.

[A person who is neither a United States citizen nor a resident alien is issued a certificate under this Part, outside the United States, only when the Administrator finds that the certificate is needed for operation of a U.S.-registered civil aircraft.]

§ 67.13 First-class medical certificate.

(a) To be eligible for a first-class medical certificate, an applicant must meet the requirements of paragraphs (b) through (f) of this section.

(b) *Eye:*

(1) Distant visual acuity of 20/20 or better in each eye separately, without correction; or of at least 20/100 in each eye separately corrected to 20/20 or better with corrective lenses (glasses or contact lenses), in which case the applicant may be qualified only on the condition that he wears those corrective lenses while exercising the privileges of his airman certificate. ☞**F76,77**

(2) Near vision of at least v = 1.00 at 18 inches with each eye separately, with or without corrective glasses.

(3) Normal color vision.

(4) Normal fields of vision.

(5) No acute or chronic pathological condition of either eye or adenexae that might interfere with its proper function, might progress to that degree, or might be aggravated by flying.

(6) Bifoveal fixation and vergencedphoria relationship sufficient to prevent a break in fusion under conditions that may reasonably occur in performing airman duties.

Tests for the factors named in subparagraph (6)

of this paragraph are not required except for applicants found to have more than one prism diopter of hyperphoria, six prism diopters of esophoria, or six prism diopters of exophoria. If these values are exceeded, the Federal Air Surgeon may require the applicant to be examined by a qualified eye specialist to determine if there is bifoveal fixation and adequate vergencephoria relationship. However, if the applicant is otherwise qualified, he is entitled to a medical certificate pending the results of the examination.

(c) *Ear, nose, throat, and equilibrium:*

(1) Ability to—

(i) Hear the whispered voice at a distance of at least 20 feet with each ear separately; or

(ii) Demonstrate a hearing acuity of at least 50 percent of normal in each ear throughout the effective speech and radio range as shown by a standard audiometer.

(2) No acute or chronic disease of the middle or internal ear.

(3) No disease of the mastoid.

(4) No unhealed (unclosed) perforation of the eardrum.

(5) No disease or malformation of the nose or throat that might interfere with, or be aggravated by, flying.

(6) No disturbance in equilibrium.

(d) *Mental and neurologic:*

(1) Mental.

(i) No established medical history or clinical diagnosis of any of the following:

(a) A personality disorder that is severe enough to have repeatedly manifested itself by overt acts.

(b) A psychosis.

[*(c)* Alcoholism, unless there is established clinical evidence, satisfactory to the Federal Air Surgeon, of recovery, including sustained total abstinence from alcohol for not less than the preceding 2 years. As used in this section, "alcoholism" means a condition in which a person's intake of alcohol is great enough to damage physical health or personal or social functioning, or when alcohol has become a prerequisite to normal functioning.]

(d) Drug dependence. As used in this section, "drug dependence" means a condition in which a person is addicted to or dependent on drugs other than alcohol, tobacco, or ordinary caffeine-containing beverages, as evidenced by habitual use or a clear sense of need for the drug.

(ii) No other personality disorder, neurosis, or mental condition that the Federal Air Surgeon finds—

(a) Makes the applicant unable to safely perform the duties or exercise the privileges of the airman certificate that he holds or for which he is applying; or

(b) May reasonably be expected, within two years after the finding, to make him unable to perform those duties or exercise those privileges;

and the findings are based on the case history and appropriate qualified, medical judgement relating to the condition involved.

(2) Neurologic.

(i) No established medical history or clinical diagnosis of either of the following:

(a) Epilepsy.

(b) A disturbance of consciousness without satisfactory medical explanation of the cause.

(ii) No other convulsive disorder, disturbance of consciousness, or neurologic condition that the Federal Air Surgeon finds—

(a) Makes the applicant unable to safely perform the duties or exercise the privileges of the airman certificate that he holds or for which he is applying; or

(b) May reasonably be expected, within two years after the finding, to make him unable to perform those duties or exercise those privileges;

and the findings are based on the case history and appropriate, qualified, medical judgement relating to the condition involved.

(e) *Cardiovascular:*

(1) No established medical history or clinical diagnosis of—

[(i) Myocardial infarction;

[(ii) Angina pectoris; or

[(iii) Coronary heart disease that has required treatment or, if untreated, that has been symptomatic or clinically significant.]

(2) If the applicant has passed his thirty-fifth birthday but not his fortieth, he must, on the first examination after his thirty-fifth birthday, show an absence of myocardial infarction on electrocardiographic examination.

(3) If the applicant has passed his fortieth birthday, he must annually show an absence of myocardial infarction on electrocardiographic examination.

(4) Unless the adjusted maximum readings apply, the applicant's reclining blood pressure may not be more than the maximum reading for his age group in the following table: ☞ **F45**

Age Group	Maximum readings (reclining blood pressure in mm)		Adjusted maximum readings (reclining blood pressure in mm)[1]	
	Systolic	Diastolic	Systolic	Diastolic
20–29	140	88	—	—
30–39	145	92	155	98
40–49	155	96	165	100
50 and over	160	98	170	100

[1]For an applicant at least 30 years of age whose reclining blood pressure is more than the maximum reading for his age group and whose cardiac and kidney conditions, after complete cardiovascular examination, are found to be normal.

(5) If the applicant is at least 40 years of age, he must show a degree of circulatory efficiency that is compatible with the safe operation of aircraft at high altitudes.

An electrocardiogram, made according to acceptable standards and techniques within the 90 days before an examination for a firstclass certificate, is accepted at the time of the physical examination as meeting the requirements of subparagraphs (2) and (3) of this paragraph.

(f) *General medical condition:*

(1) No established medical history or clinical diagnosis of diabetes mellitus that requires insulin or any other hypoglycemic drug for control.

(2) No other organic, functional, or structural disease, defect, or limitation that the Federal Air Surgeon finds— ☞**F78**

(i) Makes the applicant unable to safely perform the duties or exercise the privileges of the airman certificate that he holds or for which he is applying; or

(ii) May reasonably be expected, within two years after the finding, to make him unable to perform those duties or exercise those privileges;

and the findings are based on the case history and appropriate, qualified, medical judgment relating to the condition involved.

[(g) An applicant who does not meet the provisions of paragraphs (b) through (f) of this section may apply for the discretionary issuance of a certificate under § 67.19.]

§ 67.15 Second-class medical certificate.

(a) To be eligible for a second-class medical certificate, an applicant must meet the requirements of paragraphs (b) through (f) of this section.

(b) *Eye:*

(1) Distant visual acuity of 20/20 or better in each eye separately, without correction; or of at least 20/100 in each eye separately corrected to 20/20 or better with corrective lenses (glasses or contact lenses), in which case the applicant may be qualified only on the condition that he wears those corrective lenses while exercising the privileges of his airman certificate. ☞ **F76,77,79**

(2) Enough accommodation to pass a test prescribed by the Administrator based primarily on ability to read official aeronautical maps.

(3) Normal fields of vision.

(4) No pathology of the eye.

(5) Ability to distinguish aviation signal red, aviation signal green, and white.

(6) Bifoveal fixation and vergencephoria relationship sufficient to prevent a break in fusion under conditions that may reasonably occur in performing airman duties.

Tests for the factors named in subparagraph (6) of this paragraph are not required except for applicants found to have more than one prism diopter of hyperphoria, six prism diopters of esophoria, or six prism diopters of exophoria. If these values are exceeded, the Federal Air Surgeon may require the applicant to be examined by a qualified eye specialist to determine if there is bifoveal fixation and adequate vergencephoria relationship. However, if the applicant is otherwise qualified, he is entitled to a medical certificate pending the results of the examination.

(c) *Ear, nose, throat, and equilibrium:*

(1) Ability to hear the whispered voice at 8 feet with each ear separately.

(2) No acute or chronic disease of the middle or internal ear.

(3) No disease of the mastoid.

(4) No unhealed (unclosed) perforation of the eardrum.

(5) No disease or malformation of the nose or throat that might interfere with, or be aggravated by, flying.

(6) No disturbance in equilibrium.

(d) *Mental and neurologic:*

(1) Mental.

(i) No established medical history or clinical diagnosis of any of the following:

(a) A personality disorder that is severe enough to have repeatedly manifested itself by overt acts.

(b) A psychosis.

[(c) Alcoholism, unless there is established clinical evidence, satisfactory to the Federal Air Surgeon, of recovery, including sustained total abstinence from alcohol for not less than the preceding 2 years. As used in this section, "alcoholism" means a condition in which a person's intake of alcohol is great enough to damage physical health or personal or social functioning, or when alcohol has become a prerequisite to normal functioning.]

(d) Drug dependence. As used in this section, "drug dependence" means a condition in which a person is addicted to or dependent on drugs other than alcohol, tobacco, or ordinary caffeine-containing beverages, as evidenced by habitual use or a clear sense of need for the drug.

(ii) No other personality disorder, neurosis, or mental condition that the Federal Air Surgeon finds—

(a) Makes the applicant unable to safely perform the duties or exercise the privileges of the airman certificate that he holds or for which he is applying; or

(b) May reasonably be expected, within two years after the finding, to make him unable to perform those duties or exercise those privileges;

and the findings are based on the case history and appropriate, qualified, medical judgment relating to the condition involved.

(2) Neurologic.

(i) No established medical history or clinical diagnosis of either of the following:

(a) Epilepsy.

(b) A disturbance of consciousness without satisfactory medical explanation of the cause.

(ii) No other convulsive disorder, disturbance of consciousness, or neurologic condition that the Federal Air Surgeon finds—

(a) Makes the applicant unable to safely perform the duties or exercise the privileges of the airman certificate that he holds or for which he is applying; or

(b) May reasonably be expected, within two years after the finding, to make him unable to perform those duties or exercise those privileges;

and the findings are based on the case history and appropriate, qualified, medical judgment relating to the condition involved.

(e) *Cardiovascular:*

(1) No established medical history or clinical diagnosis of— ☞ **F45**

[(i) Myocardial infarction;

[(ii) Angina pectoris; or

[(iii) Coronary heart disease that has required treatment or, if untreated, that has been symptomatic or clinically significant.]

(f) *General medical condition:*

(1) No established medical history or clinical diagnosis of diabetes mellitus that requires insulin or any other hypoglycemic drug for control.

(2) No other organic, functional, or structural disease, defect, or limitation that the Federal Air Surgeon finds— ☞**F78**

(i) Makes the applicant unable to safely perform the duties or exercise the privileges of the airman certificate that he holds or for which he is applying; or

(ii) May reasonably be expected, within 2 years after the finding to make him unable to perform those duties or exercise those privileges;

and the findings are based on the case history and appropriate, qualified, medical judgment relating to the condition involved.

[(g) An applicant who does not meet the provisions of paragraphs (b) through (f) of this section may apply for the discretionary issuance of a certificate under § 67.19.]

§ 67.17 Third-class medical certificate.

(a) To be eligible for a third-class medical certificate, an applicant must meet the requirements of paragraphs (b) through (f) of this section.

(b) *Eye:*

(1) Distant visual acuity of 20/50 or better in each eye separately, without correction; or if the vision in either or both eyes is poorer than 20/50 and is corrected to 20/30 or better in each eye with corrective lenses (glasses or contact lenses), the applicant may be qualified on the condition that he wears those corrective lenses while exercising the privileges of his airman certificate. ☞ **F77**

(2) No serious patholoy of the eye.

(3) Ability to distinguish aviation signal red, aviation signal green, and white.

(c) *Ears, nose, throat, and equilibrium:*

(1) Ability to hear the whispered voice at 3 feet.

(2) No acute or chronic disease of the internal ear.

(3) No disease or malformation of the nose or throat that might interfere with, or be aggravated by, flying.

(4) No disturbance in equilibrium.

(d) *Mental and neurologic:*

(1) Mental.

(i) No established medical history or clinical diagnosis of any of the following:

(*a*) A personality disorder that is severe enough to have repeatedly manifested itself by overt acts.

(*b*) A psychosis.

[(*c*) Alcoholism, unless there is established clinical evidence, satisfactory to the Federal Air Surgeon, of recovery, including sustained total abstinence from alcohol for not less than the preceding 2 years. As used in this section, "alcoholism" means a condition in which a person's intake of alcohol is great enough to damage physical health or personal or social functioning, or when alcohol has become a prerequisite to normal functioning.]

(*d*) Drug dependence. As used in this section, "drug dependence" means a condition in which a person is addicted to or dependent on drugs other than alcohol, tobacco, or ordinary caffeine-containing beverages, as evidenced by habitual use or a clear sense of need for the drug.

(ii) No other personality disorder, neurosis, or mental condition that the Federal Air Surgeon finds—

(*a*) Makes the applicant unable to safely perform the duties or exercise the privileges of the airman certificate that he holds or for which he is applying; or

(*b*) May reasonably be expected, within two years after the finding, to make him unable to perform those duties or exercise those privileges;

and the findings are based on the case history and appropriate, qualified, medical judgment relating to the condition involved.

(2) Neurologic.

(i) No established medical history or clinical diagnosis of either of the following:

(*a*) Epilepsy

(*b*) A disturbance of consciousness without satisfactory medical explanation of the cause.

(ii) No other convulsive disorder, disturbance of consciousness, or neurologic

condition that the Federal Air Surgeon finds—

(*a*) Makes the applicant unable to safely perform the duties or exercise the privileges of the airman certificate that he holds or for which he is applying; or

(*b*) May reasonably be expected, within two years after the finding, to make him unable to perform those duties or exercise those privileges;

and the findings are based on the case history and appropriate, qualified, medical judgment relating to the condition involved.

(e) *Cardiovascular:* ☞ **F45, 80–82**

(1) No established medical history or clinical diagnosis of—

[(i) Myocardial infraction;

[(ii) Angina pectoris; or

[(iii) Coronary heart disease that has required treatment or, if untreated, that has been symptomatic or clinically significant.]

(f) *General medical condition:*

(1) No established medical history or clinical diagnosis of diabetes mellitus that requires insulin or any other hypoglycemic drug for control;

(2) No other organic, functional or structural disease, defect, or limitation that the Federal Air Surgeon finds— ☞ **F78**

(i) Makes the applicant unable to safely perform the duties or exercise the privileges of the airman certificate that he holds or for which he is applying; or

(ii) May reasonably be expected, within 2 years after the finding, to make him unable to perform those duties or exercise those privileges;

and the findings are based on the case history and appropriate, qualified, medical judgment relating to the condition involved.

[(g) An applicant who does not meet the provisions of paragraphs (b) through (f) of this section may apply for the discretionary issuance of a certificate under § 67.19.]

[§ **67.19 Special issue of medical certificates.**

[(a) At the discretion of the Federal Air Surgeon, a medical certificate may be issued to an applicant who does not meet the applicable provisions of §§ 67.13, 67.15, or 67.17 if the applicant shows to the satisfaction of the Federal Air Surgeon that the duties authorized by the class of medical certificate applied for can be performed without endangering air commerce during the period in which the certificate would be in force. The Federal Air Surgeon may authorize a special medical flight test, practical test, or medical evaluation for this purpose.

[(b) The Federal Air Surgeon may consider the applicant's operational experience and any medical facts that may affect the ability of the applicant to perform airman duties including:

[(1) The combined effect on the applicant of failure to meet more than one requirement of this Part; and

[(2) The prognosis derived from professional consideration of all available information regarding the airman.

[(c) In determining whether the special issuance of a third-class medical certificate should be made to an applicant, the Federal Air Surgeon considers the freedom of an airman,

exercising the privileges of a private pilot certificate, to accept reasonable risks to his or her person and property that are not acceptable in the exercise of commercial or airline transport privileges, and, at the same time, considers the need to protect the public safety of persons and property in other aircraft and on the ground.

[(d) In issuing a medical certificate under this section, the Federal Air Surgeon may do any or all of the following:

[(1) Limit the duration of the certificate.

[(2) Condition the continued effect of the certificate on the results of subsequent medical tests, examinations, or evaluations.

[(3) Impose any operational limitation on the certificate needed for safety.

[(4) Condition the continued effect of a second- or third-class medical certificate on compliance with a statement of functional limitations issued to the applicant in coordination with the Director of Flight Operations or the Director's designee.

[(e) An applicant who has been issued a medical certificate under this section based on a special medical flight or practical test need not take the test again during later physical examinations unless the Federal Air Surgeon determines that the physical deficiency has become enough more pronounced to require another special medical flight or practical test.

[(f) The authority of the Federal Air Surgeon under this section is also exercised by the Chief, Aeromedical Certification Branch, Civil Aeromedical Institute, and each Regional Flight Surgeon.]

§ **67.20 Applications, certificates, logbooks, reports, and records: falsification, reproduction, or alteration.**

(a) No person may make or cause to be made—

(1) Any fraudulent or intentionally false statement on any application for a medical certificate under this Part;

(2) Any fraudulent or intentionally false entry in any logbook, record, or report that is required to be kept, made, or used, to show compliance with any requirement for any medical certificate under this Part;

(3) Any reproduction, for fraudulent purpose, of any medical certificate under this Part; or

(4) Any alteration of any medical certificate under this Part.

(b) The commission by any person of an act prohibited under paragraph (a) of this section is a basis for suspending or revoking any airman, ground instructor, or medical certificate or rating held by that person.

Subpart B—Certification Procedures

§ **67.21 Applicability.**

This subpart prescribes the general procedures that apply to the issue of medical certificates for airmen.

§ **67.23 Medical examinations: who may give.**

(a) *First class.* Any aviation medical examiner who is specifically designated for the purpose may give the examination for the first class certificate. Any interested person may obtain a list of these aviation medical examiners, in any area, from the FAA Regional

Director of the region in which the area is located. ☞**F60**

(b) *Second class and third class.* Any aviation medical examiner may give the examination for the second or third class certificate. Any interested person may obtain a list of aviation medical examiners, in any area, from the FAA Regional Director of the region in which the area is located. ☞**F60**

§ 67.25 Delegation of authority.

(a) The authority of the Administrator, under section 602 of the Federal Aviation Act of 1958 (49 U.S.C. 1422), to issue or deny medical certificates is delegated to .the Federal Air Surgeon, to the extent necessary to—

(1) Examine applicants for and holders of medical certificates for compliance with applicable medical standards; and

(2) Issue, renew, or deny medical certificates to applicants and holders based upon compliance or noncompliance with applicable medical standards.

Subject to limitations in this chapter, the authority delegated in subparagraphs (1) and (2) of this paragraph is also delegated to aviation medical examiners and to authorized representatives of the Federal Air Surgeon within the FAA.

(b) The authority of the Administrator, under subsection 314(b) of the Federal Aviation Act of 1958 (49 U.S.C. 1355(b)), to reconsider the action of an aviation medical examiner is delegated to the Federal Air Surgeon, the Chief, Aeromedical Certification Branch, Civil Aeromedical Institute, and each Regional Flight Surgeon. [Where the applicant does not meet the standards of §§ 67.13(d) (1) (ii), (d) (2) (ii), or (f) (2), 67.15(d) (1) (ii), (d) (2) (ii), or (f) (2), or 67.17(d) (1) (ii), (d) (2) (ii), or (f) (2), any action taken under this paragraph other than by the Federal Air Surgeon is subject to reconsideration by the Federal Air Surgeon.] A certificate issued by an aviation medical examiner is considered to be affirmed as issued unless an FAA official named in this paragraph on his own initiative reverses that issuance within 60 days after the date of issuance. However, if within 60 days after the date of issuance that official requests the cer-

tificate holder to submit additional medical information, he may on his own initiative reverse the issuance within 60 days after he receives the requested information.

(c) The authority of the Administrator, under section 609 of the Federal Aviation Act of 1958 (49 U.S.C. 1429), to re-examine any civil airman, to the extent necessary to determine an airman's qualification to continue to hold an airman medical certificate, is delegated to the Federal Air Surgeon and his authorized representatives within the FAA.

§ 67.27 Denial of medical certificate.

(a) Any person who is denied a medical certificate by an aviation medical examiner may, within 30 days after the date of the denial, apply in writing and in duplicate to the Federal Air Surgeon, Attention: Chief, Aeromedical Certification Branch, Civil Aeromedical Institute, Federal Aviation Administration, P.O. Box 25082, Oklahoma City, Okla. 73125, for reconsideration of that denial. If he does not apply for reconsideration during the 30 day period after the date of the denial, he is considered to have withdrawn his application for a medical certificate.

(b) The denial of a medical certificate—

(1) By an aviation medical examiner is not a denial by the Administrator under section 602 of the Federal Aviation Act of 1958 (49 U.S.C. 1422);

(2) By the Federal Air Surgeon is considered to be a denial by the Administrator under that section of the Act; and

[(3) By the Chief, Aeromedical Certification Branch, Civil Aeromedical Institute, or a Regional Flight Surgeon is considered to be a denial by the Administrator under the Act except where the applicant does not meet the standards of §§ 67.13 (d) (1) (ii), (d) (2) (ii), or (f) (2), 67.15(d) (1) (ii), (d) (2) (ii), (f) (2), or 67.17(d) (1) (ii), (d) (2) (ii), or (f) (2).

[(c) Any action taken under § 67.25(b) that wholly or partly reverses the issue of a medical certificate by an aviation medical examiner is the denial of a medical certificate under paragraph (b) of this section.

[(d) If the issue of a medical certificate is wholly or partly reversed upon reconsideration by the Federal Air Surgeon, the Chief, Aeromedical Cer-

tification Branch, Civil Aeromedical Institute, or a Regional Flight Surgeon, the person holding that certificate shall surrender it, upon request of the FAA.]

§ 67.29 Medical certificates by senior flight surgeons of Armed Forces.

(a) The FAA has designated senior flight surgeons of the Armed Forces on specified military posts, stations, and facilities, as aviation medical examiners.

(b) An aviation medical examiner described in paragraph (a) of this section may give physical examinations to applicants for FAA medical certificates who are on active duty or who are, under Department of Defense medical programs, eligible for FAA medical certification as civil airmen. In addition, such an examiner may issue or deny an appropriate FAA medical certificate in accordance with the regulations of this chapter and the policies of the FAA.

(c) Any interested person may obtain a list of the military posts, stations and facilities at which a senior flight surgeon has been designated as an aviation medical examiner, from the Surgeon General of the Armed Force concerned or from the Chief, Aeromedical Certification Branch, AC-130, Department of Transportation, Federal Aviation Administration, Civil Aeromedical Institute, P. O. Box 25082, Oklahoma City, Oklahoma 73125.

§ 67.31 Medical records.

Whenever the Administrator finds that additional medical information or history is necessary to determine whether an applicant for or the holder of a medical certificate meets the medical standards for it, he requests that person to furnish that information or authorize any clinic, hospital, doctor, or other person to release to the Administrator any available information or records concerning that history. If the applicant, or holder, refuses to provide the requested medical information or history or to authorize the release so requested, the Administrator may suspend, modify, or revoke any medical certificate that he holds or may, in the case of an applicant, refuse to issue a medical certificate to him.

DEPARTMENT OF TRANSPORTATION

Federal Aviation Administration

14 CFR Part 67
Falsification of Airman Medical Certificate Applications; Record of Traffic Convictions

AGENCY: Federal Aviation Administration (FAA), DOT.

ACTION: Notice of Enforcement Policy.

SUMMARY: Applicants for an airman medical certificate who have failed to disclose information with respect to traffic convictions (such as convictions for driving while intoxicated) on their applications may have violated § 67.20 of the Federal Aviation Regulations (14 CFR 67.20) by making intentionally false or fraudulent statements. This notice announces the termination of a previously announced policy allowing any such applicant to avoid enforcement action against his or her certificates, based on such a falsification, by providing the FAA with corrected information before the FAA became aware of any incorrect statement.

EFFECTIVE DATE: December 1, 1988.

FOR FURTHER INFORMATION CONTACT: Peter J. Lynch, Manager, Enforcement Proceedings Branch, AGC-250, Office of the Chief Counsel, 800 Independence Avenue SW., Washington, DC 20591; telephone (202) 267-9956.

SUPPLEMENTARY INFORMATION: On October 22, 1987, the FAA issued a notice of enforcement policy with respect to persons who may have violated § 67.20 by falsifying their applications for airman medical certification with regard to their record of traffic convictions. The notice was published at 52 FR 41557 (October 29, 1987). That notice announced a policy which allowed such persons to avoid related FAA enforcement action against his or her airman, ground instructor, or medical certificates by providing corrected information before January 1, 1988, even if by that date the FAA had become aware of the apparent falsification. As to the FAA's policy which has remained in effect since January 1, 1988, the notice stated:

* * * from the date of this notice and *until further notice,* where the airman has voluntarily supplied to the FAA's Aeromedical Certification Branch information regarding a record of traffic convictions in his or her medical application prior to the FAA's becoming aware of any incorrect statement in the application, the FAA will not take action against the airman's certificates on the basis of falsification for any falsification disclosed by such voluntarily disclosed information.

(Emphasis supplied.) As the FAA stated in its notice of October 1987, the Inspector General of the United States Department of Transportation (IG) has identified some airmen who appear to have falsified their applications with regard to their record of traffic convictions. The notice further indicated that the IG was referring these cases to the FAA for appropriate action. The IG has made a large number of such referrals to the FAA.

This notice is to advise that the previously announced policy for allowing any airman to avoid possible FAA certificate action (often referred to as the FAA's "amnesty program") is terminated, effective December 1, 1988.

Availability of this Notice

Any person may obtain a copy of this notice by submitting a request to the Federal Aviation Administration, Office of Public Affairs, Attention: Public Inquiry Center, APA-230, 800 Independence Avenue, SW., Washington, DC 20591, or by calling (202) 267-3484.

Issued in Washington, DC, on October 27, 1988.

T. Allan McArtor,
Administrator.

[FR Doc. 88-25251 Filed 10-31-88; 8:45 am]

FEDERAL AVIATION REGULATIONS

Part 71—Designation of Federal Airways, Area Low Routes, Controlled Airspace, and Reporting Points

Contents

Subpart A—General

§ 71.1 Applicability.

(a) The airspace assignments described in Subparts B and C are designated as Federal airways.

(b) The airspace assignments described in Subparts B through I are designated as control areas, the continental control area, control zones, transition areas, positive control areas, and reporting points, as described in the appropriate subpart.

(c) The airspace assignments described in Subpart K of this Part are designated as terminal control areas.

(d) The airspace assignments described in Subpart J are designated as area low routes.

[(e) The airspace assignments described in Subpart L of this Part are designated as airport radar service areas.]

§ 71.3 Classification of Federal airways.

Federal airways are classified as follows:

(a) Colored Federal airways:

(1) Green Federal airways.

(2) Amber Federal airways.

(3) Red Federal airways.

(4) Blue Federal airways.

(b) VOR Federal airways.

§ 71.5 Extent of Federal airways.

(a) Each Federal airway is based on a centerline that extends from one navigational aid or intersection to another navigational aid (or through several navigational aids or intersections) specified for that airway.

(b) Unless otherwise specified in Subpart B or C—

(1) Each Federal airway includes the airspace within parallel boundary lines 4 miles each side of the centerline. Where an airway changes direction, it includes that airspace enclosed by extending the boundary lines of the airway segments until they meet;

(2) Where the changeover point for an airway segment is more than 51 miles from either of the navigational aids defining that segment, and—

(i) The changeover point is midway between the navigational aids, the airway in-cludes the airspace between lines diverging at angles of 4.5° from the centerline at each navigational aid and extending until they intersect opposite the changeover point; or

(ii) The changeover point is not midway between the navigational aids, the airway includes the airspace between lines diverging at angles of 4.5° from the centerline at the navigational aid more distant from the changeover point, and extending until they intersect with the bisector of the angle of the centerlines at the changeover point; and between lines connecting these points of intersection and the navigational aid nearer to the changeover point;

(3) Where an airway terminates at a point or intersection more than 51 miles from the closest associated navigational aid it includes the additional airspace within lines diverging at angles of 4.5° from the centerline extending from the associated navigational aid to a line perpendicular to the centerline at the termination point;

(4) Where an airway terminates, it includes

the airspace within a circle centered at the specified navigational aid or intersection having a diameter equal to the airway width at that point. However, an airway does not extend beyond the domestic/oceanic control area boundary.

(c) Unless otherwise specified in Subpart B or C—

(1) Each Federal airway includes that airspace extending upward from 1,200 feet above the surface of the earth to, but not including, 18,000 feet MSL, except that Federal airways for Hawaii have no upper limits. Variations of the lower limits of an airway are expressed in digits representing hundreds of feet above the surface (AGL) or mean sea level (MSL) and, unless otherwise specified, apply to the segment of an airway between adjoining navigational aids or intersections; and

(2) The airspace of a Federal airway within the lateral limits of a transition area has a floor coincident with the floor of the transition area.

(d) One or more alternate airways may be designated between specified navigational aids or intersections along each VOR Federal airway described in Subpart C. Unless otherwise specified, the centerline of an alternate VOR Federal airway and the centerline of the corresponding segment of the main VOR Federal airway are separated by 15°.

(e) A Federal airway does not include the airspace of a prohibited area.

§ 71.6 Extent of area low routes.

(a) Each area low route is based on a centerline that extends from one waypoint to another waypoint (or through several waypoints) specified for that area low route. An area low route does not include the airspace of a prohibited area. All mileages specified in connection with area low routes are nautical miles.

(b) Unless otherwise specified in Subpart J, the following apply:

(1) Except as provided in subparagraph (2) of this paragraph, each area low route includes, and is limited to, that airspace within parallel boundary lines 4 or more miles on each side of the route centerline as described in the middle column of the following table, plus that additional airspace outside of those parallel lines and within lines drawn outward from those parallel lines at angles of 3.25°, beginning at the distance from the tangent point specified in the right hand column of the following table:

Miles from reference facility to tangent point	Miles from centerline to parallel lines	Miles From tangent point along parallel lines to vortices of 3.25° angles
Less than 17	4	51
17 to, but not including 27	4	50
27 to, but not including 33	4	49
33 to, but not including 38	4	48
38 to, but not including 43	4	47
43 to, but not including 47	4	46
47 to, but not including 51	4	45
51 to, but not including 55	4	44
55 to, but not including 58	4	43
58 to, but not including 61	4	42
61 to, but not including 63	4	41

63 to, but not including 66	4	40
66 to, but not including 68	4	39
68 to, but not including 70	4	38
70 to, but not including 72	4	37
72 to, but not including 74	4	36
74 to, but not including 76	4	35
76 to, but not including 78	4	34
78 to, but not including 79	4	33
79 to, but not including 81	4	32
81 to, but not including 83	4	31
83 to, but not including 84	4	30
84 to, but not including 86	4	29
86 to, but not including 87	4	28
87 to, but not including 88	4	27
88 to, but not including 89	4	26
89 to, but not including 91	4	25
91 to, but not including 92	4	24
92 to, but not including 93	4	23
93 to, but not including 94	4	22
94 to, but not including 95	4	21
95 to, but not including 96	4	19
96 to, but not including 97	4	18
97 to, but not including 98	4	17
98 to, but not including 99	4	15
99 to, but not including 100	4	13
100 to, but not including 101	4	11
101 to, but not including 102	4	8
102 to, but not including 105	4	0*
105 to, but not including 115	4.25	0*
115 to, but not including 125	4.50	0*
125 to, but not including 135	4.75	0*
135 to, but not including 145	5.00	0*
145 to, but not including 150	5.25	0*

*i.e., at tangent point.

(2) Each area low route, whose centerline is at least 2 miles, and not more than 3 miles from the reference facility, includes, in addition to the airspace specified in subparagraph (1) of this paragraph, that airspace on the reference facility side of the centerline that is within lines connecting the point that is 4.9 miles from the tangent point on a perpendicular line from the centerline through the reference facility, thence to the edges of the boundary lines described in subparagraph (1) of this paragraph, intersecting those boundary lines at angles of 5.15°.

(3) Where an area low route changes direction, it includes that airspace enclosed by extending the boundary lines of the route segments until they meet.

(4) Where the widths of adjoining route segments are unequal, the following apply:

(i) If the tangent point of the narrower segment is on the route centerline, the width of the narrower segment includes that additional airspace within lines from the lateral extremity of the wider segment where the route segments join, thence toward the tangent point of the narrower route segment, until intersecting the boundary of the narrower segment.

(ii) If the tangent point of the narrower segment is on the route centerline extended, the width of the narrower segment includes that additional airspace within lines from the lateral extremity of the wider segment where the route segments join, thence toward the tangent point until reaching the point where the narrower segment terminates or changes direction, or until intersecting the boundary of the narrower segment.

(5) Where an area low route terminates, it includes that airspace within a circle whose center is the terminating waypoint and whose diameter is equal to the route

segment width at that waypoint, except that an area low route does not extend beyond the domestic/oceanic control area boundary.

(6) Each area low route includes that airspace extending upward from 1,200 feet above the surface of the earth to, but not including, 18,000 feet MSL, except that area low routes for Hawaii have no upper limits. Variations of the lower limits of an area low route are expressed in digits representing hundreds of feet above the surface (AGL) or mean sea level (MSL) and, unless otherwise specified, apply to the route segment between adjoining waypoints used in the description of the route.

(7) The airspace of an area low route within the lateral limits of a transition area has a floor coincident with the floor of the transition area.

§ 71.7 Control areas.

Control areas consist of the airspace designated in Subparts B, C, E, and J, but do not include the continental control area. Unless otherwise designated, control areas include the airspace between a segment of a main VOR Federal airway and its associated alternate segments with the vertical extent of the area corresponding to the vertical extent of the related segment of the main airway.

§ 71.9 Continental control area.

The continental control area consists of the airspace of the 48 contiguous States, the District of Columbia and Alaska, excluding the Alaska peninsula west of Long. 160°00′00′′W., at and above 14,500 feet MSL, but does not include—

(a) The airspace less than 1,500 feet above the surface of the earth; or

[(b) Prohibited and restricted areas, other than restricted areas listed in Subpart D of this Part.]

§ 71.11 Control zones.

The control zones listed in Subpart F of this Part consist of controlled airspace which extends upward from the surface of the earth and terminates at the base of the continental control area. Control zones that do not underlie the continental control area have no upper limit. A control zone may include one or more airports and is normally a circular area with a radius of 5 miles and any extensions necessary to include instrument approach and departure paths.

☞ **F130–133, 136, 137**
§ 71.12 Terminal control areas.

The terminal control areas listed in Subpart K of this part consist of controlled airspace extending upward from the surface or higher to specified altitudes, within which all aircraft are subject to operating rules and pilot and equipment requirements specified in Part 91 of this chapter. Each such location includes at least one primary airport around which the terminal control area is located.

§ 71.13 Transition areas.

The transition areas listed in Subpart G of this Part consist of controlled airspace extending upward from 700 feet or more above the surface of the earth when designated in conjunction with an airport for which an approved instrument approach procedure has been prescribed; or from 1,200 feet or more above the surface of the earth when designated in conjunction with airway route structures or segments. Unless otherwise specified, transition areas terminate at the base of the overlying controlled airspace.

[§ 71.14 Airport radar service areas.

[The airport radar service areas listed in Subpart L of this Part consist of controlled airspace extending upward from the surface or higher to specified altitudes, within which all aircraft are subject to operating rules and equipment requirements specified in Part 91 of this chapter. Each location listed includes at least one primary airport around which the airport radar service is located.]

§ 71.15 Positive control areas.

The positive control areas listed in Subpart H of this Part consist of controlled airspace within which there is positive control of aircraft.

§ 71.17 Reporting points.

(a) The reporting points listed in Subpart I of this Part consist of geographic locations, in relation to which the position of an aircraft must be reported in accordance with § 91.125 of this chapter.

(b) Unless otherwise designated, each reporting point applies to all directions of flight. In any case where a geographical location is designated as a reporting point for less than all airways passing through that point, or for a particular direction of flight along an airway only, it is so indicated by including the airways or direction of flight in the designation of geographical location.

(c) Unless otherwise specified, place names appearing in the reporting point descriptions indicate VOR or VORTAC facilities identified by those names.

§ 71.19 Bearings; radials; miles.

(a) All bearings and radials in this Part are true, and are applied from point of origin.

(b) Except as otherwise specified and except that mileages for Federal airways are stated as nautical miles, all mileages in this Part are stated as statute miles.

Subpart B—Colored Federal Airways

§ 71.101 Designation.

The airspace assignments described in this subpart are designated as colored Federal airways.*

§ 71.103 Green Federal airways.

§ 71.105 Amber Federal airways.

§ 71.107 Red Federal airways.

§ 71.109 Blue Federal airways.

Subpart C—VOR Federal Airways

§ 71.121 Designation.

The airspace assignments described in this subpart are designated as VOR Federal airways. Unless otherwise specified, place names appearing in the descriptions indicate VOR or VORTAC navigational facilities identified by those names.*

§ 71.123 Domestic VOR Federal airways.

§ 71.125 Alaskan VOR Federal airways.

§ 71.127 Hawaiian VOR Federal airways.

Subpart D—Continental Control Area

§ 71.151 Restricted areas included.

The airspace of the following restricted areas at or above 14,500 feet MSL and 1,500 feet or more above the surface of the earth is continental control area.*

Subpart E—Control Areas and Control Area Extensions

§ 71.161 Designation of control areas associated with jet routes outside the continental control area.

Unless otherwise specified, the airspace centered on each of the following jet route segments has a vertical extent identical to that of a jet route and a lateral extent identical to that of a Federal airway and is designated as a control area. Unless otherwise specified, the place names appearing in the descriptions indicate VOR or VORTAC facilities identified by those names.*

§ 71.163 Designation of additional control areas.

Unless otherwise specified, each control area designated below has a lateral extent identical to that of a Federal airway and extends upward from 700 feet (until designated from 1,200 feet or more) above the surface of the earth, except that the airspace of a control area within the lateral limits of a transition area has a floor coincident with the floor of the transition area.*

§ 71.165 Designation of control area extensions.

Unless otherwise specified, each control area extension designated below extends upward from 700 feet above the surface of the earth, except that the airspace of a control area extension within the lateral limits of a transition area has a floor coincident with that of the transition area.*

Subpart F—Control Zones

§ 71.171 Designation.

The parts of airspace described below are designated as control zones.*

Subpart G—Transition Areas

§ 71.181 Designation.

The parts of airspace described below are designated as transition areas.*

Subpart H—Positive Control Areas

§ 71.193 Designation of positive control areas.

The parts of airspace described below are designated as positive control areas.*

Subpart I—Reporting Points

§ 71.201 Designation.

The locations described in this subpart are designated as reporting points.*

§ 71.203 Domestic low altitude reporting points.

The reporting points listed below are designated at all altitudes up to but not including 18,000 feet MSL.*

§ 71.207 Domestic high altitude reporting points.

The reporting points listed below are designated at all altitudes from 18,000 feet MSL to Flight Level 450, inclusive.*

§ 71.209 Other domestic reporting points.

The reporting points listed below are designated at all altitudes.*

§ 71.211 Alaskan low altitude reporting points.

The reporting points listed below are designated up to but not including 18,000 feet MSL.*

§ 71.213 Alaskan high altitude reporting points.

The reporting points listed below are designated at 18,000 feet MSL to Flight Level 450.*

§ 71.215 Hawaiian reporting points.

The reporting points listed below are designated at all altitudes.*

Subpart J—Area Low Routes

§ 71.301 Designation.

The parts of airspace described below are designated as area low routes.**

Subpart K—Terminal Control Areas

[Subpart L—Airport Radar Service Areas

[§ 71.501 Designation.

[The airspace descriptions listed below are designated as airport radar service areas. The primary airport for each airport radar service area is also designated. Except as otherwise specified, all mileages are nautical miles and all altitudes are above mean sea level.] *

---

* The airspace descriptions in this Part are published in the Federal Register. Due to their complexity and length, they will not be included in this publication of Part 71.

** The area low route descriptions in this Part are published in the Federal Register. Due to their complexity and length, they will not be included in this Publication of Part 71.

FEDERAL AVIATION REGULATIONS

Part 73—Special Use Airspace

Contents

Subpart A—General

§ 73.1 Applicability.

The airspace that is described in Subpart B and Subpart C of this Part is designated as special use airspace. This Part prescribes the requirements for the use of that airspace.

§ 73.3 Special use airspace.

(a) Special use airspace consists of airspace of defined dimensions identified by an area on the surface of the earth wherein activities must be confined because of their nature, or wherein limitations are imposed upon aircraft operations that are not a part of those activities, or both.

(b) The vertical limits of special use airspace are measured by designated altitude floors and ceilings expressed as flight levels or as feet above mean sea level. Unless otherwise specified, the word "to" (an altitude or flight level) means "to and including" (that altitude or flight level).

(c) The horizontal limits of special use airspace are measured by boundaries described by geographic coordinates or other appropriate references that clearly define their perimeter.

(d) The period of time during which a designation of special use airspace is in effect is stated in the designation.

§ 73.5 Bearings; radials; miles.

(a) All bearings and radials in this Part are true from point of origin.

(b) Unless otherwise specified, all mileages in this Part as stated as statute miles.

Subpart B—Restricted Areas

§ 73.11 Applicability.

This subpart designates restricted areas and prescribes limitations on the operation of aircraft within them.

§ 73.13 Restrictions.

No person may operate an aircraft within a restricted area between the designated altitudes and during the time of designation, unless he has the advance permission of—

(a) The using agency described in § 73.15; or

(b) The controlling agency described in § 73.17.

§ 73.15 Using agency.

(a) For the purposes of this subpart, the following are using agencies:

(1) The agency, organization, or military command whose activity within a restricted area necessitated the area being so designated.

【(2) [Reserved]】

(b) Upon the request of the FAA, the using agency shall execute a letter establishing procedures for joint use of a restricted area by the using agency and the controlling agency, under which the using agency would notify the controlling agency whenever the controlling agency may grant permission for transit through the restricted area in accordance with the terms of the letter.

(c) The using agency shall—

(1) Schedule activities within the restricted area;

(2) Authorize transit through, or flight within, the restricted area as feasible; and

(3) Contain within the restricted area all activities conducted therein in accordance with the purpose for which it was designated.

§ 73.17 Controlling agency.

For the purposes of this Part, the controlling agency is the FAA facility that may authorize transit through or flight within a restricted area in accordance with a joint-use letter issued under § 73.15.

【§ 73.19 Reports by using agency.

【(a) Each using agency shall prepare a report on the use of each restricted area assigned thereto during any part of the preceding 12-month period ended September 30, and transmit it by the following January 31 of each year to the Chief, Air Traffic Division in the regional office of the Federal Aviation Administration having jurisdiction over the area in which the restricted area is located, with a copy to the Director, Air Traffic Service, Federal Aviation Administration, Washington, D.C. 20591.

【(b) In the report under this section the using agency shall:

【(1) State the name and number of the restricted area as published in this Part, and the period covered by the report.

【(2) State the activities (including average daily number of operations if appropriate) conducted in the area, and any other pertinent information concerning current and future electronic monitoring devices.

【(3) State the number of hours daily, the days of the week, and the number of weeks during the year that the area was used.

【(4) For restricted areas having a joint-use designation, also state the number of hours daily, the days of the week, and the number of weeks during the year that the restricted area was released to the controlling agency for public use.

【(5) State the mean sea level altitudes or flight levels (whichever is appropriate) used in aircraft operations and the maximum and average ordinate of surface firing (expressed in feet, mean sea level altitude) used on a daily, weekly, and yearly basis.

【(6) Include a chart of the area (of optional scale and design) depicting, if used, aircraft operating areas, flight patterns, ordnance delivery areas, surface firing points, and target, fan, and impact areas. After once submitting an appropriate chart, subsequent annual charts are not required unless there is a change in the area, activity or altitude (or flight levels) used, which might alter the depiction of the activities originally reported. If no change is to be submitted, a statement indicating "no change" shall be included in the report.

【(7) Include any other information not otherwise required under this Part which is considered pertinent to activities carried on in the restricted area.

【(c) If it is determined that the information submitted under paragraph (b) of this section is not sufficient to evaluate the nature and extent of the use of a restricted area, the FAA may request the using agency to submit supplementary reports. Within 60 days after receiving a request for additional information, the using agency shall submit such information as the Director of the Air Traffic Service considers appropriate. Supplementary reports must be sent to the FAA officials designated in paragraph (a) of this section.]

§§ 73.21 through 73.72 [Redesignations.] [§§ 608.21 through 608.72 of the Regulations

of the Administrator are hereby redesignated as §§ 73.21 through 73.72, respectively]*

Subpart C—Prohibited Areas

§ 73.81 Applicability.

This subpart designates prohibited areas and prescribes limitations on the operation of aircraft therein.

§ 73.83 Restrictions.

No person may operate an aircraft within a prohibited area unless authorization has been granted by the using agency.

§ 73.85 Using Agency.

For the purpose of this subpart, the using agency is the agency, organization or military command that established the requirement for the prohibited area.

§§ 73.87 through 73.99.*

These sections are reserved for descriptions of designated Prohibited Areas.

*The airspace descriptions in this Part and their subsequent changes are published in the Federal Register. Due to their complexity and length, they will not be included in this publication of Part 73.

FEDERAL AVIATION REGULATIONS

Part 91—General Operating and Flight Rules

Contents

Subpart A—General

§ 91.1 Applicability.

(a) Except as provided in paragraph (b) of this section, this Part prescribes rules governing the operation of aircraft (other than moored balloons, kites, unmanned rockets, and unmanned free balloons) within the United States.

(b) Each person operating a civil aircraft of U.S. registry outside of the United States shall—

(1) When over the high seas, comply with Annex 2 (Rules of the Air) to the Convention on International Civil Aviation and with §§ 91.70(c), 91.88, and 91.90 of Subpart B;

(2) When within a foreign country, comply with the regulations relating to the flight and maneuver of aircraft there in force;

(3) Except for §§ 91.15(b), 91.17, 91.38, and 91.43, comply with Subparts A, C, and D of this Part so far as they are not inconsistent with applicable regulations of the foreign country where the aircraft is operated or Annex 2 to the Convention on International Civil Aviation; and

(4) When over the North Atlantic within airspace designated as Minimum Naviga-tion Performance Specifications airspace, comply with § 91.20.

(c) Annex 2 to the Convention on International Civil Aviation, Sixth Edition—September 1970, with amendments through Amendment 20 effective August 1976, to which reference is made in this Part is incorporated into this Part and made a part hereof as provided in 5 U.S.C. 552 and pursuant to 1 CFR Part 51. Annex 2 (including a complete historic file of changes thereto) is available for public inspection at the Rules Docket, AGC-204, Federal Aviation Administration, 800 Independence Avenue, SW., Washington, D.C. 20591. In addition, Annex 2 may be purchased from the International Civil Aviation Organization (Attention: Distribution Officer), P.O. Box 400, Succursale; Place de L'Aviation Internationale, 1000 Sherbrooke Street West, Montreal, Quebec, Canada H3A 2R2.

§ 91.2 Certificate of authorization for certain Category II operations.

The Administrator may issue a certificate of authorization authorizing deviations from the requirements of §§ 91.6, 91.33(f), and 91.34 of this subpart for the operation of small [air-craft] identified as Category A aircraft in § 97.3 of this chapter in Category II operations, if he finds that the proposed operation can be safely conducted under the terms of the certificate. Such authorization does not permit operation of the aircraft carrying persons or property for compensation or hire.

§ 91.3 Responsibility and authority of the pilot in command.

(a) The pilot in command of an aircraft is directly responsible for, and is the final authority as to, the operation of that aircraft.

(b) In an emergency requiring immediate action, the pilot in command may deviate from any rule of this subpart or of Subpart B to the extent required to meet that emergency.

(c) Each pilot in command who deviates from a rule under paragraph (b) of this section shall, upon the request of the Administrator, send a written report of that deviation to the Administrator.

§ 91.4 Pilot in command of aircraft requiring more than one required pilot.

No person may operate an aircraft that is type certificated for more than one required pilot flight crewmember unless the pilot flight crew consists of a pilot in command who meets the requirements of § 61.58 of this chapter.

§ 91.5 Preflight action.

Each pilot in command shall, before beginning a flight, familiarize himself with all available information concerning that flight. This information must include:

(a) For a flight under IFR or a flight not in the vicinity of an airport, weather reports and forecasts, fuel requirements, alternatives available if the planned flight cannot be completed, and any known traffic delays of which he has been advised by ATC. ☞ **F83–85**

(b) For any flight, runway lengths at airports of intended use, and the following takeoff and landing distance information:

(1) For civil aircraft for which an approved airplane or rotorcraft flight manual containing takeoff and landing distance data is required, the takeoff and landing distance data contained therein; and

(2) For civil aircraft other than those specified in subparagraph (1) of this paragraph, other reliable information appropriate to the aircraft, relating to aircraft performance under expected values of airport elevation and runway slope, aircraft gross weight, and wind and temperature.

[§ 91.6 Category II and III operations: general operating rules.

[(a) No person may operate a civil aircraft in a Category II or Category III operation unless:

[(1) The flightcrew of the aircraft consists of a pilot in command and a second in command who hold the appropriate authorizations and ratings prescribed in § 61.3 of this chapter;

[(2) Each flight crewmember has adequate knowledge of, and familiarity with, the aircraft and the procedures to be used; and

[(3) The instrument panel in front of the pilot who is controlling the aircraft has appropriate instrumentation for the type of flight control guidance system that is being used.

[(b) Unless otherwise authorized by the Administrator, no person may operate a civil aircraft in a Category II or Category III operation unless each ground component required for that operation and the related airborne equipment is installed and operating.

[(c) For the purpose of this section, when the approach procedure being used provides for and requires use of a DH, the authorized decision height is the DH prescribed by the approach procedure, the DH prescribed for the pilot in command, or the DH for which the aircraft is equipped, whichever is higher.

[(d) Unless otherwise authorized by the Administrator, no pilot operating an aircraft in a Category II or Category III approach that provides and requires use of a DH may continue the approach below the authorized decision height unless the following conditions are met:

[(1) The aircraft is in a position from which a descent to a landing on the intended runway can be made at a normal rate of descent using normal maneuvers, and where that descent rate will allow touchdown to occur within the touchdown zone of the runway of intended landing.

[(2) At least one of the following visual references for the intended runway is distinctly visible and identifiable to the pilot:

[(i) The approach light system, except that the pilot may not descend below 100 feet above the touchdown zone elevation using the approach lights as a reference unless the red terminating bars or the red side row bars are also distinctly visible and identifiable.

[(ii) The threshold.

[(iii) The threshold markings.

[(iv) The threshold lights.

[(v) The touchdown zone or touchdown zone markings.

[(vi) The touchdown zone lights.

[(e) Unless otherwise authorized by the Administrator, each pilot operating an aircraft shall immediately execute an appropriate missed approach whenever prior to touchdown the requirements of paragraph (d) of this section are not met.

[(f) No person operating an aircraft using a Category III approach without decision height may land that aircraft except in accordance with the provisions of the letter of authorization issued by the Administrator.

(g) Paragraphs (a) through (f) of this section do not apply to operations conducted by the holders of certificates issued under Parts 121, 123, 125, 129, or 135 of this chapter. No person may operate a civil aircraft in a Category II or Category III operation conducted by the holder of a certificate issued under Parts 121, 123, 125, 129, or 135 of this chapter unless the operation is conducted in accordance with that certificate holder's operations specifications.

§ 91.7 Flight crewmembers at stations.

(a) During takeoff and landing, and while en route, each required flight crewmember shall—

(1) Be at his station unless his absence is necessary in the performance of his duties in connection with the operation of the aircraft or in connection with his physiological needs; and ☞ **F86**

(2) Keep his seat belt fastened while at his station.

(b) After July 18, 1978, each required flight crewmember of a U.S. registered civil airplane shall, during takeoff and landing, keep the shoulder harness fastened while at his station. This paragraph does not apply if—

(1) The seat at the crewmember's station is not equipped with a shoulder harness; or

(2) The crewmember would be unable to perform his required duties with the shoulder harness fastened.

§ 91.8 Prohibition against interference with crewmembers.

No person may assault, threaten, intimidate, or interfere with a crewmember in the performance of the crewmember's duties aboard an aircraft being operated.

§ 91.9 Careless or reckless operation.

No person may operate an aircraft in a careless or reckless manner so as to endanger the life or property of another. ☞ **F83,87**

§ 91.10 Careless or reckless operation other than for the purpose of air navigation.

No person may operate an aircraft other than for the purpose of air navigation, on any part of the surface of an airport used by aircraft for air commerce (including areas used by those aircraft for receiving or discharging persons or cargo), in a careless or reckless manner so as to endanger the life or property of another.

§ 91.11 Alcohol or drugs.

(a) No person may act or attempt to act as a crewmember of a civil aircraft—

(1) Within 8 hours after the consumption of any alcoholic beverage;

(2) While under the influence of alcohol;

(3) While using any drug that affects the person's faculties in any way contrary to safety; or ☞ **F39,88**

(4) While having .04 percent by weight or more alcohol in the blood.

(b) Except in an emergency, no pilot of a civil aircraft may allow a person who appears to be intoxicated or who demonstrates by manner or physical indications that the individual is under the influence of drugs (except a medical patient under proper care) to be carried in that aircraft.

[(c) A crewmember shall do the following:

[(1) On request of a law enforcement officer, submit to a test to indicate the percentage by weight of alcohol in the blood, when—

[(i) The law enforcement officer is authorized under State or local law to conduct the test or to have the test conducted; and

[(ii) The law enforcement officer is requesting submission to the test to investigate a suspected violation of State or local law governing the same or substantially similar conduct prohibited by paragraph (a)(1), (a)(2), or (a)(4) of this section.

[(2) Whenever the Administrator has a reasonable basis to believe that a person may have violated paragraph (a) (1), (a) (2), or (a) (4) of this section, that person shall, upon request by the Administrator, furnish the Administrator, or authorize any clinic, hospital, doctor, or other person to release to the Administrator, the results of each test taken within 4 hours after acting or attempting to act as a crewmember that indicates percentage by weight of alcohol in the blood.]

(d) Whenever the Administrator has a reasonable basis to believe that a person may have violated paragraph (a) (3) of this section, that person shall, upon request by the Administrator, furnish the Administrator, or authorize any clinic, hospital, doctor, or other person to release to the Administrator, the results of each test taken within 4 hours after acting or attempting to act as a crewmember that indicates the presence of any drugs in the body.

(e) Any test information obtained by the Administrator under paragraph (c) or (d) of this section may be evaluated in determining a person's qualifications for any airman certificate or possible violations of this chapter and may be used as evidence in any legal proceeding under Section 602, 609, or 901 of the Federal Aviation Act of 1958.

§ 91.12 Carriage of narcotic drugs, marihuana, and depressant or stimulant drugs or substances.

(a) Except as provided in paragraph (b) of this section, no person may operate a civil aircraft within the United States with knowledge that narcotic drugs, marihuana, and depressant or stimulant drugs or substances as defined in Federal or State statutes are carried in the aircraft.

(b) Paragraph (a) of this section does not apply to any carriage of narcotic drugs, marihuana, and depressant or stimulant drugs or substances authorized by or under any Federal or State statute or by any Federal or State agency.

§ 91.13 Dropping objects.

No pilot in command of a civil aircraft may allow any object to be dropped from that aircraft in flight that creates a hazard to persons or property. However, this section does not prohibit the dropping of any object if reasonable precautions are taken to avoid injury or damage to persons or property.

§ 91.14 Use of safety belts and shoulder harnesses.

(a) Unless otherwise authorized by the Administrator—

(1) No pilot may take off a U.S. registered civil aircraft (except a free balloon that incorporates a basket or gondola and an airship) unless the pilot in command of that aircraft ensures that each person on board is briefed on how to fasten and unfasten that person's safety belt and shoulder harness, if installed.

(2) No pilot may take off or land a U.S. registered civil aircraft (except free balloons that incorporate baskets or gondolas and airships) unless the pilot in command of that aircraft ensures that each person on board has been notified to fasten his safety belt and shoulder harness, if installed.

(3) During the takeoff and landing of U.S. registered civil aircraft (except free balloons that incorporate baskets or gondolas and airships) each person on board that aircraft must occupy a seat or berth with a safety belt and shoulder harnesses, if installed properly secured about him. However, a person who has not reached his second birthday may be held by an adult who is occupying a seat or berth, and a person on board for the purpose of engaging in sport parachuting may use the floor of the aircraft as a seat. ☞ **F89, 90**

(b) This section does not apply to operations conducted under Parts 121, 123, or 127 of this chapter. Subparagraphs (a) (3) of this section does not apply to persons subject to § 91.7.

§ 91.15 Parachutes and parachuting.

(a) No pilot of a civil aircraft may allow a parachute that is available for emergency use to be carried in that aircraft unless it is an approved type and—

(1) If a chair type (canopy in back), it has been packed by a certificated and appropriately rated parachute rigger within the

preceding 120 days; or

(2) If any other type, it has been packed by a certificated and appropriately rated parachute rigger—

(i) Within the preceding 120 days, if its canopy, shrouds, and harness are composed exclusively of nylon, rayon, or other similar synthetic fiber or materials that are substantially resistant to damage from mold, mildew, or other fungi and other rotting agents propagated in a moist environment; or

(ii) Within the preceding 60 days, if any part of the parachute is composed of silk, pongee, or other natural fiber, or materials not specified in subdivision (1) of this paragraph.

(b) Except in an emergency, no pilot in command may allow, and no person may make, a parachute jump from an aircraft within the United States except in accordance with Part 105.

(c) Unless each occupant of the aircraft is wearing an approved parachute, no pilot of a civil aircraft, carrying any person (other than a crewmember) may execute any intentional maneuver that exceeds— ☞ **F91**

(1) A bank of 60 degrees relative to the horizon; or

(2) A nose-up or nose-down attitude of 30 degrees relative to the horizon.

(d) Paragraph (c) of this section does not apply to—

(1) Flight tests for pilot certification or rating; or ☞ **F91**

(2) Spins and other flight maneuvers required by the regulations for any certificate or rating when given by—

(i) A certificated flight instructor; or

(ii) An airline transport pilot instructing in accordance with § 61.169 of this chapter.

(e) For the purposes of this section, "approved parachute" means—

(1) A parachute manufactured under a type certificate or a technical standard order (C–23 series); or

(2) A personnel-carrying military parachute identified by an NAF, AAF, or AN drawing number, an AAF order number, or any other military designation or specification number.

§ 91.17 Towing: gliders.

(a) No person may operate a civil aircraft towing a glider unless:

(1) The pilot in command of the towing aircraft is qualified under § 61.69 of this chapter.

(2) The towing aircraft is equipped with a towhitch of a kind, and installed in a manner, approved by the Administrator.

(3) The towline used has a breaking strength not less than 80 percent of the maximum certificated operating weight of the glider, and not more than twice this operating weight. However, the towline used may have a breaking strength more than twice the maximum certificated operating weight of the glider if—

(i) A safety link is installed at the point of attachment of the towline to the glider, with a breaking strength not less than 80 percent of the maximum certificated operating weight of the glider, and not greater than twice this operating weight; and

(ii) A safety link is installed at the point of attachment of the towline to the towing air-

craft with a breaking strength greater, but not more than 25 percent greater, than that of the safety link at the towed glider end of the towline, and not greater than twice the maximum certificated operating weight of the glider.

(4) Before conducting any towing operations within a control zone, or before making each towing flight within a control zone if required by ATC, the pilot in command notifies the control tower if one is in operation in that control zone. If such a control tower is not in operation, he must notify the FAA flight service station serving the control zone before conducting any towing operations in that control zone.

(5) The pilots of the towing aircraft and the glider have agreed upon a general course of action including takeoff and release signals, airspeeds, and emergency procedures for each pilot.

(b) No pilot of a civil aircraft may intentionally release a towline, after release of a glider, in a manner so as to endanger the life of property of another.

(c) [Reserved]

§ 91.18 Towing: other than under § 91.17.

(a) No pilot of a civil aircraft may tow anything with that aircraft (other than under § 91.17) except in accordance with the terms of a certificate of waiver issued by the Administrator.

(b) An application for a certificate of waiver under this section is made on a form and in a manner prescribed by the Administrator and must be submitted to the nearest Flight Standards District Office.

§ 91.19 Portable electronic devices.

(a) Except as provided in paragraph (b) of this section, no person may operate, nor may any operator or pilot in command of an aircraft allow the operation of, any portable electronic device on any of the following U.S. registered civil aircraft:

(1) Aircraft operated by an air carrier or commercial operator; or

(2) Any other aircraft while it is operated under IFR.

(b) Paragraph (a) of this section does not apply to:

(1) Portable voice recorders;

(2) Hearing aids;

(3) Heart pacemakers;

(4) Electric shavers; or

(5) Any other portable electronic device that the operator of the aircraft has determined will not cause interference with the navigation or communication system of the aircraft on which it is to be used.

(c) In the case of an aircraft operated by an air carrier or commercial operator, the determination required by paragraph (b) (5) of this section shall be made by the air carrier or commercial operator of the aircraft on which the particular device is to be used. In the case of other aircraft, the determination may be made by the pilot in command or other operator of the aircraft.

§ 91.20 Operations within the North Atlantic Minimum Navigation Performance Specifications airspace.

Unless otherwise authorized by the Administrator, no person may operate a civil aircraft of U.S. registry in North Atlantic (NAT) airspace designated as Minimum Navigation Performance Specifications (MNPS) airspace unless that aircraft has approved navigation performance capability which complies with the requirements of Appendix C to this Part. The Administrator authorizes deviations from the requirements of this section in accordance with Section 3 of Appendix C to this Part.

[§ 91.21 Flight instruction; simulated instrument flight and certain flight tests.

[(a) No person may operate a civil aircraft (except a manned free balloon) that is being used for flight instruction unless that aircraft has fully functioning, dual controls. However, instrument flight instruction may be given in a single-engine airplane equipped with a single, functioning throwover control wheel, in place of fixed, dual controls of the elevator and ailerons, when:

[(1) The instructor has determined that the flight can be conducted safely; and

(2) The person manipulating the controls has at least a private pilot certificate with appropriate category and class ratings.]

(b) No person may operate a civil aircraft in simulated instrument flight unless—

☞ **F28–30, 32, 36, 92–94**

(1) An appropriately rated pilot occupies the other control seat as safety pilot;

(2) The safety pilot has adequate vision forward and to each side of the aircraft, or a competent observer in the aircraft adequately supplements the vision of the safety pilot; and

(3) Except in the case of lighter-than-air aircraft, that aircraft is equipped with fully functioning dual controls. However, simulated instrument flight may be conducted in a single-engine airplane, equipped with a single, functioning, throwover control wheel, in place of fixed, dual controls of the elevator and ailerons, when—

(i) The safety pilot has determined that the flight can be conducted safely; and

(ii) The person manipulating the control has at least a private pilot certificate with appropriate category and class ratings.

(c) No person may operate a civil aircraft that is being used for a flight test for an airline transport pilot certificate or a class or type rating on that certificate, or for a Federal Aviation Regulation Part 121 proficiency flight test, unless the pilot seated at the controls, other than the pilot being checked, is fully qualified to act as pilot in command of the aircraft.

§ 91.22 Fuel requirements for flight under VFR.

(a) No person may begin a flight in an airplane under VFR unless (considering wind and forecast weather conditions) there is enough fuel to fly to the first point of intended landing and, assuming normal cruising speed—

(1) During the day, to fly after that for at least 30 minutes; or

(2) At night, to fly after that for at least 45 minutes.

(b) No person may begin a flight in a rotorcraft under VFR unless (considering wind and forecast weather conditions) there is enough fuel to fly to the first point of intended landing and, assuming normal cruising speed, to fly after that for at least 20 minutes.

§ 91.23 Fuel requirements for flight in IFR conditions.

(a) Except as provided in paragraph (b) of this section, no person may operate a civil aircraft in IFR conditions unless it carries enough fuel (considering weather reports and forecasts, and weather conditions) to—

☞ **F84, 95**

(1) Complete the flight to the first airport of intended landing;

(2) Fly from that airport to the alternate airport; and

(3) Fly after that for 45 minutes at normal cruising speed or, for helicopters, fly after that for 30 minutes at normal cruising speed.

(b) Paragraph (a) (2) of this section does not apply if—

(1) Part 97 of this subchapter prescribes a standard instrument approach procedure for the first airport of intended landing; and

(2) For at least 1 hour before and 1 hour after the estimated time of arrival at the airport, the weather reports or forecasts or any combination of them, indicate—

(i) The ceiling will be at least 2,000 feet above the airport elevation; and

(ii) Visibility will be at least 3 miles.

§ 91.24 ATC transponder and altitude reporting equipment and use.

[(a) *All airspace: U. S. registered civil aircraft.* For operations not conducted under Parts 121, 123, 127, or 135 of this chapter, ATC transponder equipment installed within the time periods indicated below must meet the performance and environmental requirements of the following TSO's.

[(1) *Through January 1, 1992:*

[(i) Any class of TSO-C74b or any class of TSO-C74c as appropriate, provided that the equipment was manufactured before January 1, 1990; or

[(ii) The appropriate class of TSO-C112 (Mode S).

[(2) *After January 1, 1992: The appropriate class of TSO-C112 (Mode S).* For purposes of paragraph (a)(2) of this section, "installation" does not include—

[(i) Temporary installation of TSO-C74b or TSO-C74c substitute equipment, as

appropriate, during maintenance of the permanent equipment;

[(ii) Reinstallation of equipment after temporary removal for maintenance; or

[(iii) For fleet operations, installation of equipment in a fleet aircraft after removal of the equipment for maintenance from another aircraft in the same operators's fleet.

[(b) *All Airspace:* No person may operate an aircraft in the airspace described in paragraphs (b)(1) through (b)(5) of this section, unless that aircraft is equipped with an operating coded radar beacon transponder having either a Mode 3/A 4096 code capability, replying to Mode 3/A interrogations with the code specified by ATC, or a Mode S capability, replying to Mode 3/A interrogations with the code specified by ATC and intermode and Mode S interrogations in accordance with the applicable provisions specified in TSO-C112, and that aircraft is equipped with automatic pressure altitude reporting equipment having a Mode C capability that automatically replies to Mode C interrogations by transmitting pressure altitude information in 100-foot increments. This requirement applies—

[(1) *All aircraft.* In terminal control areas and positive control areas;

[(2) *Effective July 1, 1989. All aircraft.* In all airspace within 30 nautical miles of a terminal control area primary airport, from the surface upward to 10,000 feet MSL;

[(3) *Effective July 1, 1989.* Notwithstanding paragraph (b)(2) of this section, any aircraft which was not originally certificated with an engine-driven electrical system or which has not subsequently been certified with such a system installed, balloon, or glider may conduct operations in the airspace within 30 nautical miles of a terminal control area primary airport provided such operations are conducted—

[(i) Outside any terminal conbrol area and positive control areas; and

[(ii) Below the altitude of the terminal control area ceiling or 10,000 feet MSL, whichever is lower; and

[(4) *Effective December 30, 1990. All aircraft.*

[(i) In the airspace of an airport radar service area, and

[(ii) In all airspace above the ceiling and within the lateral boundaries of an airport radar service area upward to 10,000 feet MSL; and

[(5) All aircraft except any aircraft which was not originally certificated with an engine-driven electrical system or which has not subsequently been certified with such a system installed, balloon, or glider.

[(i) In all airspace of the 48 contiguous states and the District of Columbia:

[(A) *Through June 30, 1989.* Above 12,500 feet MSL and below the floor of a positive control area, excluding the airspace at and below 2,500 feet AGL.

[(B) *Effective July 1, 1989.* At and above 10,000 feet MSL and below the floor of a positive control area, excluding the airspace at and below 2,500 feet AGL; and

〖(ii) *Effective December 30, 1990.* In the airspace from the surface to 10,000 feet MSL within a 10-nautical-mile radius of any airport listed in Appendix D of this part excluding the airspace below 1,200 feet AGL outside of the airport traffic area for that airport.

〖(c) *Transponder-on operation.* While in the airspace as specified in paragraph (b) of this section or in all controlled airspace, each person operating an aircraft equipped with an operable ATC transponder maintained in accordance with § 91.172 of this part shall operate the transponder, including Mode C equipment if installed, and shall reply on the appropriate code or as assigned by ATC.〗 ☞**F96,**

(d) *ATC authorized deviations.* ATC may authorize deviations from paragraph (b) of this section—

(1) Immediately, to allow an aircraft with an inoperative transponder to continue to the airport of ultimate destination, including any intermediate stops, or to proceed to a place where suitable repairs can be made, or both;

(2) Immediately, for operations of aircraft with an operating transponder but without operating automatic pressure altitude reporting equipment having a Mode C capability; and

(3) On a continuing basis, or for individual flights, for operations of aircraft without a transponder, in which case the request for a deviation must be submitted to the ATC facility having jurisdiction over the airspace concerned at least one hour before the proposed operation.

§ 91.25 VOR equipment check for IFR operations.

(a) No person may operate a civil aircraft under IFR using the VOR system of radio navigation unless the VOR equipment of that aircraft—

(1) Is maintained, checked, and inspected under an approved procedure; or

(2) Has been operationally checked within the preceding 30 days and was found to be within the limits of the permissible indicated bearing error set forth in paragraph (b) or (c) of this section. ☞**F97**

(b) Except as provided in paragraph (c) of this section, each person conducting a VOR check under subparagraph (a)(2) of this section, shall—

(1) Use, at the airport of intended departure, an FAA operated or approved test signal or a test signal radiated by a certificated and appropriately rated radio repair station or, outside the United States, a test signal operated or approved by appropriate authority, to check the VOR equipment (the maximum permissible indicated bearing error is plus or minus 4 degrees).

(2) If a test signal is not available at the airport of intended departure, use a point on an airport surface designated as a VOR system checkpoint by the Administrator or, outside the United States, by appropriate authority (the maximum permissible bearing error is plus or minus 4 degrees);

(3) If neither a test signal nor a designated checkpoint on the surface is available, use an airborne checkpoint designated by the Administrator or, outside the United States, by appropriate authority (the maximum permissible bearing error is plus or minus 6 degrees); or

(4) If no check signal or point is available, while in flight— ☞**F98**

(i) Select a VOR radial that lies along the centerline of an established VOR airway;

(ii) Select a prominent ground point along the selected radial preferably more than 20 miles from the VOR ground facility and maneuver the aircraft directly over the point at a reasonably low altitude; and

(iii) Note the VOR bearing indicated by the receiver when over the ground point (the maximum permissible variation between the published radial and the indicated bearing is 6 degrees).

(c) If dual system VOR (units independent of each other except for the antenna) is installed in the aircraft, the person checking the equipment may check one system against the other in place of the check procedures specified in paragraph (b) of this section. He shall tune both systems to the same VOR ground facility and note the indicated bearings to that station. The maximum permissible variation between the two indicated bearings is 4 degrees.

(d) Each person making the VOR operational check as specified in paragraph (b) or (c) of this section shall enter the date, place, bearing error, and sign the aircraft log or other record. In addition, if a test signal radiated by a repair station, as specified in paragraph (b)(1) of this section, is used, an entry must be made in the aircraft log or other record by the repair station certificate holder or the certificate holder's representative certifying to the bearing transmitted by the repair station for the check and the date of transmission.

§ 91.27 Civil aircraft: certifications required.

(a) Except as provided in § 91.28, no person may operate a civil aircraft unless it has within it the following:

(1) An appropriate and current airworthiness certificate. Each U.S. airworthiness certificate used to comply with this subparagraph (except a special flight permit, a copy of the applicable operations specifications issued under § 21.197(c) of this chapter, appropriate sections of the air carrier manual required by Parts 121 and 127 of this chapter containing that portion of the operations specifications issued under § 21.197(c), or an authorization under § 91.45), must have on it the registration number assigned to the aircraft under Part 47 of this chapter. However, the airworthiness certificate need not have on it an assigned special identification number before 10 days after that number is first affixed to the aircraft. A revised airworthiness certificate having on it an assigned special identification number, that has been affixed to an aircraft, may only be obtained upon application to an FAA Flight Standards District Office.

(2) A registration certificate issued to its owner.

(b) No person may operate a civil aircraft unless the airworthiness certificate required by paragraph (a) of this section or a special flight authorization issued under § 91.28 is displayed at the cabin or cockpit entrance so that it is legible to passengers or crew.

(c) No person may operate an aircraft with a fuel tank installed within the passenger compartment or a baggage compartment unless the installation was accomplished pursuant to Part 43 of this chapter, and a copy of FAA Form 337 authorizing that installation is on board the aircraft.

§ 91.28 Special flight authorizations for foreign civil aircraft.

(a) Foreign civil aircraft may be operated without the airworthiness certificates required under § 91.27 if a special flight authorization for that operation is issued under this section. Application for a special flight authorization must be made to the Regional Director of the FAA Region in which the applicant is located, or to the region within which the U.S. point of entry is located. However, in the case of an aircraft to be operated in the U.S. for the purpose of demonstration at an air show, the application may be made to the Regional Director of the FAA region in which the air show is located.

(b) The Administrator may issue a special flight authorization for a foreign civil aircraft, subject to any conditions and limitations that the Administrator considers necessary for safe operation in the U.S. airspace.

(c) No person may operate a foreign civil aircraft under a special flight authorization unless that operation also complies with Part 375 of the Special Regulations of the Civil Aeronautics Board (14 CFR 375).

§ 91.29 Civil aircraft airworthiness.

(a) No person may operate a civil aircraft unless it is in an airworthy condition. ☞**F72**

(b) The pilot in command of a civil aircraft is responsible for determining whether that aircraft is in condition for safe flight. He shall discontinue the flight when unairworthy mechanical or structural conditions occur.

§ 91.30 Inoperative instruments and equipment.

(a) Except as provided in paragraphs (d) of this section, no person may takeoff an aircraft with inoperative instruments or equipment installed unless the following conditions are met:

〖(1) An approved Minimum Equipment List exists for that aircraft.

〖(2) The aircraft has within it a letter of authorization, issued by the FAA Flight Standards Office having jurisdiction over the area in which the operator is located, authorizing operation of the aircraft under the Minimum Equipment List. The letter of authorization may be obtained by written request of the airworthiness certificate holder. The Minimum Equipment List and the letter of authorization constitute a supplemental type certificate for the aircraft.

⟦(3) The approved Minimum Equipment List must:

⟦(i) Be prepared in accordance with the limitations specified in paragraph (b) of this section.

⟦(ii) Provide for the operation of the aircraft with the instruments and equipment in an inoperable condition.

⟦(4) The aircraft records available to the pilot must include an entry describing the inoperable instruments and equipment.

⟦(5) The aircraft is operated under all applicable conditions and limitations contained in the Minimum Equipment List and the letter authorizing the use of the list.

⟦(b) The following instruments and equipment may not be included in a Minimum Equipment List:

⟦(1) Instruments and equipment that are either specifically or otherwise required by the airworthiness requirements under which the aircraft is type certificated and which are essential for safe operations under all operating conditions.

⟦(2) Instruments and equipment required by an airworthiness directive to be in operable condition unless the airworthiness directive provides otherwise.

⟦(3) Instruments and equipment required for specific operations by this Part.

(c) A person authorized to use an approved Minimum Equipment List issued for a specific aircraft under Part 121, 125, or 135 of this chapter shall use that Minimum Equipment List in connection with operations conducted with that aircraft under this part without additional approval requirements.

(d) Except for operations conducted in accordance with paragraphs (a) or (c) of this section, a person may takeoff an aircraft in operations conducted under this part with inoperative instruments and equipment without an approved Minimum Equipment List provided—

(1) The flight operation is conducted in a—

(i) Rotorcraft, nonturbine-powered airplane, glider, or lighter-than-air aircraft for which a Master Minimum Equipment List has not been developed; or

(ii) Small rotorcraft, nonturbine-powered small airplane, glider, or lighter-than-air aircraft for which a Master Minimum Equipment List has been developed; and

(2) The inoperative instruments and equipment are not—

(i) Part of the VFR-day type certification instruments and equipment prescribed in the applicable airworthiness regulations under which the aircraft was type certificated;

(ii) Indicated as required on the aircraft's equipment list, or on the Kinds of Operations Equipment List for the kind of flight operation being conducted;

(iii) Required by § 91.33 or any other rule of this part for the specific kind of flight operation being conducted; or

(iv) Required to be operational by an airworthiness directive; and

(3) The inoperative instruments and equipment are—

(i) Removed from the aircraft, the cockpit control placarded, and the maintenance recorded in accordance with § 43.9 of this chapter; or

(ii) Deactivated and placarded "Inoperative." If deactivation of the inoperative instrument or equipment involves maintenance, it must be accomplished and recorded in accordance with Part 43 of this chapter; and

(4) A determination is made by a pilot, who is certificated and appropriately rated under Part 61, or by a person, who is certificated and appropriately rated to perform maintenance on the aircraft, that the inoperative instrument or equipment does not constitute a hazard to the aircraft.

An aircraft with inoperative instruments or equipment as provided in paragraph (d) of this section is considered to be in a properly altered condition acceptable to the Administrator.

(e) Notwithstanding any other provision of this section, an aircraft with inoperable instruments or equipment may be operated under a special flight permit issued in accordance with §§ 21.197 and 21.199 of this chapter.

§ 91.31 Civil aircraft flight manual marking, and placard requirements.

(a) Except as provided in paragraph (d) of this section, no person may operate a civil aircraft without complying with the operating limitations specified in the approved Airplane or Rotorcraft Flight Manual, markings, and placards, or as otherwise prescribed by the certificating authority of the country of registry.
☞F99

(b) No person may operate a U.S. registered civil aircraft—

(1) For which an Airplane or Rotorcraft Flight Manual is required by § 21.5 unless there is available in the aircraft a current approved Airplane or Rotorcraft Flight Manual or the manual provided for in § 121.141(b); and

(2) For which an Airplane or Rotorcraft Flight Manual is not required by § 21.5, unless there is available in the aircraft a current approved Airplane or Rotorcraft Flight

Manual, approved manual material, markings, and placards, or any combination thereof.

(c) No person may operate a U.S. registered civil aircraft unless that aircraft is identified in accordance with Part 45 of this chapter.

(d) Any person taking off or landing a helicopter certificated Under Part 29 of this chapter at a heliport constructed over water may make such momentary flight as is necessary for takeoff or landing through the prohibited range of the limiting height-speed envelope established for that helicopter if that flight through the prohibited range takes place over water on which a safe ditching can be accomplished, and if the helicopter is amphibious or is equipped with floats or other emergency flotation gear adequate to accomplish a safe emergency ditching on open water.

⟦(e) [Removed]⟧

§ 91.32 Supplemental oxygen.   ☞F121

(a) *General.*   No person may operate a civil aircraft of U.S. registry—

(1) At cabin pressure altitudes above 12,500 feet (MSL) up to and including 14,000 feet (MSL), unless the required minimum flight crew is provided with and uses supplemental oxygen for that part of the flight at those altitudes that is of more than 30 minutes duration;

(2) At cabin pressure altitudes above 14,000 feet (MSL), unless the required minimum flight crew is provided with and uses supplemental oxygen during the entire flight time at those altitudes; and

(3) At cabin pressure altitudes above 15,000 feet (MSL), unless each occupant of the aircraft is provided with supplemental oxygen.

(b) *Pressurized cabin aircraft.*

(1) No person may operate a civil aircraft of U.S. registry with a pressurized cabin—

(i) At flight altitudes above flight level 250, unless at least a 10-minute supply of supplemental oxygen, in addition to any oxygen required to satisfy paragraph (a) of this section, is available for each occupant of the aircraft for use in the event that a descent is necessitated by loss of cabin pressurization; and

(ii) At flight altitudes above flight level 350, unless one pilot at the controls of the airplane is wearing and using an oxygen mask that is secured and sealed, and that either supplies oxygen at all times or automatically supplies oxygen whenever the cabin pressure altitude of the airplane exceeds 14,000 feet (MSL), except that the one pilot need not wear and use an oxygen mask while at or below flight level 410 if there are two pilots at the controls and each pilot has a quick-donning type of oxygen mask that can be placed on the face with one hand from the ready position within five seconds, supplying oxygen and properly secured and sealed.

(2) Notwithstanding subparagraph (1) (ii) of this paragraph, if for any reason at any time it is necessary for one pilot to leave his station at the controls of the aircraft when operating at flight altitudes above flight level 350, the remaining pilot at the controls shall put on and use his oxygen mask until the other pilot has returned to his station.

§ 91.33 Powered civil aircraft with standard category U.S. airworthiness certificates; instrument and equipment requirements.   ☞ F100

(a) *General.* Except as provided in paragraphs (c) (3) and (e) of this section, no person may operate a powered civil aircraft with a standard category U.S. airworthiness certificate in any operation described in paragraphs (b) through (f) of this section unless that aircraft contains the instruments and equipment specified in those paragraphs (or FAA approved equivalents) for that type of operation, and those instruments and items of equipment are in operable condition.

(b) *Visual flight rules (day).* For VFR flight during the day the following instruments and equipment are required:

(1) Airspeed indicator.

(2) Altimeter.

(3) Magnetic direction indicator.

(4) Tachometer for each engine.

(5) Oil pressure gauge for each engine using pressure system.

(6) Temperature gauge for each liquid-cooled engine.

(7) Oil temperature gauge for each air-cooled engine.

(8) Manifold pressure gauge for each altitude engine.

(9) Fuel gauge indicating the quantity of fuel in each tank.

(10) Landing gear position indicator, if the aircraft has a retractable landing gear.

(11) If the aircraft is operated for hire over water and beyond power-off gliding distance from shore, approved flotation gear readily available to each occupant, and at least one pyrotechnic signaling device.

(12) Except as to airships, an approved safety belt for all occupants who have reached their second birthday. After December 4, 1981, each safety belt must be equipped with an approved metal to metal latching device. The rated strength of each safety belt shall not be less than that corresponding with the ultimate load factors specified in the current applicable aircraft airworthiness requirements considering the dimensional characteristics of the safety belt installation for the specific seat or berth arrangement. The webbing of each safety belt shall be replaced as required by the Administrator.

(13) For small civil airplanes manufactured after July 18, 1978, an approved shoulder harness for each front seat. The shoulder harness must be designed to protect the occupant from serious head injury when the occupant experiences the ultimate inertia forces specified in § 23.561(b) (2) of this chapter. Each shoulder harness installed at a flight crewmember station must permit the crewmember, when seated and with his safety belt and shoulder harness fastened, to perform all functions necessary for flight operations. For purposes of this paragraph—

(i) The date of manufacture of an airplane is the date the inspection acceptance records reflect that the airplane is complete and meets the FAA Approved Type Design Data; and

(ii) A front seat is a seat located at a flight crewmember station or any seat located alongside such a seat.

〖(14) For normal, utility, and acrobatic category airplanes with a seating configuration, excluding pilot seats, of nine or less, manufactured after December 12, 1986, a shoulder harness for—

〖(i) Each front seat that meets the requirements of § 23.785(g) and (h) of this chapter in effect on December 12, 1985;

〖(ii) Each additional seat that meets the requirements of § 23.785(g) of this chapter in effect on December 12, 1985.〗

(c) *Visual flight rules (night).* For VFR flight at night the following instruments and equipment are required:

(1) Instruments and equipment specified in paragraph (b) of this section.

(2) Approved position lights.

(3) An approved aviation red or aviation white anticollision light system on all U.S. registered civil aircraft. Anticollision light systems initially installed after August 11, 1971, on aircraft for which a type certificate was issued or applied for before August 11, 1971, must at least meet the anticollision light standards of Parts 23, 25, 27, or 29, as applicable, that were in effect on August 10, 1971, except that the color may be either aviation red or aviation white. In the event of failure of any light of the anticollision light system, operations with the aircraft may be continued to a stop where repairs or replacement can be made.

(4) If the aircraft is operated for hire, one electric landing light.

(5) An adequate source of electrical energy for all installed electrical and radio equipment.

(6) One spare set of fuses, or three spare fuses of each kind required.

(d) *Instrument flight rules.* For IFR flight the following instruments and equipment are required:

(1) Instruments and equipment specified in paragraph (b) of this section and for night flight, instruments and equipment specified in paragraph (c) of this section.

(2) Two-way radio communications system and navigational equipment appropriate to the ground facilities to be used.

(3) Gyroscopic rate-of-turn indicator, except on the following aircraft:

(i) Large airplanes with a third attitude instrument system useable through flight attitudes of 360 degrees of pitch and roll and installed in accordance with § 121.305(j) of this chapter; and

(ii) Rotorcraft, type certificate under Part 29 of this chapter, with a third attitude instrument system useable through flight attitudes of ±80 degrees of pitch and ±120 degrees of roll and installed in accordance with § 29.1303(g) of this chapter.

(4) Slip-skid indicator.

(5) Sensitive altimeter adjustable for barometric pressure.

(6) A clock displaying hours, minutes, and seconds with a sweep-second pointer or digital presentation.

(7) Generator of adequate capacity.

(8) Gyroscopic bank and pitch indicator (artificial horizon).

(9) Gyroscopic direction indicator (directional gyro or equivalent).

(e) *Flight at and above 24,000 feet MSL.* IF VOR navigational equipment is required under paragraph (d) (2) of this section, no person may operate a U.S. registered civil aircraft within the 50 states, and the District of Columbia, at or above 24,000 feet MSL unless that aircraft is equipped with approved distance measuring equipment (DME). When DME required by this paragraph fails at and above 24,000 feet MSL, the pilot in command of the aircraft shall notify ATC immediately, and may then continue operations at and above 24,000 feet MSL to the next airport of intended landing at which repairs or replacement of the equipment can be made.

(f) *Category II operations.* For Category II operations the instruments and equipment specified in paragraph (d) of this section and Appendix A to this Part are required. This paragraph does not apply to operations conducted by the holder of a certificate issued under Part 121 of this chapter.

§ 91.34 Category II manual.

(a) No person may operate a civil aircraft of United States registry in a Category II operation unless—

(1) There is available in the aircraft a current approved Category II manual for that aircraft;

(2) The operation is conducted in accordance with the procedures, instructions, and limitations in that manual; and

(3) The instruments and equipment listed in the manual that are required for a particular Category II operation have been inspected and maintained in accordance with the maintenance program contained in that manual.

(b) Each operator shall keep a current copy of the approved manual at its principal base of operations and shall make it available for inspection upon request of the Administrator.

(c) This section does not apply to operations conducted by the holder of a certificate issued under Part 121 of this chapter.

§ 91.35 Flight recorders and cockpit voice recorders.

[(a)] No holder of an air carrier or commercial operator certificate may conduct any operation under this Part with an aircraft listed in his operations specifications or current list of aircraft used in air transportation unless that aircraft complies with any applicable flight recorder and cockpit voice recorder requirements of the Part under which its certificate is issued; except that it may—

[(1)] Ferry an aircraft with an inoperative flight recorder or cockpit voice recorder from a place where repair or replacement cannot be made to a place where they can be made;

[(2)] Continue a flight as originally planned, if the flight recorder or cockpit voice recorder becomes inoperative after the aircraft has taken off;

[(3)] Conduct an airworthiness flight test, during which the flight recorder or cockpit voice recorder is turned off to test it or to test any communications or electrical equipment installed in the aircraft; or

[(4)] Ferry a newly acquired aircraft from the place where possession of it was taken to a place where the flight recorder or cockpit voice recorder is to be installed.]

[(b) No person may operate a U.S. civil registered, multiengine, turbine-powered airplane or rotorcraft having a passenger seating configuration, excluding any pilot seats of 10 or more that has been manufactured after October 11, 1991, unless it is equipped with one or more approved flight recorders that utilize a digital method of recording and storing medium, that are capable of recording the date specified in Appendix E, for an airplane, or Appendix F, for a rotorcraft, of this part within the range, accuracy, and recording interval specified, and that are capable of retaining no less than 8 hours of aircraft operation.]

[(c) Whenever a flight recorder, required by this section, is installed, it must be operated continuously from the instant the airplane begins the takeoff roll or the rotorcraft begins lift-off until the airplane has completed the landing roll or the rotorcraft has landed at its destination.]

[(d) Unless otherwise authorized by the Administrator, after Ocotober 11, 1991, no person may operate a U.S. civil registered, multiengine, turbine-powered airplane or rotorcraft having a passenger seating configuration of six passengers or more and for which two pilots are required by type certification or operating rule unless it is equipped with an approved cockpit voice recorder that—

[(1)] Is installed in compliance with § 23.1457(a)(1) and (2), (b), (c), (d), (e), (f), and (g), § 25.1457(a)(1) and (2), (b), (c), (d), (e), (f), and (g); § 29.1457(a)(1) and (2), (b), (c), (d), (e), (f), and (g) of this chapter, as applicable; and

[(2)] Is operated continuously from the use of the check list before the flight to completion of the final check list at the end of the flight.]

[(e) In complying with this section, an approved cockpit voice recorder having an erasure feature may be used, so that at any time during the operation of the recorder, information recorded more than 15 minutes earlier may be erased or otherwise obliterated.]

[(f) In the event of an accident or occurrence requiring immediate notification to the National Transportation Safety Board under Part 830 of its regulations that results in the termination of the flight, any operator who has installed approved flight recorders and approved cockpit voice recorders shall keep the recorded information for at least 60 days or, if requested by the Administrator or the Board, for a longer period. Information obtained from the record is used to assist in determining the cause of accidents or occurrences in connection with investigation under Part 830. The Administrator does not use the cockpit voice recorder record in civil penalty or certification action.

§ 91.36 Data correspondence between automatically reported pressure altitude data and the pilot's altitude reference.

No person may operate any automatic pressure altitude reporting equipment associated with a radar beacon transponder—

(a) When deactivation of that equipment is directed by ATC;

(b) Unless, as installed, that equipment was tested and calibrated to transmit altitude data corresponding with 125 feet (on a 95 percent probability basis) of the indicated or calibrated datum of the altimeter normally used to maintain flight altitude, with that altimeter referenced to 29.92 inches of mercury for altitudes from sea level to the maximum operating altitude of the aircraft; or

(c) After September 1, 1979, unless the altimeters and digitizers in that equipment meet the standards in TSO-C10b and TSO-C88, respectively.

§ 91.37 Transport category civil airplane weight limitations.

(a) No person may take off any transport category airplane (other than a turbine engine powered airplane certificated after September 30, 1985) unless—

(1) The takeoff weight does not exceed the authorized maximum takeoff weight for the elevation of the airport of takeoff;

(2) The elevation of the airport of takeoff is within the altitude range for which maximum takeoff weights have been determined;

(3) Normal consumption of fuel and oil in flight to the airport of intended landing will leave a weight on arrival not in excess of the authorized maximum landing weight for the elevation of that airport; and

(4) The elevation of the airport of intended landing and of all specified alternate airports are within the altitude range for which maximum landing weights have been determined.

(b) No person may operate a turbine engine powered transport category airplane certificated after September 30, 1958 contrary to the Airplane Flight Manual, nor takeoff that airplane unless—

(1) The takeoff weight does not exceed the takeoff weight specified in the Airplane Flight Manual for the elevation of the airport and for the ambient temperature existing at the time of takeoff;

(2) Normal consumption of fuel and oil in flight to the airport of intended landing and to the alternate airports will leave a weight not in excess of the landing weight specified in the Airplane Flight Manual for the elevation of each of the airports involved and for the ambient temperatures expected at the time of landing:

(3) The takeoff weight does not exceed the weight shown in the Airplane Flight Manual to correspond with the minimum distances required for takeoff considering the elevation of the airport, the runway to be used, the effective runway gradient, and the ambient temperature and wind component existing at the time of takeoff; and

(4) Where the takeoff distance includes a clearway, the clearway distance is not greater than one-half of—

(i) The takeoff run, in the case of airplanes certificated after September 30, 1958 and before August 30, 1959; or

(ii) The runway length, in the case of airplanes certificated after August 29, 1959.

No person may take off a turbine engine powered transport category airplane certificated after August 29, 1959 unless, in addition to the requirements of paragraph (b) of this section—

(1) The accelerate-stop distance is no greater than the length of the runway plus the length of the stopway (if present);

(2) The takeoff distance is no greater than the length of the runway plus the length of the clearway (if present); and

(3) The takeoff run is no greater than the length of the runway.

§ 91.38 Increased maximum certificated weights for certain airplanes operated in Alaska.

(a) Notwithstanding any other provision of the Federal Aviation Regulations, the Administrator will, as provided in this section, approve an increase in the maximum certificated weight of an airplane type certificated under Aeronautics Bulletin No. 7-A of the U.S. Department of Commerce dated January 1, 1931, as amended, or under the normal category of Part 4a of the former Civil Air Regulations, (14 CFR 4a, 1964 ed.) if that airplane is operated in the State of Alaska by—

(1) An air taxi operator or other air carrier; or

(2) The U.S. Department of Interior in conducting its game and fish law enforcement activities or its management, fire detection, and fire suppression activities concerning public lands.

(b) The maximum certificated weight approved under this section may not exceed—

(1) 12,500 pounds;

(2) 115 percent of the maximum weight listed in the FAA Aircraft Specifications;

(3) The weight at which the airplane meets the positive maneuvering load factor requirement for the normal category specified in § 23.337 of this chapter; or

(4) The weight at which the airplane meets the climb performance requirements under which it was type certificated.

(c) In determining the maximum certificated weight the Administrator considers the structural soundness of the airplane and the terrain to be traversed.

(d) The maximum certificated weight determined under this section is added to the airplane's operation limitations and is identified as the maximum weight authorized for operations within the State of Alaska.

§ 91.39 Restricted category civil aircraft; operating limitations.

(a) No person may operate a restricted category civil aircraft—

(1) For other than the special purpose for which it is certificated; or

(2) In an operation other than one necessary for the accomplishment of the work activity directly associated with that special purpose.

For the purposes of this paragraph, the operation of a restricted category civil aircraft to provide flight crewmember training in a special purpose operation for which the aircraft is certificated is considered to be an operation for that special purpose.

(b) No person may operate a restricted category civil aircraft carrying persons or property for compensation or hire. For the purposes of this paragraph, a special purpose operation involving the carriage of persons or materials necessary for the accomplishment of that operation such as crop dusting, seeding, spraying, and banner towing (including the carrying of required persons or materials to the location of that operation), and an operation for the purpose of providing flight crewmember training in a special purpose operation, are not considered to be the carrying of persons or property for compensation or hire.

(c) No person may be carried on a restricted category civil aircraft unless—

(1) He is a flight crewmember;

(2) He is a flight crewmember trainee;

(3) He performs an essential function in connection with a special purpose operation for which the aircraft is certificated; or

(4) He is necessary for the accomplishment of the work activity directly associated with that special purpose.

(d) Except when operating in accordance with the terms and conditions of a certificate of waiver or special operating limitations issued by the Administrator, no person may operate a restricted category civil aircraft within the United States—

(1) Over densely populated area;

(2) In a congested airway; or

(3) Near a busy airport where passenger transport operations are conducted.

(e) An application for a certificate of waiver under paragraph (d) of this section is made on a form and in a manner prescribed by the Administrator and must be submitted to the Flight Standards District Office having jurisdiction over the area in which the applicant is located.

(f) After December 9, 1977, this section does not apply to non-passenger-carrying civil rotorcraft external-load operations conducted under Part 133 of this chapter.

(g) No person may operate a small restricted category civil airplane, manufactured after July 18, 1978, unless an approved shoulder harness is installed for each front seat. The shoulder harness must be designed to protect each occupant from serious head injury when the occupant experiences the ultimate inertia forces specified in § 23.561(b)(2) of this chapter. The shoulder harness installation at each flight crewmember station must permit the crewmember, when seated and with his safety belt and shoulder harness fastened, to perform all functions necessary for flight operations. For purposes of this paragraph—

(1) The date of manufacture of an airplane is the date the inspection acceptance records reflect that the airplane is complete and meets the FAA Approved Type Design Data; and

(2) A front seat is a seat located at a flight crewmember station or any seat located alongside such a seat.

§ 91.40 Limited category civil aircraft; operating limitations.

No person may operate a limited category civil aircraft carrying persons or property for compensation or hire.

§ 91.41 Provisionally certificated civil aircraft; operating limitations.

(a) No person may operate a provisionally certificated civil aircraft unless he is eligible for a provisional airworthiness certificate under § 21.213 of this chapter.

(b) No person may operate a provisionally certificated civil aircraft outside the United States unless he has specific authority to do so from the Administrator and each foreign country involved.

(c) Unless otherwise authorized by the [Director of Airworthiness], no person may operate a provisionally certificated civil aircraft in air transportation.

(d) Unless otherwise authorized by the Administrator, no person may operate a provisionally certificated civil aircraft except—

(1) In direct conjunction with the type or supplemental type certification of that aircraft;

(2) For training flight crews, including simulated air carrier operations;

(3) Demonstration flights by the manufacturer for prospective purchasers;

(4) Market surveys by the manufacturer;

(5) Flight checking of instruments, accessories, and equipment, that do not affect the basic airworthiness of the aircraft; or

(6) Service testing of the aircraft.

(e) Each person operating a provisionally certificated civil aircraft shall operate within the prescribed limitations displayed in the aircraft or set forth in the provisional aircraft flight manual or other appropriate document. However, when operating in direct conjunction with the type or supplemental type certification of the aircraft, he shall operate under the experimental aircraft limitations of § 21.191 of this chapter and when flight testing, shall operate under the requirements of § 91.93 of this chapter.

(f) Each person operating a provisionally certificated civil aircraft shall establish approved procedures for—

(1) The use and guidance of flight and ground personnel in operating under this section; and

(2) Operating in and out of airports where takeoffs or approaches over populated areas are necessary.

No person may operate that aircraft except in compliance with the approved procedures.

(g) Each person operating a provisionally certificated civil aircraft shall ensure that each flight crewmember is properly certificated and has adequate knowledge of, and familiarity with, the aircraft and procedures to be used by that crewmember.

(h) Each person operating a provisionally certificated civil aircraft shall maintain it as required by applicable regulations and as may be specially prescribed by the Administrator.

(i) Whenever the manufacturer, or the Administrator, determines that a change in design, construction or operation is necessary to ensure safe operation, no person may operate a provisionally certificated civil aircraft until that change has been made and approved. Section 21.99 of this chapter applies to operations under this section.

(j) Each person operating a provisionally certificated civil aircraft—

(1) May carry in that aircraft only persons who have a proper interest in the operations allowed by this section or who are specifically authorized by both the manufacturer and the Administrator; and

(2) Shall advise each person carried that the aircraft is provisionally certificated.

(k) The Administrator may prescribe additional limitations or procedures that he considers necessary, including limitations on the number of persons who may be carried in the aircraft.

§ 91.42 Aircraft having experimental certificates; operating limitations.

(a) No person may operate an aircraft that has an experimental certificate—

(1) For other than the purpose for which the certificate was issued; or

(2) Carrying persons or property for compensation or hire. ☞ **F65**

(b) No person may operate an aircraft that has an experimental certificate outside of an area assigned by the Administrator until it is shown that—

(1) The aircraft is controllable throughout its normal range of speeds and throughout all the maneuvers to be executed; and

(2) The aircraft has no hazardous operating characteristics or design features.

(c) Unless otherwise authorized by the Administrator in special operating limitations, no person may operate an aircraft that has an experimental certificate over a densely populated area or in a congested airway. The Administrator may issue special operating limitations for particular aircraft to permit takeoffs and landings to be conducted over a densely populated area or in a congested airway, in accordance with terms and conditions specified in the authorization in the interest of safety in air commerce.

(d) Each person operating an aircraft that has an experimental certificate shall—

(1) Advise each person carried of the experimental nature of the aircraft;

(2) Operate under VFR, day only, unless otherwise specifically authorized by the Administrator; and

(3) Notify the control tower of the experimental nature of the aircraft when operating the aircraft into or out of airports with operating control towers.

(e) The Administrator may prescribe additional limitations that he considers necessary, including limitations on the persons that may be carried in the aircraft.

§ 91.43 Special rules for foreign civil aircraft.

(a) *General.* In addition to the other applicable regulations of this Part, each person operating a foreign civil aircraft within the United States shall comply with this section.

[(b) *VFR.* No person may conduct VFR operations which require two-way radio communications under this Part, unless at least one crewmember of that aircraft is able to conduct two-way radio communications in the English language and is on duty during that operation.]

(c) *IFR.* No person may operate a foreign civil aircraft under IFR unless—

(1) That aircraft is equipped with—

(i) Radio equipment allowing two-way radio communication with ATC when it is operated in a control zone or control area; and

(ii) Radio navigational equipment appropriate to the navigational facilities to be used;

(2) Each person piloting the aircraft—

(i) Holds a current United States instrument rating or is authorized by his foreign airman certificate to pilot under IFR; and

(ii) Is thoroughly familiar with the United States en route, holding, and letdown procedures; and

(3) At least one crewmember of that aircraft is able to conduct two-way radiotelephone communications in the English language and that crewmember is on duty while the aircraft is approaching, operating within, or leaving the United States.

(d) *Overwater.* Each person operating a foreign civil aircraft over water off the shores of the United States shall give flight notification or file a flight plan, in accordance with the Supplementary Procedures for the ICAO region concerned.

(e) *Flight at and above 24,000 feet MSL.* If VOR navigation equipment is required under paragraph (c)(1)(ii) of this section, no person may operate a foreign civil aircraft within the 50 states and the District of Columbia at or above 24,000 feet MSL, unless the aircraft is equipped with distance measuring equipment (DME) capable of receiving and indicating distance information from the VORTAC facilities to be used. When DME required by this paragraph fails at and above 24,000 feet MSL, the pilot in command of the aircraft shall notify ATC immediately, and may then continue operations at and above 24,000 feet MSL to the next airport of intended landing at which repairs or replacement of the equipment can be made.

However, paragraph (e) does not apply to foreign civil aircraft that are not equipped with DME when operated for the following purposes, and if ATC is notified prior to each takeoff:

(1) Ferry flights to and from a place in the United States where repairs or alterations are to be made.

(2) Ferry flights to a new country of registry.

(3) Flight of a new aircraft of U.S. manufacture for the purpose of—

(i) Flight testing the aircraft;

(ii) Training foreign flight crews in the operation of the aircraft; or

(iii) Ferrying the aircraft for export delivery outside the United States.

(4) Ferry, demonstration, and test flights of an aircraft brought to the United States for the purpose of demonstration or testing the whole or any part thereof.

§ 91.45 [Authorization for ferry flights with one engine inoperative.]

[(a) *General.* The holder of an air carrier operating certificate, an operating certificate issued under Part 125, or until January 1, 1983, an operating certificate issued under Part 121, may conduct a ferry flight of a four-engine airplane or a turbine-engine-powered airplane equipped with three engines, with one engine inoperative, to a base for the purpose of repairing that engine subject to the following:]

(1) The airplane model has been test flown and found satisfactory for safe flight in accordance with paragraph (b) or (c) of this section, as appropriate. However, each operator who before November 19, 1966, has shown that a model of airplane with an engine inoperative is satisfactory for safe flight by a test flight conducted in accordance with performance data contained in the applicable Airplane Flight Manual under § 91.45(a)(2) need not repeat the test flight for that model.

(2) The approved Airplane Flight Manual contains the following performance data and the flight is conducted in accordance with that data:

(i) Maximum weight.

(ii) Center of gravity limits.

(iii) Configuration of the inoperative propeller (if applicable).

(iv) Runway length for takeoff (including temperature accountability).

(v) Altitude range.

(vi) Certificate limitations.

(vii) Ranges of operational limits.

(viii) Performance information.

(ix) Operating procedures.

(3) The operator's manual contains operating procedures for the safe operation of the airplane, including specific requirements for—

(i) A limitation that the operating weight on any ferry flight must be the minimum necessary therefor with the necessary reserve fuel load;

(ii) A limitation that takeoffs must be made from dry runways unless, based on a showing of actual operating takeoff techniques on wet runways with one engine inoperative, takeoffs with full controllability from wet runways have been approved for the specific model aircraft and included in the Airplane Flight Manual;

(iii) Operations from airports where the runways may require a takeoff or approach over populated areas; and

(iv) Inspection procedures for determining the operating condition of the operative engines.

(4) No person may take off an airplane under this section if—

(i) The initial climb is over thickly populated areas; or

(ii) Weather conditions at the takeoff or destination airport are less than those required for VFR flight.

[(5) Persons other than required flight crewmembers shall not be carried during the flight.]

[(6) No person may use a flight crewmember for flight under this section unless that crewmember is thoroughly familiar with the operating procedures for one-engine inoperative ferry flight contained in the certificate holder's manual, and the limitations and performance information in the Airplane Flight Manual.]

(b) *Flight tests: reciprocating engine*

powered airplanes. The airplane performance of a reciprocating engine powered airplane with one engine inoperative must be determined by flight test as follows:

(1) A speed not less than 1.3 V_{S_1} must be chosen at which the airplane may be controlled satisfactorily in a climb with the critical engine inoperative (with its propeller removed or in a configuration desired by the operator) and with all other engines operating at the maximum power determined in subparagraph (3) of this paragraph.

(2) The distance required to accelerate to the speed listed in subparagraph (1) of this paragraph and to climb to 50 feet must be determined with—

(i) The landing gear extended;

(ii) The critical engine inoperative and its propeller removed or in a configuration desired by the operator; and

(iii) The other engines operating at not more than the maximum power established under subparagraph (3) of this paragraph.

(3) The takeoff, flight, and landing procedures such as the approximate trim settings, method of power application, maximum power, and speed must be established.

(4) The performance must be determined at a maximum weight not greater than the weight that allows a rate of climb of at least 400 feet a minute in the en route configuration set forth in § 25.67(d) of this chapter at an altitude of 5,000 feet.

(5) The performance must be determined using temperature accountability for the takeoff field length, computed in accordance with § 25.61 of this chapter.

(c) *Flight tests: turbine engine powered airplanes.* The airplane performance of a turbine engine powered airplane with one engine inoperative must be determined in accordance with the following, by flight tests including at least three takeoff tests:

(1) Takeoff speeds V_R and V_2, not less than the corresponding speeds under which the airplane was type certificated under § 25.107 of this chapter, must be chosen at which the airplane may be controlled satisfactorily with the critical engine inoperative (with its propeller removed or in a configuration desired by the operator, if applicable) and with all other engines operating at not more than the power selected for type certification, as set forth in § 25.101 of this chapter.

(2) The minimum takeoff field length must be the horizontal distance required to accelerate, and climb to the 35-foot height at V_2 speed (including any additional speed increment obtained in the tests), multiplied by 115 percent, and determined with—

(i) The landing gear extended;

(ii) The critical engine inoperative and its propeller removed or in a configuration desired by the operator (if applicable); and

(iii) The other engines operating at not more than the power selected for type certification, as set forth in § 25.101 of this chapter.

(3) The takeoff, flight, and landing procedures such as the approximate trim settings, method of power application, maximum power, and speed, must be established. The airplane must be satisfactorily controllable during the entire takeoff run when operated according to these procedures.

(4) The performance must be determined at a maximum weight not greater than the weight determined under § 25.121(c) of this chapter, but with—

(i) The actual steady gradient of the final takeoff climb requirement not less than 1.2 percent at the end of the takeoff path with two critical engines inoperative; and

(ii) The climb speed not less than the two-engine inoperative trim speed for the actual steady gradient of the final takeoff climb prescribed by subdivision (i) of this subparagrah.

(5) The airplane must be satisfactorily controllable in a climb with two critical engines inoperative. Climb performance may be shown by calculations based on, and equal in accuracy to, the results of testing.

(6) The performance must be determined using temperature accountability for takeoff distance and final takeoff climb, computed in accordance with § 25.101 of this chapter.

For the purposes of subparagraphs (4) and (5), "two critical engines" means two adjacent engines on one side of an airplane with four engines, and the center engine and one outboard engine on an airplane with three engines.

§ 91.47 Emergency exits for airplanes carrying passengers for hire.

(a) Notwithstanding any other provision of this chapter, no person may operate a large airplane (type certificated under the Civil Air Regulations effective before April 9, 1957) in passenger-carrying operations for hire, with more than the number of occupants—

(1) Allowed under Civil Air Regulation § 4b.362(a), (b), and (c), as in effect on December 20, 1951; or

(2) Approved under Special Civil Air Regulations SR–387, SR–389, SR–389A, or SR–389B, or under this section as in effect.

However, an airplane type listed in the following table may be operated with up to the listed number of occupants (including crewmembers) and the corresponding number of exits (including emergency exits and doors) approved for the emergency exit of passengers or with an occupant-exit configuration approved under paragraph (b) or (c) of this section:

Airplane Type	Maximum number of occupants including all crewmembers	Corresponding number of exits authorized for passenger use
B–307	61	4
B–377	96	9
C–46	67	4
CV–240	53	6
CV–340 and CV–440	53	6
DC–3	35	4
DC–3(Super)	39	5
DC–4	86	5
DC–6	87	7
DC–6B	112	11
L–18	17	3
L–049, L–649, L–749	87	7
L–1049 series	96	9
M–202	53	6
M–404	53	7
Viscount 700 series	53	7

(b) Occupants in addition to those authorized under paragraph (a) of this section may be carried as follows:

(1) For each additional floor-level exit at least 24 inches wide by 48 inches high, with an unobstructed 20-inch wide access aisleway between the exit and the main passenger aisle: Twelve additional occupants.

(2) For each additional window exit located over a wing that meets the requirements of the airworthiness standards under which the airplane was type certificated or that is large enough to inscribe an ellipse 19 x 26 inches: Eight additional occupants.

(3) For each additional window exit that is not located over a wing but that otherwise complies with subparagraph (2) of this paragraph: Five additional occupants.

(4) For each airplane having a ratio (as computed from the table in paragraph (a) of this section) of maximum number of occupants to number of exits greater than 14:1, and for each airplane that does not have at least one full-size door-type exit in the side of the fuselage in the rear part of the cabin, the first additional exit must be a floor-level exit that complies with subparagraph (1) of this paragraph and must be located in the rear part of the cabin on the opposite side of the fuselage from the main entrance door.

However, no person may operate an airplane under this section carrying more than 115 occupants unless there is such an exit on each side of the fuselage in the rear part of the cabin.

(c) No person may eliminate any approved exit except in accordance with the following:

(1) The previously authorized maximum number of occupants must be reduced by the same number of additional occupants authorized for that exit under this section.

(2) Exits must be eliminated in accordance with the following priority schedule: First, non-over-wing window exits; second, over-wing-window exits; third, floor-level exits located in the forward part of the cabin; fourth, floor-level exits located in the rear of the cabin.

(3) At least one exit must be retained on each side of the fuselage regardless of the number of occupants.

(4) No person may remove any exit that would result in a ratio of maximum number of occupants to approved exits greater than 14:1.

(d) This section does not relieve any person operating under Part 121 of this chapter from complying with § 121.291 of this chapter.

§ 91.49 Aural speed warning device.

No person may operate a transport category airplane in air commerce unless that airplane is equipped with an aural speed warning device that complies with § 25.1303(c) (1).

[§ 91.50 [Reserved.]]

§ 91.51 Altitude alerting system or device; turbojet powered civil airplanes.

(a) Except as provided in paragraph (d) of this section, no person may operate a turbojet powered U.S. registered civil airplane unless that airplane is equipped with an approved altitude alerting system or device that is in operable condition and meets the requirements of paragraph (b) of this section.

(b) Each altitude alerting system or device required by paragraph (a) of this section must be able to:

(1) Alert the pilot:

(i) Upon approaching a preselected altitude in either ascent or descent, by a sequence of both aural and visual signals in sufficient time to establish level flight at that preselected altitude; or

(ii) Upon approaching a preselected altitude in either ascent or descent, by a sequence of visual signals in sufficient time to establish level flight at that preselected altitude, and when deviating above and below that preselected altitude, by an aural signal;

(2) Provide the required signals from sea level to the highest operating altitude approved for the airplane in which it is installed;

(3) Preselect altitudes in increments that are commensurate with the altitudes at which the aircraft is operated;

(4) Be tested without special equipment to determine proper operation of the alerting signals; and

(5) Accept necessary barometric pressure settings if the system or device operates on barometric pressure.

However, for operations below 3,000 feet AGL, the system or device need only provide one signal, either visual or aural, to comply with this paragraph. A radio altimeter may be included to provide the signal, if the operator has an approved procedure for its use to determine DH or MDA, as appropriate.

(c) Each operator to which this section applies must establish and assign procedures for the use of the altitude alerting system or device and each flight crewmember must comply with those procedures assigned to him.

(d) Paragraph (a) of this section does not apply to any operation of an airplane that has an experimental certificate or to the operation of an airplane for the following purposes:

(1) Ferrying a newly acquired airplane from the place where possession of it was taken to a place where the altitude alerting system or device is to be installed.

(2) Continuing a flight as originally planned, if the altitude alerting system or device becomes inoperative after the airplane has taken off; however, the flight may not depart from a place where repair or replacement can be made.

(3) Ferrying an airplane with an inoperative altitude alerting system or device from a place where repair or replacement cannot be made to a place where they can be made.

(4) Conducting an airworthiness flight test of the airplane.

(5) Ferrying an airplane to a place outside the United States for the purpose of registering it in a foreign country.

(6) Conducting a sales demonstration of the operation of the airplane.

(7) Training foreign flight crews in the operation of the airplane prior to ferrying it to a place outside the United States for the purpose of registering it in a foreign country.

§ 91.52 Emergency locator transmitters.

(a) Except as provided in paragraphs (e) and (f) of this section, no person may operate a U.S. registered civil airplane unless it meets the applicable requirements of paragraphs (b), (c), and (d) of this section.

(b) To comply with paragraph (a) of this section, each U.S. registered civil airplane must be equipped as follows:

(1) For operations governed by the supplemental air carrier and commercial operator rules of Part 121 of this chapter, or the air travel club rules of Part 123 of this chapter, there must be attached to the airplane an automatic type emergency locator transmitter that is in operable condition and meets the applicable requirements of TSO-C91;

(2) For charter flights governed by the domestic and flag air carrier rules of Part 121 of this chapter, there must be attached to the airplane an automatic type emergency locator transmitter that is in operable condition and meets the applicable requirements of TSO-C91;

(3) For operations governed by Part 135 of this chapter, there must be attached to the airplane an automatic type emergency locator transmitter that is in operable condition and meets the applicable requirements of TSO-C91; and

(4) For operations other than those specified in subparagraphs (1), (2), and (3) of this paragraph, there must be attached to the airplane a personnel type or an automatic type emergency locator transmitter that is in operable condition and meets the applicable requirements of TSO-C91.

(c) Each emergency locator transmitter required by paragraphs (a) and (b) of this section must be attached to the airplane in such a manner that the probability of damage to the transmitter, in the event of crash impact, is minimized. Fixed and deployable automatic type transmitters must be attached to the airplane as far aft as practicable.

(d) Batteries used in the emergency locator transmitters required by paragraphs (a) and (b) of this section must be replaced (or recharged, if the battery is rechargeable)—

(1) When the transmitter has been in use for more than one cumulative hour; or

(2) When 50 percent of their useful life (or, for rechargeable batteries, 50 percent of their useful life of charge), as established by the transmitter manufacturer under TSO-C91, paragraph (g)(2), has expired.

The new expiration date for the replacement (or recharge) of the battery must be legibly marked on the outside of the transmitter and entered in the aircraft maintenance record. Subparagraph (d) (2) of this paragraph does not apply to batteries such as water-activated batteries) that are essentially unaffected during probable storage intervals.

(e) Notwithstanding paragraphs (a) and (b) of this section, a person may—

(1) Ferry a newly acquired airplane from the place where possession of it was taken to a place where the emergency locator transmitter is to be installed; and

(2) Ferry an airplane with an inoperative emergency locator transmitter from a place where repairs or replacement cannot be made to a place where they can be made.

No person other than required crewmembers may be carried board an airplane being ferried pursuant to paragraph (e) of this section.

(f) Paragraphs (a) and (b) of this section do not apply to—

(1) Turbojet-powered aircraft;

(2) Aircraft while engaged in scheduled flights by scheduled air carriers certificated by the Civil Aeronautics Board;

(3) Aircraft while engaged in training operations conducted entirely within a 50-mile radius of the airport from which such local flight operations began;

(4) Aircraft while engaged in flight operations incident to design and testing;

(5) New aircraft while engaged in flight operations incident to their manufacture, preparation, and delivery;

(6) Aircraft while engaged in flight operations incident to the aerial application of chemicals and other substances for agricultural purposes;

(7) Aircraft certificated by the Administrator for research and development purposes;

(8) Aircraft while used for showing compliance with regulations, crew training, exhibition, air racing, or market surveys;

(9) Aircraft equipped to carry not more than one person; and

(10) An aircraft during any period for which the transmitter has been temporarily removed for inspection, repair, modification or replacement, subject to the following:

(i) No person may operate the aircraft unless the aircraft records contain an entry which includes the date of initial removal, the make, model, serial number and reason for removal of the transmitter, and a placard is located in view of the pilot to show "ELT not installed."

(ii) No person may operate the aircraft more than 90 days after the ELT is initially removed from the aircraft.

§ 91.53 [Reserved.]

§ 91.54 Truth in leasing clause requirement in leases and conditional sales contracts.

(a) Except as provided in paragraph (b) of this section, the parties to a lease or contract of conditional sale involving a United States registered large civil aircraft and entered into after January 2, 1973, shall execute a written lease or contract and include therein a written truth in leasing clause as a concluding paragraph in large print, immediately preceding the space for the signature of the parties, which contains the following with respect to each such aircraft:

(1) Identification of the Federal Aviation Regulations under which the aircraft has been maintained and inspected during the 12 months preceding the execution of the lease or contract of conditional sale; and certification by the parties thereto regarding the aircraft's status of compliance with applicable maintenance and inspection requirements in this Part for the operation to be conducted under the lease or contract of conditional sale.

(2) The name and address (printed or typed) and the signature of the person responsible for operational control of the aircraft under the lease or contract of conditional sale, and certification that each person understands that person's responsibilities for compliance with applicable Federal Aviation Regulations.

(3) A statement that an explanation of factors bearing on operational control and pertinent Federal Aviation Regulations can be obtained from the nearest FAA Flight Standards District Office, General Aviation District Office, or Air Carrier District Office.

(b) The requirements of paragraph(a) of this section do not apply—

(1) To a lease or contract of conditional sale when:

(i) The party to whom the aircraft is furnished is a foreign air carrier or certificate holder under Part 121, 123, [125,] 127, 135, or 141 of this chapter; or

(ii) The party furnishing the aircraft is a foreign air carrier, certificate holder under Part 121, 123, [125,] 127, or 141 of this

chapter, or a certificate holder under Part 135 of this chapter having appropriate authority to engage in air taxi operations with large aircraft.

(2) To a contract of conditional sale, when the aircraft involved has not been registered anywhere prior to the execution of the contract, except as a new aircraft under a dealer's aircraft registration certificate issued in accordance with § 47.61 of this chapter.

(c) No person may operate a large civil aircraft of U.S. registry that is subject to a lease or contract of conditional sale to which paragraph (a) of this section applies, unless—

(1) The lessee or conditional buyer, or the registered owner if the lessee is not a citizen of the United States, has mailed a copy of the lease or contract that complies with the requirements of paragraph (a) of this section, within 24 hours of its execution, to the Flight Standards Technical Division, P.O. Box 25724, Oklahoma City, Oklahoma 73125;

(2) A copy of the lease or contract that complies with the requirements of paragraph (a) of this section is carried in the aircraft. The copy of the lease or contract shall be made available for review upon request by the Administrator; and

(3) The lessee or conditional buyer, or the registered owner if the lessee is not a citizen of the United States, has notified by telephone or in person, the FAA Flight Standards District Office, General Aviation District Office, Air Carrier District Office, or International Field Office nearest the airport where the flight will originate. Unless otherwise authorized by that office, the notification shall be given at least forty-eight hours prior to takeoff in the case of the first flight of that aircraft under that lease or contract and inform the FAA of—

(i) The location of the airport of departure;

(ii) The departure time; and

(iii) The registration number of the aircraft involved.

(d) The copy of the lease or contract furnished to the FAA under paragraph (c) of this section is commercial or financial information obtained from a person. It is, therefore, privileged and confidential, and will not be made available by the FAA for public inspection or copying under 5 U.S.C. 552(b)(4), unless recorded with the FAA under Part 49 of this chapter.

(e) For the purpose of this section, a lease means any agreement by a person to furnish an aircraft to another person for compensation or hire, whether with or without flight crewmembers, other than an agreement for the sale of an aircraft and a contract of conditional sale under section 101 of the Federal Aviation Act of 1958. The person furnishing the aircraft is referred to as the lessor and the person to whom it is furnished the lessee.

§ 91.55 Civil aircraft sonic boom.

(a) No person may operate a civil aircraft in

the United States at a true flight Mach number greater than 1 except in compliance with conditions and limitations in an authorization to exceed Mach 1 issued to the operator under Appendix B of this Part.

(b) In addition, no person may operate a civil aircraft, for which the maximum operating limit speed M_{mo} exceeds a Mach number of 1, to or from an airport in the United States unless—

(1) Information available to the flight crew includes flight limitations that ensure that flights entering or leaving the United States will not cause a sonic boom to reach the surface within the United States; and

(2) The operator complies with the flight limitations prescribed in paragraph (b)(1) of this section or complies with conditions and limitations in an authorization to exceed Mach 1 issued under Appendix B of this Part.

[§ 91.56 Agricultural and fire fighting airplanes; noise operating limitations.

[(a) This section applies to propeller-driven, small airplanes having standard airworthiness certificates, that are designed for "agricultural aircraft operations" (as defined in § 137.3 of this chapter, as effective on January 1, 1966) or for dispensing fire fighting materials.

[(b) If the Airplane Flight Manual, or other approved manual material, information, markings, or placards for the airplane indicate that the airplane has not been shown to comply with the noise limits under Part 36 of this chapter, no person may operate that airplane, except—

[(1) To the extent necessary to accomplish the work activity directly associated with the purpose for which it is designed;

[(2) To provide flight crewmember training in the special purpose operation for which the airplane is designed; and

[(3) To conduct "nondispensing aerial work operations" in accordance with the requirements under § 137.29(c) of this chapter.]

§ 91.57 Aviation Safety Reporting Program; prohibition against use of reports for enforcement purposes.

The Administrator the the FAA will not use reports submitted to the National Aeronautics and Space Administration under the Aviation Safety Reporting Program (or information derived therefrom) in any enforcement action, except information concerning criminal offenses or accidents which are wholly excluded from the Program.

§ 91.58 Materials for compartment interiors.

No person may operate an airplane that conforms to an amended or supplemental type certificate issued in accordance with SFAR No. 41 for a maximum certificated takeoff weight in excess of 12,500 pounds, unless within one year after issuance of the initial airworthiness certificate under that SFAR, the airplane meets the compartment interior requirements set forth in § 25.853(a), (b), (b-1), (b-2), and (b-3) of this chapter in effect on September 26, 1978.

§ 91.59 Carriage of candidates in Federal elections.

(a) An aircraft operator, other than one

operating an aircraft under the rules of Part 121, 127, or 135 of this chapter, may receive payment for the carriage of a candidate in a Federal election, an agent of the candidate, or a person traveling on behalf of the candidate, if—

(1) That operator's primary business is not as an air carrier or commercial operator;

(2) The carriage is conducted under the rules of Part 91; and

(3) The payment for the carriage is required, and does not exceed the amount required to be paid, by regulations of the Federal Election Commission (11 CFR et seq.).

(b) For the purposes of this section, the terms "candidate" and "election" have the same meaning as that set forth in the regulations of the Federal Election Commission.

Subpart B—Flight Rules

GENERAL

§ 91.61 Applicability.

This subpart prescribes flight rules governing the operation of aircraft within the United States.

§ 91.63 Waivers.

(a) The Administrator may issue a certificate of waiver authorizing the operation of aircraft in deviation of any rule of this subpart if he finds that the proposed operation can be safely conducted under the terms of that certificate of waiver.

(b) An application for a certificate of waiver under this section is made on a form and in a manner prescribed by the Administrator and may be submitted to any FAA office.

(c) A certificate of waiver is effective as specified in that certificate.

§ 91.65 Operating near other aircraft.

(a) No person may operate an aircraft so close to another aircraft as to create a collision hazard.

(b) No person may operate an aircraft in formation flight except by arrangement with the pilot in command of each aircraft in the formation.

(c) No person may operate an aircraft, carrying passengers for hire, in formation flight.

(d) Unless otherwise authorized by ATC, no person operating an aircraft may operate his aircraft in accordance with any clearance or instruction that has been issued to the pilot of another aircraft for radar Air Traffic Control purposes.

§ 91.67 Right-of-way rules; except water operations.   ☞F101

(a) *General.* When weather conditions permit, regardless of whether an operation is conducted under Instrument Flight Rules or Visual Flight Rules, vigilance shall be maintained by each person operating an aircraft so as to see and avoid other aircraft in compliance with this section. When a rule of this section gives another aircraft the right of way, he shall give

way to that aircraft and may not pass over, under, or ahead of it, unless well clear.

(b) *In distress.* An aircraft in distress has the right of way over all other air traffic.

(c) *Converging.* When aircraft of the same category are converging at approximately the same altitude (except head-on, or nearly so) the aircraft to the other's right has the right of way. If the aircraft are of different categories—

(1) A balloon has the right of way over any other category of aircraft;

(2) A glider has the right of way over an airship, airplane or rotorcraft; and

(3) An airship has the right of way over an airplane or rotorcraft.

However, an aircraft towing or refueling other aircraft has the right of way over all other engine-driven aircraft.

(d) *Approaching head-on.* When aircraft are approaching each other head-on, or nearly so, each pilot of each aircraft shall alter course to the right.

(e) *Overtaking.* Each aircraft that is being overtaken has the right of way and each pilot of an overtaking aircraft shall alter course to the right to pass well clear.

(f) *Landing.* Aircraft, while on final approach to land, or while landing, have the right of way over other aircraft in flight or operating on the surface. When two or more aircraft are approaching an airport for the purpose of landing, the aircraft at the lower altitude has the right of way, but it shall not take advantage of this rule to cut in front of another which is on final approach to land, or to overtake that aircraft.

(g) *Inapplicability.* This section does not apply to the operation of an aircraft on water.

§ 91.69 Right-of-way rules; water operations.

(a) *General.* Each person operating an aircraft on the water shall, insofar as possible, keep clear of all vessels and avoid impeding their navigation, and shall give way to any vessel or other aircraft that is given the right of way by any rule of this section.

(b) *Crossing.* When aircraft, or an aircraft and a vessel are on crossing courses, the aircraft or vessel to the other's right has the right of way.

(c) *Approaching head-on.* When aircraft, or an aircraft and a vessel, are approaching head-on or nearly so, each shall alter its course to the right to keep well clear.

(d) *Overtaking.* Each aircraft or vessel that is being overtaken has the right of way, and the one overtaking shall alter course to keep well clear.

(e) *Special circumstances.* When aircraft, or an aircraft and a vessel, approach so as to involve risk of collision, each aircraft or vessel shall proceed with careful regard to existing circumstances, including the limitations of the respective craft.

§ 91.70 Aircraft speed.

(a) Unless otherwise authorized by the Administrator, no person may operate an aircraft below 10,000 feet MSL at an indicated airspeed of more than 250 knots (288 m.p.h.).

(b) Unless otherwise authorized or required by ATC, no person may operate an aircraft within an airport traffic area at an indicated airspeed of more than—

(1) In the case of a reciprocating engine aircraft, 156 knots (180 m.p.h.); or

(2) In the case of a turbine-powered aircraft, 200 knots (230 m.p.h.).

Paragraph (b) does not apply to any operations within a Terminal Control Area. Such operations shall comply with paragraph (a) of this section.

(c) No person may operate an aircraft in the airspace underlying a terminal control area, or in a VFR corridor designated through a terminal control area, at an indicated airspeed of more than 200 knots (230 m.p.h.).

However, if the minimum safe airspeed for any particular operation is greater than the maximum speed prescribed in this section, the aircraft may be operated at that minimum speed.

§ 91.71 Acrobatic flight.   ☞F102,103

No person may operate an aircraft in acrobatic flight—

(a) Over any congested area of a city, town, or settlement:

(b) Over an open air assembly of persons;

(c) Within a control zone or Federal airway:

(d) Below an altitude of 1,500 feet above the surface: or

(e) When flight visibility is less than three miles.

For the purposes of this section, acrobatic flight means an intentional maneuver involving an abrupt change in an aircraft's attitude, an abnormal attitude, or abnormal acceleration, not necessary for normal flight.

§ 91.73 Aircraft lights.

No person may, during the period from sunset to sunrise (or, in Alaska, during the period a prominent unlighted object cannot be seen from a distance of three statute miles or the sun is more than six degrees below the horizon)—

(a) Operate an aircraft unless it has lighted position lights;

(b) Park or move an aircraft in, or in dangerous proximity to, a night flight operations area of an airport unless the aircraft—

(1) Is clearly illuminated;

(2) Has lighted position lights; or

(3) Is in an area which is marked by obstruction lights[.]

(c) Anchor an aircraft unless the aircraft—

(1) Has lighted anchor lights; or

(2) Is in an area where anchor lights are not required on vessels[; or

[(d) Operate an aircraft, required by § 91.-33(c)(3) to be equipped with an anticollision light system, unless it has approved and lighted aviation red or aviation white anti-collision lights. However, the anticollision lights need not be lighted when the pilot in command determines that, because of operating conditions, it would be in the interest of safety to turn the lights off.]

§ 91.75 Compliance with ATC clearances and instructions.

(a) When an ATC clearance has been obtained, no pilot in command may deviate from that clearance, except in an emergency, unless he obtains an amended clearance. However, except in positive controlled airspace, this paragraph does not prohibit him from cancelling an IFR flight plan if he is operating in VFR weather conditions. If a pilot is uncertain of the meaning of an ATC clearance, he shall immediately request clarification from ATC.

(b) Except in an emergency, no person may, in an area in which air traffic control is exercised, operate an aircraft contrary to an ATC instruction.

(c) Each pilot in command who deviates, in an emergency, from an ATC clearance or instruction shall notify ATC of that deviation as soon as possible.

(d) Each pilot in command who (though not deviating from a rule of this subpart) is given priority by ATC in an emergency, shall, if requested by ATC, submit a detailed report of that emergency within 48 hours to the chief of that ATC facility.

§ 91.77 ATC light signals.

ATC light signals have the meaning shown in the following table.

Color and type of signal	Meaning with respect to aircraft on the surface	Meaning with respect to aircraft in flight
Steady green..........	Cleared for takeoff....	Cleared to land.
Flashing green.........	Cleared to taxi.........	Return for landing (to be followed by steady green at proper time).
Steady red............	Stop....................	Give way to other aircraft and continue circling.
Flashing red..........	Taxi clear of runway in use.	Airport unsafe—do not land.
Flashing white........	Return to starting point on airport.	Not applicable.
Alternating red and green.	Exercise extreme caution.	Exercise extreme caution.

§ 91.79 Minimum safe altitudes; general.

Except when necessary for takeoff or landing, no person may operate an aircraft below the following altitudes:

(a) *Anywhere.* An altitude allowing, if a power unit fails, an emergency landing without undue hazard to persons or property on the surface.

(b) *Over congested areas.* Over any congested area of a city, town, or settlement, or over any open air assembly of persons, an altitude of 1,000 feet above the highest obstacle within a horizontal radius of 2,000 feet of the aircraft.

(c) *Over other than congested areas.* An altitude of 500 feet above the surface, except over open water or sparsely populated areas. In that case, the aircraft may not be operated closer than 500 feet to any person, vessel, vehicle, or structure.

(d) *Helicopters.* Helicopters may be operated at less than the minimums prescribed in paragraph (b) or (c) of this section if the operation is conducted without hazard to persons or property on the surface. In addition, each person operating a helicopter shall comply with routes or altitudes specifically prescribed for helicopters by the Administrator.

§ 91.81 Altimeter settings.

(a) Each person operating an aircraft shall maintain the cruising altitude or flight level of that aircraft, as the case may be, by reference to an altimeter that is set, when operating—

(1) Below 18,000 feet MSL, to—

(i) The current reported altimeter setting of a station along the route and within 100 nautical miles of the aircraft;

(ii) If there is no station within the area prescribed in subdivision (i) of this subparagraph, the current reported altimeter setting of an appropriate available station; or

(iii) In the case of an aircraft not equipped with a radio, the elevation of the departure airport or an appropriate altimeter setting available before departure; or

(2) At or above 18,000 feet MSL, to 29.92″ Hg.

(b) The lowest usable flight level is determined by the atmospheric pressure in the area of operation, as shown in the following table:

Current altimeter setting	Lowest usable flight level
29.92 (or higher) _____	180
29.91 thru 29.42 _____	185
29.41 thru 28.92 _____	190
28.91 thru 28.42 _____	195
28.41 thru 27.92 _____	200
27.91 thru 27.42 _____	205
27.41 thru 26.92 _____	210

(c) To convert minimum altitude prescribed under §§ 91.79 and 91.119 to the minimum flight level, the pilot shall take the flight-level equivalent of the minimum altitude in feet and add the appropriate number of feet specified below, according to the current reported altimeter setting:

Current altimeter setting	Adjustment factor
29.92 (or higher) _____	None
29.91 thru 29.42 _____	500 feet
29.41 thru 28.92 _____	1000 feet
28.91 thru 28.42 _____	1500 feet
28.41 thru 27.92 _____	2000 feet
27.91 thru 27.42 _____	2500 feet
27.41 thru 26.92 _____	3000 feet

§ 91.83 Flight plan; information required.

(a) Unless otherwise authorized by ATC, each person filing an IFR or VFR flight plan shall include in it the following information:

(1) The aircraft identification number and, if necessary, its radio call sign.

(2) The type of the aircraft or, in the case of a formation flight, the type of each aircraft and the number of aircraft, in the formation.

(3) The full name and address of the pilot in command or, in the case of a formation flight, the formation commander.

(4) The point and proposed time of departure.

(5) The proposed route, cruising altitude (or flight level), and true airspeed at that altitude.

(6) The point of first intended landing and the estimated elapsed time until over that point. ☞ F126, 127

(7) The radio frequencies to be used.

(8) The amount of fuel on board (in hours).

(9) In the case of an IFR flight plan, an alternate airport, except as provided in paragraph (b) of this section.

[(10) The number of persons in the aircraft, except where that information is otherwise readily available to the FAA.]

(11) Any other information the pilot in command or ATC believes is necessary for ATC purposes.

[(b) *Exceptions to applicability of paragraph (a)(9) of this section.* Paragraph (a)(9) of this section does not apply if Part 97 of this subchapter prescribes a standard instrument approach procedure for the first airport of intended landing and, for at least one hour before and one hour after the estimated time of arrival, the weather reports or forecasts or any combination of them, indicate—

[(1) The ceiling will be at least 2,000 feet above the airport elevation; and

[(2) Visibility will be at least 3 miles.]

(c) *IFR alternate airport weather minimums.* Unless otherwise authorized by the Administrator, no person may include an alternate airport in an IFR flight plan unless current weather forecasts indicate that, at the estimated time of arrival at the alternate airport, the ceiling and visibility at that airport will be at or above the following alternate airport weather minimums: ☞ F84, 95

(1) If an instrument approach procedure has been published in Part 97 for that airport, the alternate airport minimums specified in that procedure or, if none are so specified, the following minimums:

(i) Precision approach procedure: ceiling 600 feet and visibility 2 statute miles.

(ii) Non-precision approach procedure: ceiling 800 feet and visibility 2 statute miles.

(2) If no instrument approach procedure has been published in Part 97 for that airport, the ceiling and visibility minimums are those allowing descent from the MEA, approach, and landing, under basic VFR.

(d) *Cancellation.* When a flight plan has been activated, the pilot in command, upon cancelling or completing the flight under the flight plan, shall notify an FAA Flight Service Station or ATC facility.

§ 91.84 Flights between Mexico or Canada and the United States.

Unless otherwise authorized by ATC, no person may operate a civil aircraft between Mexico or Canada and the United States without filing an IFR or VFR flight plan, as appropriate.

§ 91.85 Operating on or in the vicinity of an airport; general rules.

(a) Unless otherwise required by Part 93 of this chapter, each person operating an aircraft on or in the vicinity of an airport shall comply with the requirements of this section and of §§ 91.87 and 91.89.

(b) Unless otherwise authorized or required by ATC, no person may operate an aircraft within an airport traffic area except for the purpose of landing at, or taking off from, an airport within that area. ATC authorizations may be given as individual approval of specific operations or may be contained in written agreements between airport users and the tower concerned.

(c) After March 28, 1977, except when necessary for training or certification, the pilot in command of a civil turbojet-powered airplane shall use, as a final landing flap setting, the minimum certificated landing flap setting set forth in the approved performance information in the Airplane Flight Manual for the applicable conditions. However, each pilot in command has the final authority and responsibility for the safe operation of his airplane, and he may use a different flap setting approved for that airplane if he determines that it is necessary in the interest of safety.

§ 91.87 Operation at airports with operating control towers.

(a) *General.* Unless otherwise authorized or required by ATC, each person operating an aircraft to, from, or on an airport with an operating control tower shall comply with the applicable provisions of this section.

(b) *Communications with control towers operated by the United States.* No person may, within an airport traffic area, operate an aircraft to, from, or on an airport having a control tower operated by the United States unless two-way radio communications are maintained between that aircraft and the control tower. However, if the aircraft radio fails in flight, he may operate that aircraft and land if weather conditions are at or above basic VFR weather minimums, he maintains visual contact with the tower, and he receives a clearance to land. If the aircraft radio fails while in flight under IFR, he must comply with § 91.127.

(c) *Communications with other control towers.* No person may, within an airport traffic area, operate an aircraft to, from, or on an airport having a control tower that is operated by any person other than the United States unless—

(1) If that aircraft's radio equipment so allows, two-way radio communications are maintained between the aircraft and the tower; or

(2) If that aircraft's radio equipment allows only reception from the tower, the pilot has the tower's frequency monitored.

(d) *Minimum altitudes.* When operating to an airport with an operating control tower, each pilot of—

(1) A turbine-powered airplane or a large airplane shall, unless otherwise required by the applicable distance from cloud criteria, enter the airport traffic area at an altitude of at least 1,500 feet above the surface of the airport and maintain at least 1,500 feet within the airport traffic area, including the traffic pattern, until further descent is required for a safe landing;

(2) A turbine-powered airplane or a large airplane approaching to land on a runway being served by an ILS shall, if the airplane is ILS equipped, fly that airplane at an altitude at or above the glide slope between the outer marker (or the point of interception with the glide slope, if compliance with the applicable distance from clouds criteria requires interception closer in) and the middle marker; and,

(3) An airplane approaching to land on a runway served by a visual approach slope indicator, shall maintain an altitude at or above the glide slope until a lower altitude is necessary for a safe landing.

However, subparagraphs (2) and (3) of this paragraph do not prohibit normal bracketing maneuvers above or below the glide slope that are conducted for the purpose of remaining on the glide slope.

(e) *Approaches.* When approaching to land at an airport with an operating control tower, each pilot of—

(1) An airplane, shall circle the airport to the left; and

(2) A helicopter, shall avoid the flow of fixed-wing aircraft.

(f) *Departures.* No person may operate an aircraft taking off from an airport with an operating control tower except in compliance with the following:

(1) Each pilot shall comply with any departure procedures established for that airport by the FAA.   ☞ **F105,106**

(2) Unless otherwise required by the departure procedure or the applicable distance from clouds criteria, each pilot of a turbine-powered airplane and each pilot of a large airplane shall climb to an altitude of 1,500 feet above the surface as rapidly as practicable.

(g) *Noise abatement runway system.* When landing or taking off from an airport with an operating control tower, and for which a formal runway use program has been established by the FAA, each pilot of a turbine-powered airplane and each pilot of a large airplane, assigned a noise abatement runway by ATC, shall use that runway. **[**However, consistent with the final authority of the pilot in command concerning the safe operation of the aircraft as prescribed in § 91.3(a), ATC may assign a different runway if requested by the pilot in the interest of safety.**]**

(h) Clearances required. No person may, at an airport with an operating control tower, operate an aircraft on a runway or taxiway, or takeoff or land an aircraft, unless an appropriate clearance is received from ATC. A clearance to "taxi to" the takeoff runway assigned to the aircraft is not a clearance to cross that assigned takeoff runway, or to taxi on that runway at any point, but is a clearance to cross other runways that intersect the taxi route to that assigned takeoff runway. A clearance to "taxi to" any point other than an assigned takeoff runway is a clearance to cross all runways that intersect the taxi route to that point.

[91.88 Airport radar service areas.

☞ **F107,108**

[(a) *General.* For the purposes of this section, the primary airport is the airport designated in Part 71, Subpart L, for which the airport radar service area is designated. A satellite airport is any other airport within the airport radar service area.

[(b) *Deviations.* An operator may deviate from any provision of this section under the provisions of an ATC authorization issued by the ATC facility having jurisdiction of the airport radar service area. ATC may authorize a deviation on a continuing basis or for an individual flight, as appropriate.

[(c) *Arrivals and Overflights.* No person may operate an aircraft in an airport radar service area unless two-way radio communication is established with ATC prior to entering that area and is thereafter maintained with ATC while within that area.

[(d) *Departures.* No person may operate an aircraft within an airport radar service area unless two-way radio communication is maintained with ATC while within that area, except that for aircraft departing a satellite airport, two-way radio communication is established as soon as practicable and thereafter maintained with ATC while within that area.

[(e) *Traffic Patterns.* No person may takeoff or land an aircraft at a satellite airport within an airport radar service area except in compliance with FAA arrival and departure traffic patterns.**]**

[(f)] *Equipment requirements.* Unless otherwise authorized by ATC, no person may operate an aircraft within an airport radar service area unless that aircraft is equipped with the applicable equipment specified in § 91.24.**]**

§ 91.89   Operation at airports without control towers.

(a) Each person operating an aircraft to or from an airport without an operating control tower shall—

(1) In the case of an airplane approaching to land, make all turns of that airplane to the left unless the airport displays approved light signals or visual markings indicating that turns should be made to the right, in which case the pilot shall make all turns to the right;

☞ **F109**

(2) In the case of a helicopter approaching to land, avoid the flow of fixed-wing aircraft; and

(3) In the case of an aircraft departing the airport, comply with any FAA traffic pattern for that airport.

§ 91.90   Terminal control areas.

(a) *Operating rules.* No person may operate an aircraft within a terminal control area designated in Part 71 of this chapter except in compliance with the following rules:

(1) No person may operate an aircraft within a terminal control area unless that person has received an appropriate authorization from ATC prior to operation of that aircraft in that area.

(2) Unless otherwise authorized by ATC, each person operating a large turbine engine-powered airplane to or from a primary airport shall operate at or above the designated floors while within the lateral limits of the terminal control area.

(3) Any person conducting pilot training operations at an airport within a terminal control area shall comply with any procedures established by ATC for such operations in the terminal control area.

(b) *Pilot requirements.* (1) No person may takeoff or land a civil aircraft at an airport within a terminal control area or operate a civil aircraft within a terminal control area unless:

(i) The pilot-in-command holds at least a private pilot certificate; or,

(ii) The aircraft is operated by a student pilot who has met the requirements of § 61.95.

(2) Notwithstanding the provisions of (b)(1)(ii) of this section, at the following TCA primary airports, no person may takeoff or land a civil aircraft unless the pilot-in-command holds at least a private pilot certificate:

(i) Atlanta Hartsfield Airport, GA.
(ii) Boston Logan Airport, MA.
(iii) Chicago O'Hare International Airport IL.
(iv) Dallas/Fort Worth International Airport, TX.
(v) Los Angeles International Airport, CA.
(vi) Miami International Airport, FL.
(vii) Newark International Airport, NJ.
(viii) New York Kennedy Airport, NY.

(ix) New York La Guardia Airport, NY.
(x) San Francisco International Airport, CA.
(xi) Washington National Airport, DC.
(xii) Andrews Air Force Base, MD.

(c) *Communications and navigation equipment requirements.* Unless otherwise authorized by ATC, no person may operate an aircraft within a terminal control area unless that aircraft is equipped with—

(1) An operable VOR or TACAN receiver (except for helicopter operations prior to July 1, 1989); and

(2) An operable two-way radio capable of communications with ATC on appropriate frequencies for that terminal control area.

(d) *Transponder requirement.* No person may operate an aircraft in a terminal control area unless the aircraft is equipped with the applicable operating transponder and automatic altitude reporting equipment specified in paragraph (a) of § 91.24, except as provided in paragraph (d) of that section.

91.91   Temporary flight restrictions.

(a) The Administrator will issue a Notice to Airmen (NOTAM) designating an area within which temporary flight restrictions apply and specifying the hazard or condition requiring their imposition, whenever he determines it is necessary in order to—

(1) Protect persons and property on the surface or in the air from a hazard associated with an incident on the surface;

(2) Provide a safe environment for the operation of disaster relief aircraft; or

(3) Prevent an unsafe congestion of sightseeing and other aircraft above an incident or event which may generate a high degree of public interest.

The Notice to Airmen will specify the hazard or condition that requires the imposition of temporary flight restrictions.

[(b) When a NOTAM has been issued under paragraph (a) (1) of this section, no person may operate an aircraft within the designated area unless that aircraft is participating in the hazard relief activities and is being operated under the direction of the official in charge of on scene emergency response activities.

[(c) When a NOTAM has been issued under paragraph (a) (2) of this section, no person may operate an aircraft within the designated area unless at least one of the following conditions are met:

[(1) The aircraft is participating in hazard relief activities and is being operated under the direction of the official in charge of on scene emergency response activities.

[(2) The aircraft is carrying law enforcement officials.

[(3) The aircraft is operating under an ATC approved IFR flight plan.

[(4) The operation is conducted directly to or from an airport within the area, or is necessitated by the impracticability of VFR flight above or around the area due to weather, or terrain; information is given to the Flight Service Station (FSS) or ATC facility specified in the NOTAM to receive ad-

visories concerning disaster relief aircraft operations; and the operation does not hamper or endanger relief activities and is not conducted for the purpose of observing the disaster.

[(5) The aircraft is carrying properly accredited news representatives, and, prior to entering the area, a flight plan is filed with the appropriate FAA or ATC facility specified in the Notice to Airmen and the operation is conducted above the altitude used by the disaster relief aircraft, unless otherwise authorized by the official in charge of on scene emergency response activities.

[(d) When a NOTAM has been issued under paragraph (a) (3) of this section, no person may operate an aircraft within the designated area unless at least one of the following conditions is met:

[(1) The operation is conducted directly to or from an airport within the area, or is necessitated by the impracticability of VFR flight above or around the area due to weather or terrain, and the operation is not conducted for the purpose of observing the incident or event.

[(2) The aircraft is operating under an ATC approved IFR flight plan.

[(3) The aircraft is carrying incident or event personnel, or law enforcement officials.

[(4) The aircraft is carrying properly accredited news representatives and, prior to entering that area, a flight plan is filed with the appropriate FSS or ATC facility specified in the NOTAM.

[(e) Flight plans filed and notifications made with an FSS or ATC facility under this section shall include the following information:

[(1) Aircraft identification, type and color.

[(2) Radio communications frequencies to be used.

[(3) Proposed times of entry of, and exit from, the designated area.

[(4) Name of news media or organization and purpose of flight.

[(5) Any other information requested by ATC.**]**

§ 91.93   Flight test areas.

No person may flight test an aircraft except over open water, or sparsely populated areas, having light air traffic.

§ 91.95   Restricted and prohibited areas.

(a) No person may operate an aircraft within a restricted area (designated in Part 73) contrary to the restrictions imposed, or within a prohibited area, unless he has the permission of the using or controlling agency, as appropriate.

(b) Each person conducting, within a restricted area, an aircraft operation (approved by the using agency) that creates the same hazards as the operations for which the restricted area was designated, may deviate from the rules of this subpart that are not compatible with his operation of the aircraft.

§ 91.97   Positive control areas and route segments.

(a) Except as provided in paragraph (b) of this section, no person may operate an aircraft within a positive control area or positive control route segment, designated in Part 71 of this chapter, unless that aircraft is—

(1) Operated under IFR at a specific flight level assigned by ATC;

(2) Equipped with instruments and equipment required for IFR operations;

(3) Flown by a pilot rated for instrument flight; and

(4) Equipped, when in a positive control area, with—

(i) The applicable equipment specified in § 91.24; and

(ii) A radio providing direct pilot controller communication on the frequency specified by ATC for the area concerned.

(b) ATC may authorize deviations from the requirements of paragraph (a) of this section. In the case of an inoperative transponder, ATC may immediately approve an operation within a positive control area allowing flight to continue, if desired, to the airport of ultimate destination, including any intermediate stops, or to proceed to a place where suitable repairs can be made, or both. A request for authorization to deviate from a requirement of paragraph (a) of this section, other than for operation with an inoperative transponder as outlined above, must be submitted at least four days before the proposed operation, in writing, to the ATC center having jurisdiction over the positive control area concerned. ATC may authorize a deviation on a continuing basis or for an individual flight, as appropriate.

§ 91.99 [Reserved]

§ 91.100 Emergency air traffic rules.

(a) This section prescribes a process for utilizing Notices to Airmen (NOTAM) to advise of the issuance and operations under emergency air traffic rules and regulations and designates the official who is authorized to issue NOTAMs on behalf of the Administrator in certain matters under this section.

(b) Whenever the Administrator determines that an emergency condition exists, or will exist, relating to the FAA's ability to operate the Air Traffic Control System and during which normal flight operations under this chapter cannot be conducted consistent with the required levels of safety and efficiency—

(1) The Administrator issues an immediately effective Air Traffic rule or regulation in response to that emergency condition.

(2) The Administrator, or the Director, Air Traffic Service, may utilize the Notice to Airmen (NOTAMs) system to provide notification of the issuance of the rule or regulation. Those NOTAMs communicate information concerning the rules and regulations that govern flight operations, the use of navigation facilities, and designation of that airspace in which the rules and regulations apply.

(c) When a NOTAM has been issued under this section, no person may operate an aircraft, or other device governed by the regulation concerned, within the designated airspace, except in accordance with the authorizations, terms, and conditions prescribed in the regulation covered by the NOTAM.

§ 91.101 Operations to Cuba.

No person may operate a civil aircraft from the United States to Cuba unless—

(a) Departure is from an international airport of entry designated in § 6.13 of the Air Commerce Regulations of the Bureau of Customs (19 CFR 6.13); and

(b) In the case of departure from any of the 48 contiguous States or the District of Columbia, the pilot in command of the aircraft has filed—

(1) A DVFR or IFR flight plan as prescribed in § 99.11 or 99.13 of this chapter; and

(2) A written statement, within one hour before departure, with the office of Immigration and Naturalization Service at the airport of departure, containing—

(i) All information in the flight plan;

(ii) The name of each occupant of the aircraft;

(iii) The number of occupants of the aircraft; and

(iv) A description of the cargo, if any.

This section does not apply to the operation of aircraft by a scheduled air carrier over routes authorized in operations specifications issued by the Administrator.

§ 91.102 Flight limitation in the proximity of space flight operations.

[No person may operate any aircraft of United States registry, or pilot any aircraft under the authority of an airman certificate issued by the Federal Aviation Administration within areas designated in a NOTAM for space flight operations except when authorized by ATC, or operated under the control of the Department of Defense Manager for Space Transportation System Contingency Support Operations.]

§ 91.103 Operation of civil aircraft of Cuban registry.

No person may operate a civil aircraft of Cuban registry except in controlled airspace and in accordance with air traffic clearances or air traffic control instructions that may require use of specific airways or routes and landings at specific airports.

§ 91.104 Flight restrictions in the proximity of the Presidential and other parties.

No person may operate an aircraft over or in the vicinity of any area to be visited or traveled by the President, the Vice President, or other public figures contrary to the restrictions established by the Administrator and published in a Notice to Airmen (NOTAM).

VISUAL FLIGHT RULES
§ 91.105 Basic VFR weather minimums.

(a) Except as provided in § 91.107, no person may operate an aircraft under VFR when the flight visibility is less, or at a distance from clouds that is less, than that prescribed for the corresponding altitude in the following table: ☞ **F3**

See table at bottom of page

(b) When the visibility is less than one mile, a helicopter may be operated outside controlled airspace at 1,200 feet or less above the surface if operated at a speed that allows the pilot adequate opportunity to see any air traffic or other obstruction in time to avoid a collision.

(c) Except as provided in § 91.107, no person may operate an aircraft, under VFR, within a control zone beneath the ceiling when the ceiling is less than 1,000 feet. ☞ **F132–134**

(d) Except as provided in § 91.107, no person may take off or land an aircraft, or enter the traffic pattern of an airport, under VFR, within a control zone—

(1) Unless ground visibility at that airport is at least three statute miles; or

(2) If ground visibility is not reported at that airport, unless flight visibility during landing or take off, or while operating in the traffic pattern, is at least three statute miles. ☞ **F134, 135**

(e) For the purposes of this section, an aircraft operating at the base altitude of a transition area or control area is considered to be within the airspace directly below that area.

Altitude	Flight visibility	Distance from clouds
1,200 feet or less above the surface (regardless of MSL altitude)-		
Within controlled airspace_____	3 statute miles_____	500 feet below. 1,000 feet above. 2,000 feet horizontal.
Outside controlled airspace_____	1 statute mile except as provided in § 91.105(b).	Clear of clouds.
More than 1,200 feet above the surface but less than 10,000 feet MSL—		
Within controlled airspace_____	3 statute miles_____	500 feet below. 1,000 feet above. 2,000 feet horizontal.
Outside controlled airspace_____	1 statue mile_____	500 feet below. 1,000 feet above. 2,000 feet horizontal.
More than 1,200 feet above the surface and at or above 10,000 feet MSL.	5 statute miles_____	1,000 feet below. 1,000 feet above. 1 mile horizontal.

§ 91.107 Special VFR weather minimums.

(a) Except as provided in § 93.113, when a person has received an appropriate ATC clearance, the special weather minimums of this section instead of those contained in § 91.105 apply to the operation of an aircraft by that person in a control zone under VFR.
☞ **F128, 129**

(b) No person may operate an aircraft in a control zone under VFR except clear of clouds.

(c) No person may operate an aircraft (other than a helicopter) in a control zone under VFR unless flight visibility is at least one statute mile.

(d) No person may take off or land an aircraft (other than a helicopter) at any airport in a control zone under VFR—

(1) Unless ground visibility at that airport is at least one statute mile; or

(2) If ground visibility is not reported at that airport, unless flight visibility during landing or takeoff is at least one statute mile.

(e) No person may operate an aircraft (other than a helicopter) in a control zone under the special weather minimums of this section, between sunset and sunrise (or in Alaska, when the sun is more than six degrees below the horizon) unless:

(1) That person meets the applicable requirements for instrument flight under Part 61 of this chapter; and

(2) The aircraft is equipped as required in § 91.33(d).

§ 91.109 VFR cruising altitude or flight level.

Except while holding in a holding pattern of 2 minutes or less, or while turning, each person operating an aircraft under VFR in level cruising flight more than 3,000 feet above the surface shall maintain the appropriate altitude or flight level prescribed below, unless otherwise authorized by ATC:

(a) When operating below 18,000 feet MSL and—

(1) On a magnetic course of zero degrees through 179 degrees, any odd thousand foot MSL altitude + 500 feet (such as 3,500, 5,500, or 7,500); or

(2) On a magnetic course of 180 degrees through 359 degrees, any even thousand foot MSL altitude + 500 feet (such as 4,500, 6,500, or 8,500).

(b) When operating above 18,000 feet MSL to flight level 290 (inclusive), and—

(1) On a magnetic course of zero degrees through 179 degrees, any odd flight level + 500 feet (such as 195, 215, or 235); or

(2) On a magnetic course of 180 degrees through 359 degrees, any even flight level + 500 feet (such as 185, 205, or 225).

(c) When operating above flight level 290 and—

(1) On a magnetic course of zero degrees through 179 degrees, any flight level, at 4,000-foot intervals, beginning at and including flight level 300 (such as flight level 300, 340, or 380); or

(2) On a magnetic course of 180 degrees through 359 degrees, any flight level, at 4,000-foot intervals, beginning at and including flight level 320 (such as flight level 320, 360, or 400).

INSTRUMENT FLIGHT RULES

§ 91.115 ATC clearance and flight plan required.

No person may operate an aircraft in controlled airspace under IFR unless—

(a) He has filed an IFR flight plan; and

(b) He has received an appropriate ATC clearance.
☞ **F3, 87**

§ 91.116 [Takeoff and landing under IFR: General].

(a) *Instrument approaches to civil airports.* Unless otherwise authorized by the Administrator for paragraphs (a) through (k) of this section, when an instrument letdown to a civil airport is necessary, each person operating an aircraft, except a military aircraft of the United States, shall use a standard instrument approach procedure prescribed for the airport in Part 97 of this chapter.

(b) *Authorized DH or MDA.* For the purpose of this section, when the approach procedure being used provides for and requires use of a DH or MDA, the authorized decision height or authorized minimum descent altitude is the DH or MDA prescribed by the approach procedure, the DH or MDA prescribed for the pilot in command, or the DH or MDA for which the aircraft is equipped, whichever is higher. ☞ **F110**

(c) *Operation below DH or MDA.* Where a DH or MDA is applicable, no pilot may operate an aircraft, except a military aircraft of the United States, at any airport below the authorized MDA or continue an approach below the authorized DH unless— ☞ **F111**

(1) The aircraft is continuously in a position from which a descent to a landing on the intended runway can be made at a normal rate of descent using normal maneuvers, and for operations conducted under Part 121 or Part 135 unless that descent rate will allow touchdown to occur within the touchdown zone of the runway of intended landing;

(2) The flight visibility is not less than the visibility prescribed in the standard instrument approach being used;

(3) Except for a Category II or Category III approach where any necessary visual reference requirements are specified by the Administrator, at least one of the following visual references for the intended runway is distinctly visible and identifiable to the pilot:

(i) The approach light system, except that the pilot may not descend below 100 feet above the touchdown zone elevation using the approach lights as a reference unless the red terminating bars or the red side row bars are also distinctly visible and identifiable. ☞ **F119**

(ii) The threshold.

(iii) The threshold markings.

(iv) The threshold lights.

(v) The runway end identifier lights.

(vi) The visual approach slope indicator.

(vii) The touchdown zone or touchdown zone markings.

(viii) The touchdown zone lights.

(ix) The runway or runway markings.

(x) The runway lights; and ☞ **F112**

(4) When the aircraft is on a straight-in nonprecision approach procedure which incorporates a visual descent point, the aircraft has reached the visual descent point, except where the aircraft is not equipped for or capable of establishing that point or a descent to the runway cannot be made using normal procedures or rates of descent if descent is delayed until reaching the point.

(d) *Landing.* No pilot operating an aircraft, except a military aircraft of the United States, may land that aircraft when the flight visibility is less than the visibility prescribed in the standard instrument approach procedure being used.

(e) *Missed approach procedures.* Each pilot operating an aircraft, except a military aircraft of the United States, shall immediately execute an appropriate missed approach procedure when either of the following conditions exist:

(1) Whenever the requirements of paragraph (c) of this section are not met at either of the following times:

(i) When the aircraft is being operated below MDA; or

(ii) Upon arrival at the missed approach point, including a DH where a DH is specified and its use is required, and at any time after that until touchdown.

(2) Whenever an identifiable part of the airport is not distinctly visible to the pilot during a circling maneuver at or above MDA, unless the inability to see an identifiable part of the airport results only from a normal bank of the aircraft during the circling approach.

[(f) *Civil airport takeoff minimums.* Unless otherwise authorized by the Administrator, no person operating an aircraft under Part 121, 125, 127, 129, or 135 of this chapter] may take off from a civil airport under IFR unless weather conditions are at or above the weather minimums for IFR takeoff prescribed for that airport under Part 97 of this chapter. If takeoff minimums are not prescribed under Part 97 of this chapter for a particular airport, the following minimums apply to takeoffs under IFR for aircraft operating under those Parts:

[(1) For aircraft, other than helicopters, having two engines or less—1 statute mile visibility.]

(2) For aircraft having more than two engines—½ statute mile visibility.

(3) For helicopters—½ statute mile visibility.]

(g) *Military airports.* Unless otherwise prescribed by the Administrator, each person operating a civil aircraft under IFR into or out of a military airport shall comply with the instrument approach procedures and the takeoff and landing minimums prescribed by the military authority having jurisdiction of that airport.

(h) *Comparable values of RVR and ground visibility.*

(1) Except for Category II or Category III minimums, if RVR minimums for takeoff or landing are prescribed in an instrument approach procedure, but RVR is not reported for the runway of intended operation, the RVR minimum shall be converted to ground visibility in accordance with the table in paragraph (h)(2) of this section and shall be the visibility minimum for takeoff or landing on that runway.

[(2) RVR (feet)	Visibility (statute miles)
1,600	¼
2,400	½
3,200	⁵/₈
4,000	¾
4,500	⁷/₈
5,000	1
6,000	1¼

[(i) *Operations on unpublished routes and use of radar in instrument approach procedures.* When radar is approved at certain locations for ATC purposes, it may be used not only for surveillance and precision radar approaches, as applicable, but also may be used in conjunction with instrument approach procedures predicted on other types of radio navigational aids. Radar vectors may be authorized to provide course guidance through the segments of an approach to the final course or fix. When operating on an unpublished route or while being radar vectored, the pilot, when an approach clearance is received, shall, in addition to complying with § 91.119, maintain the last altitude assigned to that pilot until the aircraft is established on a segment of a published route or instrument approach procedure unless a different altitude is assigned by ATC. After the aircraft is so established, published altitudes apply to descent within each succeeding route or approach segment unless a different altitude is assigned by ATC. Upon reaching the final approach course or fix, the pilot may either complete the instrument approach in accordance with a procedure approved for the facility or continue a surveillance or precision radar approach to a landing. ☞ F113

[(j) *Limitation on procedure turns.* In the case of a radar vector to a final approach course or fix, a timed approach from a holding fix, or an approach for which the procedure specifies "No PT", no pilot may make a procedure turn unless cleared to do so by ATC.

[(k) *ILS components.* The basic ground components of an ILS are the localizer, glide slope, outer marker, middle marker, and, when installed for use with Category II or Category III instrument approach procedures, an inner marker. A compass locator or precision radar may be substituted for the outer or middle

marker. DME, VOR, or nondirectional beacon fixes authorized in the standard instrument approach procedure or surveillance radar may be substituted for the outer marker. Applicability of, and substitution for, the inner marker for Category II or III approaches is determined by the appropriate Part 97 approach procedure, letter of authorization, or operations specification pertinent to the operations.]

§ 91.117 [[Reserved]]

§§ 91.119 Minimum altitudes for IFR operations.

(a) Except when necessary for takeoff or landing, or unless otherwise authorized by the Administrator, no person may operate an aircraft under IFR below—

(1) The applicable minimum altitudes prescribed in Parts 95 and 97 of this chapter; or

(2) If no applicable minimum altitude is prescribed in those Parts—

(i) In the case of operations over an area designated as a mountainous area in Part 95, an altitude of 2,000 feet above the highest obstacle within a horizontal distance of five statute miles from the course to be flown; or

(ii) In any other case, an altitude of 1,000 feet above the highest obstacle within a horizontal distance of five statute miles from the course to be flown.

However, if both a MEA and a MOCA are prescribed for a particular route or route segment, a person may operate an aircraft below the MEA down to, but not below, the MOCA, when within 25 statute miles of the VOR concerned (based on the pilot's reasonable estimate of that distance). ☞ F117

(b) *Climb.* Climb to a higher minimum IFR altitude shall begin immediately after passing the point beyond which that minimum altitude applies, except that, when ground obstructions intervene, the point beyond which the higher minimum altitude applies shall be crossed at or above the applicable MCA.

§ 91.121 IFR cruising altitude or flight level.

(a) *In controlled airspace.* Each person operating an aircraft under IFR in level cruising flight in controlled airspace shall maintain the altitude or flight level assigned that aircraft by ATC. However, if the ATC clearance assigns "VFR conditions-on-top," he shall maintain an altitude or flight level as prescribed by § 91.109. ☞ F3

(b) *In uncontrolled airspace.* Except while holding in a holding pattern of two minutes or less, or while turning, each person operating an aircraft under IFR in level cruising flight, in uncontrolled airspace, shall maintain an appropriate altitude as follows: ☞ F87

(1) When operating below 18,000 feet MSL and—

(i) On a magnetic course of zero degrees through 179 degrees, any odd thou-

sand foot MSL altitude (such as 3,000, 5,000, or 7,000); or

(ii) On a magnetic course of 180 degrees through 359 degrees, any even thousand foot MSL altitude (such as 2,000, 4,000, or 6,000).

(2) When operating at or above 18,000 feet MSL but below flight level 290, and—

(i) On a magnetic course of zero degrees through 179 degrees, any odd flight level (such as 190, 210, or 230); or

(ii) On a magnetic course of 180 degrees through 359 degrees, any even flight level (such as 180, 200, or 220).

(3) When operating at flight level 290 and above, and—

(i) On a magnetic course of zero degrees through 179 degrees, any flight level, at 4,000-foot intervals, beginning at and including flight level 290 (such as flight level 290, 330, or 370); or

(ii) On a magnetic course of 180 degrees through 359 degrees, any flight level, at 4,000-foot intervals, beginning at and including flight level 310 (such as flight level 310, 350, or 390).

§ 91.123 Course to be flown.

Unless otherwise authorized by ATC, no person may operate an aircraft within controlled airspace, under IFR, except as follows:

(a) On a Federal airway, along the centerline of that airway.

(b) On any other route, along the direct course between the navigational aids or fixes defining that route.

However, this section does not prohibit maneuvering the aircraft to pass well clear of other air traffic or the maneuvering of the aircraft, in VFR conditions, to clear the intended flight path both before and during climb or descent.

§ 91.125 IFR radio communications.

The pilot in command of each aircraft operated under IFR in controlled airspace shall have a continuous watch maintained on the appropriate frequency and shall report by radio as soon as possible—

(a) The time and altitude of passing each designated reporting point, or the reporting points specified by ATC, except that while the aircraft is under radar control, only the passing of those reporting points specifically requested by ATC need be reported;

(b) Any unforecast weather conditions encountered; and

(c) Any other information relating to the safety of flight.

§ 91.127 IFR operations; two-way radio communications failure.

(a) *General.* Unless otherwise authorized by ATC, each pilot who has two-way radio communications failure when operating under IFR shall comply with the rules of this section.

(b) *VFR conditions.* If the failure occurs in VFR conditions, or if VFR conditions are encountered after the failure, each pilot shall continue the flight under VFR and land as soon as practicable.

(c) *IFR conditions.* If the failure occurs in IFR conditions, or if paragraph (b) of this section cannot be complied with, each pilot shall continue the flight according to the following:

(1) *Route.*

(i) By the route assigned in the last ATC clearance received;

(ii) If being radar vectored, by the direct route from the point of radio failure to the fix, route, or airway specified in the vector clearance;

(iii) In the absence of an assigned route, by the route that ATC has advised may be expected in a further clearance; or

(iv) In the absence of an assigned route or a route that ATC has advised may be expected in a further clearance, by the route filed in the flight plan.

(2) *Altitude.* At the highest of the following altitudes or flight levels for the route segment being flown:

(i) The altitude or flight level assigned in the last ATC clearance received;

(ii) The minimum altitude (converted, if appropriate, to minimum flight level as prescribed in § 91.81(c)) for IFR operations; or

(iii) The altitude or flight level ATC has advised may be expected in a further clearance.

[(3) *Leave clearance limit.* (i) When the clearance limit is a fix from which an approach begins, commence descent or descent and approach as close as possible to the expect further clearance time if one has been received, or if one has not been received, as close as possible to the estimated time of arrival as calculated from the filed or amended (with ATC) estimated time en route.

[(ii) If the clearance limit is not a fix from which an approach begins, leave the clearance limit at the expect further clearance time if one has been received, or if none has been received, upon arrival over the clearance limit, and proceed to a fix from which an approach begins and commence descent or descent and approach as close as possible to the estimated time of arrival as calculated from the filed or amended (with ATC) estimated time en route.]

§ 91.129 Operation under IFR in controlled airspace; malfunction reports.

(a) The pilot in command of each aircraft operated in controlled airspace under IFR,

shall report immediately to ATC any of the following malfunctions of equipment occurring in flight:

(1) Loss of VOR, TACAN, ADF, or low frequency navigation receiver capability.

(2) Complete or partial loss of ILS receiver capability.

(3) Impairment of air/ground communications capability.

(b) In each report required by paragraph (a) of this section, the pilot in command shall include the—

(1) Aircraft identification;

(2) Equipment affected;

(3) Degree to which the capability of the pilot to operate under IFR in the ATC system is impaired; and

(4) Nature and extent of assistance he desires from ATC.

Subpart C—Maintenance, Preventive Maintenance, and Alterations

§ 91.161 Applicability.

(a) This subpart prescribes rules governing the maintenance, preventive maintenance, and alterations of U.S. registered civil aircraft operating within or without the United States.

[(b) Sections 91.165, 91.169, 91.171, 91.173, and 91.174 of this subpart do not apply to an aircraft maintained in accordance with a continuous airworthiness maintenance program as provided in Part 121, 127, 129, or § 135.411(a) (2) of this chapter.]

(c) Sections 91.165, 91.169, 91.171, and Subpart D of this Part do not apply to an airplane inspected in accordance with Part 125 of this chapter.

§ 91.163 General.

(a) The owner or operator of an aircraft is primarily responsible for maintaining that aircraft in an airworthy condition, including compliance with Part 39 of this chapter.

(b) No person may perform maintenance, preventive maintenance, or alterations on an aircraft other than as prescribed in this subpart and other applicable regulations, including Part 43.

(c) No person may operate an aircraft for which a manufacturer's maintenance manual or Instructions for Continued Airworthiness has been issued that contains an Airworthiness Limitations section unless the mandatory replacement times, inspection intervals, and related procedures specified in that section or alternative inspection intervals and related procedures set forth in an operations specification approved by the Administrator under Parts 121, 123, 127, or 135, or in accordance with an inspection program approved under § 91.217(e), have been complied with.

§ 91.165 Maintenance required.

Each owner or operator of an aircraft—

(a) Shall have that aircraft inspected as prescribed in Subpart C of this part and shall between required inspections, except as provided in paragraph (c) of this section, have discrepancies repaired as prescribed in Part 43 of this chapter;

(b) Shall ensure that maintenance personnel make appropriate entries in the aircraft maintenance records indicating the aircraft has been approved for return to service;

(c) Shall have any inoperative instrument or item of equipment, permitted to be inoperative by § 91.30(d)(2) of this part, repaired, replaced, removed, or inspected at the next required inspection; and

(d) When listed discrepancies include inoperative instruments or equipment, shall ensure that a placard has been installed as required by § 43.11 of this chapter.

§ 91.167 Operation after maintenance, preventive maintenance, rebuilding, or alteration.

(a) No person may operate any aircraft that has undergone maintenance, preventive maintenance, rebuilding, or alteration unless—

(1) It has been approved for return to service by a person authorized under § 43.7 of this chapter; and

(2) The maintenance record entry required by § 43.9 or § 43.11, as applicable, of this chapter has been made.

(b) No person may carry any person (other than crewmembers) in an aircraft that has been maintained, rebuilt, or altered in a manner that may have appreciably changed its flight characteristics or substantially affected its operation in flight until an appropriately rated pilot with at least a private pilot certificate flies the aircraft, makes an operational check of the maintenance performed or alteration made, and logs the flight in the aircraft records.

(c) The aircraft does not have to be flown as required by paragraph (b) of this section if, prior to flight, ground tests, inspections, or both show conclusively that the maintenance, preventive maintenance, rebuilding, or alteration has not appreciably changed the flight characteristics or substantially affected the flight operation of the aircraft.

§ 91.169 Inspections.

(a) Except as provided in paragraph (c) of this section, no person may operate an aircraft unless, within the preceding 12 calendar months, it has had—

(1) An annual inspection in accordance with Part 43 of this chapter and has been approved for return to service by a person authorized by § 43.7 of this chapter; or

(2) An inspection for the issue of an airworthiness certificate.

No inspection performed under paragraph (b) of the section may be substituted for any inspection required by this paragraph unless it is performed by a person authorized to perform annual inspections, and is entered as an 'annual' inspection in the required maintenance records.

(b) Except as provided in paragraph (c) of this section, no person may operate an aircraft carrying any person (other than a crewmember) for hire, and no person may give flight instruction for hire in an aircraft which that person provides, unless within the preceding 100 hours of time in service it has received an annual or 100-hour inspection and been approved for return to service in accordance with Part 43 of this chapter, or received an inspection for the issuance of an airworthiness certificate in accordance with Part 21 of this chapter. The 100-hour limitation may be exceeded by not more than 10 hours if necessary to reach a place at which the inspection can be done. The excess time, however, is included in computing the next 100 hours of time in service.   ☞ F11

(c) Paragraphs (a) and (b) of this section do not apply to—

(1) An aircraft that carries a special flight permit, a current experimental certificate, or a provisional airworthiness certificate;

(2) An aircraft inspected in accordance with an approved aircraft inspection program under Part 123, 125, or 135 of this chapter and so identified by the registration number in the operations specifications of the certificate holder having the approved inspection program; or

(3) An aircraft subject to the requirements of paragraph (d) or (e) of this section.

(d) *Progressive inspection.* Each registered owner or operator of an aircraft desiring to use a progressive inspection program must submit a written request to the FAA Flight Standards district office having jurisdiction over the area in which the applicant is located, and shall provide—

(1) A certificated mechanic holding an inspection authorization, a certificated airframe repair station, or the manufacturer of the aircraft to supervise or conduct the progressive inspection;

(2) A current inspection procedures manual available and readily understandable to pilot and maintenance personnel containing, in detail—

(i) An explanation of the progressive inspection, including the continuity of inspection responsibility, the making of reports, and the keeping of records and technical reference material;

(ii) An inspection schedule, specifying the intervals in hours or days when routine and detailed inspections will be performed and including instructions for exceeding an inspection interval by not more than 10 hours while en route and for changing an inspection interval because of service experience;

(iii) Sample routine and detailed inspection forms and instructions for their use; and

(iv) Sample reports and records and instructions for their use;

(3) Enough housing and equipment for necessary disassembly and proper inspection of the aircraft; and

(4) Appropriate current technical information for the aircraft.

[The frequency and detail of the progressive inspection shall provide for the complete inspection of the aircraft within each 12-calendar months and be consistent with the manufacturer's recommendations, field service experience, and the kind of operation in which the aircraft is engaged. The progressive inspection schedule must ensure that the aircraft, at all times, will be airworthy and will conform to all applicable FAA aircraft specifications, type certificate data sheets, airworthiness directives, and other approved data. If the progressive inspection is discontinued, the owner or operator shall immediately notify the local FAA Flight Standards district office, in writing, of the discontinuance. After the discontinuance, the first annual inspection under § 91.169(a) is due within 12 calendar months after the last complete inspection of the aircraft under the progressive inspection. The 100-hour inspection under § 91.169(b) is due within 100 hours after that complete inspection. A complete inspection of the aircraft, for the purpose of determining when the annual and 100-hour inspections are due, requires a detailed inspection of the aircraft and all its components in accordance with the progressive inspection. A routine inspection of the aircraft and a detailed inspection of several components is not considered to be a complete inspection.

(e) *Large airplanes (to which Part 125 is not applicable), turbojet multiengine airplanes, and turbopropeller-powered multiengine airplanes.* No person may operate a large airplane, turbojet multiengine airplane, or turbopropeller-powered multiengine airplane unless the replacement times for life-limited parts specified in the aircraft specifications, type data sheets, or other documents approved by the Administrator are complied with and the airplane, including the airframe engines, propellers, appliances, survival equipment, and emergency equipment, is inspected in accordance with an inspection program selected under the provisions of paragraph (f) of this section.

(f) *Selection of inspection programs under paragraph (e) of this section.* The registered owner or operator of each airplane described in paragraph (e) of this section must select, identify in the aircraft maintenance records, and use one of the following programs for the inspection of that airplane:

(1) A continuous airworthiness inspection program that is part of a continuous airworthiness maintenance program currently in use by a person holding an air carrier operating certificate or an operating certificate issued under Part 121, 127, or 135 of

this chapter and operating that make and model airplane under Part 121 or 127 or operating that make and model under Part 135 and maintaining it under § 135.411(a) (2).

(2) An approved aircraft inspection program approved under § 135.419 of this chapter and currently in use by a person holding an operating certificate issued under Part 135.

(3) An approved continuous inspection program currently in use by a person certificated under Part 123 of this chapter.

(4) A current inspection program recommended by the manufacturer.

(5) Any other inspection program established by the registered owner or operator of that airplane and approved by the Administrator under paragraph (g) of this section. However, the Administrator may require revision to this inspection program in accordance with the provisions of § 91.170.

[Each operator shall include in the selected program the name and address of the person responsible for scheduling the inspections required by the program and make a copy of that program available to the person performing inspections on the airplane and, upon request, to the Administrator.

(g) *Inspection program approval under paragraph (e) of this section.* Each operator of an airplane desiring to establish or change an approved inspection program under paragraph (f) (5) of this section must submit the program for approval to the local FAA Flight Standards district office having jurisdiction over the area in which the airplane is based. The program must be in writing and include at least the following information:

(1) Instruction and procedures for the conduct of inspections for the particular make and model airplane, including necessary tests and checks. The instructions and procedures must set forth in detail the parts and areas of the airframe, engines, propellers, and appliances, including survival and emergency equipment required to be inspected.

(2) A schedule for performing the inspections that must be performed under the program expressed in terms of the time in service, calendar time, number of system operations, or any combination of these.

(h) *Changes from one inspection program to another.* When an operator changes from one inspection program under paragraph (f) of this section to another, the time in service, calendar times, or cycles of operation accumulated under the previous program must be applied in determining inspection due times under the new program.]

§ 91.170 Changes to aircraft inspection programs.

(a) Whenever the Administrator finds that revisions to an approved aircraft inspection program under § 91.169(f) (5) are necessary for the continued adequacy of the program, the owner or operator shall, after notification by the Administrator, make any changes in the program found to be necessary by the Administrator.

(b) The owner or operator may petition the Administrator to reconsider the notice to make any changes in a program in accordance with paragraph (a) of this section.

(c) The petition must be filed with the FAA Flight Standards district office which requested the change to the program within 30 days after the certificate holder receives the notice.

(d) Except in the case of an emergency requiring immediate action in the interest of safety, the filing of the petition stays the notice pending a decision by the Administrator.

§ 91.171 Altimeter system and altitude reporting equipment tests and Inspections.

(a) No person may operate an airplane [or helicopter] in controlled airspace under IFR unless—

(1) Within the preceding 24 calendar months, each static pressure system, each altimeter instrument, and each automatic pressure altitude reporting system has been tested and inspected and found to comply with Appendices E and F of Part 43 of this chapter;

(2) Except for the use of system drain and alternate static pressure valves, following any opening and closing of the static pressure system, that system has been tested and inspected and found to comply with paragraph (a), Appendix E, of Part 43 of this chapter; and

(3) Following installation or maintenance on the automatic pressure altitude reporting system or the ATC transponder where data correspondence error could be introduced, the integrated system has been tested, inspected, and found to comply with paragraph (c), Appendix E, of Part 43 of this chapter.

(b) The tests required by paragraph (a) of this section must be conducted by—

(1) The manufacturer of the airplane [or helicopter] on which the tests and inspections are to be performed;

(2) A certificated repair station properly equipped to perform those functions and holding—

(i) An instrument rating, Class I;

(ii) A limited instrument rating appropriate to the make and model of appliance to be tested;

(iii) A limited rating appropriate to the test to be performed;

(iv) An airframe rating appropriate to the airplane [or helicopter] to be tested; or

(v) A limited rating for a manufacturer issued for the appliance in accordance with § 145.101(b)(4) of this chapter; or

(3) A certificated mechanic with an airframe rating (static pressure system tests and inspections only).

(c) Altimeter and altitude reporting equipment approved under Technical Standard Orders are considered to be tested and inspected as of the date of their manufacture.

(d) No person may operate an airplane [or helicopter] in controlled airspace under IFR at an altitude above the maximum altitude at which all altimeters and the automatic altitude reporting system of that airplane [or helicopter] have been tested.

§ 91.172 ATC transponder tests and Inspections.

(a) No person may use an ATC transponder that is specified in Part 125, § 91.24(a), § 121.345(c), § 127.123(b) or § 135.143(c) of this chapter unless, within the preceding 24 calendar months, that ATC transponder has been tested and inspected and found to comply with Appendix F of Part 43 of this chapter; and

(b) Following any installation or maintenance on an ATC transponder where data correspondence error could be introduced, the integrated system has been tested, inspected, and found to comply with paragraph (c), Appendix E, of Part 43 of this chapter.

(c) The tests and inspections specified in this section must be conducted by—

(1) A certificated repair station properly equipped to perform those functions and holding—

(i) A radio rating, Class III;

(ii) A limited rating appropriate to the make and model transponder to be tested;

(iii) A limited rating appropriate to the test to be performed;

(iv) A limited rating for a manufacturer issued for the transponder in accordance with § 145.101(b) (4) of this chapter; or

(2) A holder of a continuous airworthiness maintenance program as provided in Part 121, 127, or [135.411(a) (2) of this chapter; or

(3) The manufacturer of the aircraft on which the transponder to be tested is installed, if the transponder was installed by that manufacturer.

§ 91.173 Maintenance records.

(a) Except for work performed in accordance with 91.171, each registered owner or operator shall keep the following records for the periods specified in paragraph (b) of this section:

(1) Records of the maintenance [preventive maintenance,] and alteration, and records of the 100-hour, annual, progressive, and other required or approved inspections, as appropriate, for each aircraft (including the airframe) and each engine, propeller, rotor, and appliance of an aircraft. The records must include— ☞ **F73–75**

(i) A description (or reference to data acceptable to the Administrator) of the work performed;

(ii) The date of completion of the work performed; and

(iii) The signature and certificate number of the person approving the aircraft for return to service.

(2) Records containing the following information:

(i) The total time in service of the airframe, each engine and each propeller.

(ii) The current status of life-limited parts of each airframe, engine, propeller, rotor, and appliance.

(iii) The time since last overhaul of all items installed on the aircraft which are required to be overhauled on a specified time basis.

(iv) The identification of the current inspection status of the aircraft, including the times since the last inspections required by the inspection program under which the aircraft and its appliances are maintained.

(v) The current status of applicable airworthiness directives (AD) including, for each, the method of compliance, the AD number, and revision date. If the AD involves recurring action, the time and date when the next action is required. ☞**F72**

(vi) Copies of the forms prescribed by § 43.9(a) of this chapter for each major alteration to the airframe and currently installed engines, rotors, propellers, and appliances.

(b) The owner and operator shall retain the following records for the periods prescribed:

(1) The records specified in paragraph (a) (1) of this section shall be retained until the work is repeated or superseded by other work or for one year after the work is performed.

(2) The records specified in paragraph (a) (2) of this section shall be retained and transferred with the aircraft at the time the aircraft is sold.

(3) A list of defects furnished to a registered owner or operator under § 43.11 of this chapter, shall be retained until the defects are repaired and the aircraft is approved for return to service.

(c) [The owner or operator shall make all maintenance records required to be kept by this section available for inspection by the Administrator or any authorized representative of the National Transportation Safety Board (NTSB). In addition, the owner or operator shall present the Form 337 described in paragraph (d) of this section for inspection upon request of any law enforcement officer.]

[(d) When a fuel tank is installed within the passenger compartment or a baggage compartment pursuant to Part 43, a copy of the FAA Form 337 shall be kept on board the modified aircraft by the owner or operator.]

§ 91.174 Transfer of maintenance records.

Any owner or operator who sells a U.S. registered aircraft shall transfer to the purchaser, at the time of sale, the following records of that aircraft, in plain language form or in coded form at the election of the purchaser, if the coded form provides for the preservation and retrieval of information in a manner acceptable to the Administrator:

(a) The records specified in § 91.173(a) (2).

(b) The records specified in § 91.173(a) (1)

which are not included in the records covered by paragraph (a) of this section, except that the purchaser may permit the seller to keep physical custody of such records. However, custody of records in the seller does not relieve the purchaser of his responsibility under § 91.173(c), to make the records available for inspection by the Administrator or any authorized representative of the National Transportation Safety Board (NTSB).

§ 91.175 Rebuilt engine maintenance records.

(a) The owner or operator may use a new maintenance record, without previous operating history, for an aircraft engine rebuilt by the manufacturer or by an agency approved by the manufacturer.

(b) Each manufacturer or agency that grants zero time to an engine rebuilt by it shall enter, in the new record—

(1) A signed statement of the date the engine was rebuilt;

(2) Each change made as required by Airworthiness Directives; and

(3) Each change made in compliance with manufacturer's service bulletins, if the entry is specifically requested in that bulletin.

(c) For the purposes of this section, a rebuilt engine is a used engine that has been completely disassembled, inspected, repaired as necessary, reassembled, tested, and approved in the same manner and to the same tolerances and limits as a new engine with either new or used parts. However, all parts used in it must conform to the production drawing tolerances and limits for new parts or be of approved oversized or undersized dimensions for a new engine.

⟦§ 91.177 [Reserved.]⟧

Subpart D—Large and Turbine-Powered Multiengine Airplanes

§ 91.181 Applicability.

(a) Sections 91.181 through 91.215 prescribe operating rules, in addition to those prescribed in other subparts of this Part, governing the operation of large and of turbojet-powered multiengine civil airplanes of U.S. registry. The operating rules in this subpart do not apply to those airplanes when they are required to be operated under Parts 121, 123, 125, 129, 135, and 137 of this chapter. ⟦Section 91.169 prescribes an inspection program for large and for turbine-powered (turbojet and turboprop) multiengine airplanes of U.S. registry when they are operated under this Part or Parts 129 or 137.⟧

(b) Operations that may be conducted under the rules in this subpart instead of those in Parts 121, 123, 129, 135, and 137 when common carriage is not involved, include—

(1) Ferry or training flights;

(2) Aerial work operations such as aerial photography or survey, or pipeline patrol, but not including fire fighting operations;

(3) Flights for the demonstration of an airplane to prospective customers when no charge is made except for those specified in paragraph (d) of this section;

(4) Flights conducted by the operator of an airplane for his personal transportation, or the transportation of his guests when no charge, assessment, or fee is made for the transportation;

(5) The carriage of officials, employees, guests, and property of a company on an airplane operated by that company, or the parent or a subsidiary of that company or a subsidiary of the parent, when the carriage is within the scope of, and incidental to, the business of the company (other than transportation by air) and no charge, assessment or fee is made for the carriage in excess of the cost of owning, operating, and maintaining the airplane, except that no charge of any kind may be made for the carriage of a guest of a company, when the carriage is not within the scope of, and incidental to, the business of that company;

(6) The carriage of company officials, employees, and guests of the company on an airplane operated under a time sharing, interchange, or joint ownership agreement as defined in paragraph (c) of this section;

(7) The carriage of property (other than mail) on an airplane operated by a person in the furtherance of a business or employment (other than transportation by air) when the carriage is within the scope of, and incidental to, that business or employment and no charge, assessment, or fee is made for the carriage other than those specified in paragraph (d) of this section;

(8) The carriage on an airplane of an athletic team, sports group, choral group, or similar group having a common purpose or objective when there is no charge, assessment, or fee of any kind made by any person for that carriage; and

(9) The carriage of persons on an airplane operated by a person in the furtherance of a business other than transportation by air for the purpose of selling to them land, goods, or property, including franchises or distributorships, when the carriage is within the scope of, and incidental to, that business and no charge, assessment, or fee is made for that carriage.

(c) As used in this section—

(1) A "time sharing agreement" means an arrangement whereby a person leases his airplane with flight crew to another person, and no charge is made for the flights conducted under that arrangement other than those specified in paragraph (d) of this section;

(2) An "interchange agreement" means an arrangement whereby a person leases his airplane to another person in exchange for equal time, when needed, on the other person's airplane, and no charge, assessment, or fee is made, except that a charge may be made not to exceed the difference between the cost of owning, operating, and maintaining the two airplanes;

(3) A "joint ownership agreement" means an arrangement whereby one of the registered joint owners of an airplane employs and furnishes the flight crew for that airplane and each of the registered joint owners pays a share of the charges specified in the agreement.

(d) The following may be charged, as expenses of a specific flight, for transportation as authorized by subparagraphs (b) (3) and (7) and (c) (1) of this section:

(1) Fuel, oil, lubricants, and other additives.

(2) Travel expenses of the crew, including food, lodging, and ground transportation.

(3) Hangar and tie-down costs away from the aircraft's base of operations.

(4) Insurance obtained for the specific flight.

(5) Landing fees, airport taxes, and similar assessments.

(6) Customs, foreign permit, and similar fees directly related to the flight.

(7) In flight food and beverages.

(8) Passenger ground transportation.

(9) Flight planning and weather contract services.

(10) An additional charge equal to 100 percent of the expenses listed in subparagraph (1) of this paragraph.

§ 91.183 Flying equipment and operating information.

(a) The pilot in command of an airplane shall insure that the following flying equipment and aeronautical charts and data, in current and appropriate form, are accessible for each flight at the pilot station of the airplane:

(1) A flashlight having at least two size "D" cells, or the equivalent, that is in good working order.

(2) A cockpit checklist containing the procedures required by paragraph (b) of this section.

(3) Pertinent aeronautical charts.

(4) For IFR, VFR over-the-top, or night operations, each pertinent navigational en route, terminal area, and approach and letdown chart.

(5) In the case of multiengine airplanes, one-engine inoperative climb performance data.

(b) Each cockpit checklist must contain the following procedures and shall be used by the flight crewmembers when operating the airplane:

(1) Before starting engines.

(2) Before takeoff.

(3) Cruise.

(4) Before landing.

(5) After landing.

(6) Stopping engines.

(7) Emergencies.

(c) Each emergency cockpit checklist procedure required by paragraph (b) (7) of this section must contain the following procedures, as appropriate:

(1) Emergency operation of fuel, hydraulic, electrical, and mechanical systems.

(2) Emergency operation of instruments and controls.

(3) Engine inoperative procedures.

(4) Any other procedures necessary for safety.

(d) The equipment, charts, and data prescribed in this section shall be used by the pilot in command and other members of the flight crew, when pertinent.

§ 91.185 Familiarity with operating limitations and emergency equipment.

(a) Each pilot in command of an airplane shall, before beginning a flight, familiarize himself with the airplane flight manual for that airplane, if one is required, and with any placards, listings, instrument markings, or any combination thereof, containing each operating limitation prescribed for that airplane by the Administrator, including those specified in § 91.31(b).

(b) Each required member of the crew shall, before beginning a flight, familiarize himself with the emergency equipment installed on the airplane to which he is assigned and with the procedures to be followed for the use of that equipment in an emergency situation.

§ 91.187 Equipment requirements: over-the-top, or night VFR operations.

No person may operate an airplane over-the-top, or at night under VFR unless that airplane is equipped with the instruments and equipment required for IFR operations under § 91.33(d) and one electric landing light for night operations. Each required instrument and item of equipment must be in operable condition.

§ 91.189 Survival equipment for overwater operations.

(a) No person may takeoff an airplane for a flight over water more than 50 nautical miles from the nearest shoreline, unless that airplane is equipped with a life preserver or an approved flotation means for each occupant of the airplane.

(b) No person may take off an airplane for a flight over water more than 30 minutes flying time or 100 nautical miles from the nearest shoreline, unless it has on board the following survival equipment:

(1) A life preserver equipped with an approved survivor locator light, for each occupant of the airplane.

(2) Enough liferafts (each equipped with an approved survival locator light) of a rated capacity and buoyancy to accommodate the occupants of the airplane.

(3) At least one pyrotechnic signaling device for each raft.

(4) One self-buoyant, water-resistant, portable emergency radio signaling device, that is capable of transmission on the appropriate emergency frequency or frequencies, and not

dependent upon the airplane power supply.

(5) After June 26, 1979, a lifeline stored in accordance with § 25.1411(g) of this chapter.

(c) The required liferafts, life preservers, and signaling devices must be installed in conspicuously marked locations and easily accessible in the event of a ditching without appreciable time for preparatory procedures.

(d) A survival kit, appropriately equipped for the route to be flown, must be attached to each required liferaft.

§ 91.191 Radio equipment for overwater operations.

(a) Except as provided in paragraphs (c) and (d) of this section, no person may takeoff an airplane for a flight over water more than 30 minutes flying time or 100 nautical miles from the nearest shoreline, unless it has at least the following operable radio communication and navigational equipment appropriate to the facilities to be used and able to transmit to, and receive from, at any place on the route, at least one surface facility:

(1) Two transmitters.

(2) Two microphones.

(3) Two headsets or one headset and one speaker.

(4) Two independent receivers for navigation.

(5) Two independent receivers for communications.

However, a receiver that can receive both communications and navigational signals may be used in place of a separate communications receiver and a separate navigational signal receiver.

(b) For the purposes of paragraphs (a) (4) and (5) of this section, a receiver is independent if the function of any part of it does not depend on the functioning of any part of another receiver.

(c) Notwithstanding the provisions of paragraph (a) of this section, a person may operate an airplane on which no passengers are carried from a place where repairs or replacement cannot be made to a place where they can be made, if not more than one of each of the dual items of radio communication and navigation equipment specified in subparagraphs (1)–(5) of paragraph (a) of this section malfunctions or becomes inoperative.

(d) Notwithstanding the provisions of paragraph (a) of this section, when both VHF and HF communications equipment are required for the route and the airplane has two VHF transmitters and two VHF receivers for communications, only one HF transmitter and one HF receiver is required for communications.

§ 91.193 Emergency equipment.

(a) No person may operate an airplane unless it is equipped with the emergency equipment listed in this section:

(b) Each item of equipment—

(1) Must be inspected in accordance with § 91.217 to ensure its continued serviceability

and immediate readiness for its intended purposes;

(2) Must be readily accessible to the crew;

(3) Must clearly indicate its method of operation: and

(4) When carried in a compartment or container, must have that compartment or container marked as to contents and date of last inspection.

(c) Hand fire extinguishers must be provided for use in crew, passenger, and cargo compartments in accordance with the following:

(1) The type and quantity of extinguishing agent must be suitable for the kinds of fires likely to occur in the compartment where the extinguisher is intended to be used.

(2) At least one hand fire extinguisher must be provided and located on or near the flight deck in a place that is readily accessible to the flight crew.

(3) At least one hand fire extinguisher must be conveniently located in the passenger compartment of each airplane accommodating more than six but less than 31 passengers, and at least two hand fire extinguishers must be conveniently located in the passenger compartment of each airplane accommodating more than 30 passengers.

[(4) Hand fire extinguishers must be installed and secured in such a manner that they will not interfere with the safe operation of the airplane or adversely affect the safety of the crew and passengers. They must be readily accessible, and unless the locations of the fire extinguishers are obvious, their stowage provisions must be properly identified.]

(d) First aid kits for treatment of injuries likely to occur in flight or in minor accidents must be provided.

(e) Each airplane accommodating more than 19 passengers must be equipped with a crash ax.

(f) Each passenger-carrying airplane must have a portable battery-powered megaphone or megaphones readily accessible to the crewmembers assigned to direct emergency evacuation, installed as follows:

(1) One megaphone on each airplane with a seating capacity of more than 60 but less than 100 passengers, at the most rearward location in the passenger cabin where it would be readily accessible to a normal flight attendant seat. However, the Administrator may grant a deviation from the requirements of this subparagraph if he finds that a different location would be more useful for evacuation of persons during an emergency.

(2) Two megaphones in the passenger cabin on each airplane with a seating capacity of more than 99 passengers, one installed at the forward end and the other at the most rearward location where it would be readily accessible to a normal flight attendant seat

§ 91.195 Flight altitude rules.

(a) Notwithstanding § 91.179, and except as provided in paragraph (b) of this section, no person may operate an airplane under VFR at less than—

(1) One thousand feet above the surface, or 1,000 feet from any mountain, hill, or other obstruction to flight, for day operations; and

(2) The altitudes prescribed in § 91.119, for night operations.

(b) This section does not apply—

(1) During takeoff or landing;

(2) When a different altitude is authorized by a waiver to this section under § 91.63; or

(3) When a flight is conducted under the special VFR weather minimums of § 91.107 with an appropriate clearance from ATC.

§ 91.197 Smoking and safety belt signs.

(a) Except as provided in paragraph (b) of this section, no person may operate an airplane carrying passengers unless it is equipped with signs that are visible to passengers and cabin attendants to notify them when smoking is prohibited and when safety belts should be fastened. The signs must be so constructed that the crew can turn them on and off. They must be turned on for each takeoff and each landing and when otherwise considered to be necessary by the pilot in command.

(b) The pilot in command of an airplane that is not equipped as provided in paragraph (a) of this section shall insure that the passengers are orally notified each time that it is necessary to fasten their safety belts and when smoking is prohibited.

§ 91.199 Passenger briefing.

(a) Before each takeoff the pilot in command of an airplane carrying passengers shall ensure that all passengers have been orally briefed on:

(1) Smoking;

(2) Use of safety belts;

(3) Location and means for opening the passenger entry door and emergency exits;

(4) Location of survival equipment;

(5) Ditching procedures and the use of flotation equipment required under § 91.189 for a flight over water; and

(6) The normal and emergency use of oxygen equipment installed on the airplane.

(b) The oral briefing required by paragraph (a) of this section shall be given by the pilot in command or a member of the crew, but need not be given when the pilot in command determines that the passengers are familiar with the contents of the briefing. It may be supplemented by printed cards for the use of each passenger containing—

(1) A diagram of, and methods of operating, the emergency exits; and

(2) Other instructions necessary for use of emergency equipment.

Each card used under this paragraph must be carried in convenient locations on the airplane for use of each passenger and must contain information that is pertinent only to the type and model airplane on which it is used.

[§ 91.200 Shoulder harness.

[(a) No person may operate a transport category airplane that was type certificated after January 1, 1958, unless it is equipped at each seat at a flight-deck station with a combined safety belt and shoulder harness that meets the applicable requirements in § 25.785 of this chapter, except that—

[(1) Shoulder harnesses and combined safety belt and shoulder harnesses that were approved and installed before March 6, 1980, may continue to be used; and

[(2) Safety belt and shoulder harness restraint systems may be designed to the inertia load factors established under the certification basis of the airplane.

[(b) No person may operate a transport category airplane unless it is equipped at each required flight attendant seat in the passenger compartment with a combined safety belt and shoulder harness that meets the applicable requirements specified in § 25.785 of this chapter, except that—

[(1) Shoulder harnesses and combined safety belt and shoulder harnesses that were approved and installed before March 6, 1980, may continue to be used; and

[(2) Safety belt and shoulder harness restraint systems may be designed to the inertia load factors established under the certification basis of the airplane.]

§ 91.201 Carry-on-baggage.

No pilot in command of an airplane having a seating capacity of more than 19 passengers may permit a passenger to stow his baggage aboard that airplane except—

(a) In a suitable baggage or cargo storage compartment, or as provided in § 91.203; or

(b) Under a passenger seat in such a way that it will not slide forward under crash impacts severe enough to induce the ultimate inertia forces specified in § 25.561(b) (3) of this chapter, or the requirements of the regulations under which the airplane was type certificated. After December 4, 1979 restraining devices must also limit sideward motion of under-seat baggage and be designed to withstand crash impacts severe enough to induce sideward forces specified in § 25.561(b) (3) of this chapter.

§ 91.203 Carriage of cargo.

(a) No pilot in command may permit cargo to be carried in any airplane unless—

(1) It is carried in an approved cargo rack, bin, or compartment installed in the airplane;

(2) It is secured by means approved by the Administrator; or

(3) It is carried in accordance with each of the following:

(i) It is properly secured by a safety belt or other tiedown having enough strength to eliminate the possibility of shifting under all normally anticipated flight and ground conditions.

(ii) It is packaged or covered to avoid possible injury to passengers.

(iii) It does not impose any load on seats or on the floor structure that exceeds the load limitation for those components.

(iv) It is not located in a position that restricts the access to or use of any required emergency or regular exit, or the use of the aisle between the crew and the passenger compartment.

(v) It is not carried directly above seated passengers.

(b) When cargo is carried in cargo compartments that are designed to require the physical entry of a crewmember to extinguish any fire that may occur during flight, the cargo must be loaded so as to allow a crewmember to effectively reach all parts of the compartment with the contents of a hand fire extinguisher.

§ 91.205 Transport category airplane weight limitations.

No person may takeoff a transport category airplane, except in accordance with the weight limitations prescribed for that airplane in § 91.37 of this Part.

§ 91.207 Deleted.

§ 91.209 Operating in icing conditions.

(a) No pilot may take off an airplane that has— ☞F83

(1) Frost, snow, or ice adhering to any propeller, windshield, or power plant installation, or to an airspeed, altimeter, rate of climb, or flight attitude instrument system;

(2) Snow or ice adhering to the wings, or stabilizing or control surfaces; or

(3) Any frost adhering to the wings, or stabilizing or control surfaces, unless that frost has been polished to make it smooth.

(b) Except for an airplane that has ice protection provisions that meet the requirements in section 34 of Special Federal Aviation Regulation No. 23, or those for transport category airplane type certification, no pilot may fly—

(1) Under IFR into known or forecast moderate icing conditions; or

(2) Under VFR into known light or moderate icing conditions; unless the aircraft has functioning de-icing or anti-icing equipment protecting each propeller, windshield, wing, stabilizing or control surface, and each airspeed, altimeter, rate of climb, or flight attitude instrument system.

(c) Except for an airplane that has ice protection provisions that meet the requirements in section 34 of Special Federal Aviation Regulation No. 23, or those for transport category airplane type certification, no pilot may fly an airplane into known or forecast severe icing conditions.

(d) If current weather reports and briefing information relied upon by the pilot in command indicate that the forecast icing conditions that would otherwise prohibit the flight will not be encountered during the flight because of changed weather conditions since the forecast, the restrictions in paragraph (b) and (c) of this section based on forecast conditions do not apply.

§ 91.211   Flight engineer requirements.

(a) No person may operate the following airplanes without a flight crewmember holding a current flight engineer certificate:

(1) An airplane for which a type certificate was issued before January 2, 1964, having a maximum certificated takeoff weight of more than 80,000 pounds.

(2) An airplane type certificated after January 1, 1964, for which a flight engineer is required by the type certification requirements.

(b) No person may serve as a required flight engineer on an airplane unless, within the preceding 6 calendar months, he has had at least 50 hours of flight time as a flight engineer on that type of airplane, or the Administrator has checked him on that type airplane and determined that he is familiar and competent with all essential current information and operating procedures.

§ 91.213   Second in command requirements.

(a) Except as provided in paragraph (b) of this section, no person may operate the following airplanes without a pilot who is designated as second in command of that airplane:

(1) A large airplane, except that a person may operate an airplane certificated under SFAR 41 without a pilot who is designated as second in command if that airplane is certificated for operation with one pilot.

(2) A turbojet-powered multiengine airplane for which two pilots are required under the type certification requirements for that airplane.

[(3) A commuter category airplane, except that a person may operate a commuter category airplane notwithstanding paragraph (a)(1) of this section, that has a passenger seating configuration, excluding pilot seats, of nine or less without a pilot who is designated as second in command if that airplane is type certificated for operations with one pilot.]

(b) The Administrator may issue a letter of authorization for the operation of an airplane without compliance with the requirements of paragraph (a) of this section if that airplane is designed for and type certificated with only one pilot station. The authorization contains any conditions that the Administrator finds necessary for safe operation.

(c) No person may designate a pilot to serve as second in command nor may any pilot serve as second in command of an airplane required under this section to have two pilots, unless that pilot meets the qualifications for second in command prescribed in § 61.55 of this chapter.

§ 91.215   Flight attendant requirements.

(a) No person may operate an airplane unless at least the following number of flight attendants are on board the airplane:

(1) For airplanes having more than 19 but less than 51 passengers on board—one flight attendant.

(2) For airplanes having more than 50 but less than 101 passengers on board—two flight attendants.

(3) For airplanes having more than 100 passengers on board—two flight attendants plus one additional flight attendant for each unit (or part of a unit) of 50 passengers above 100.

(b) No person may serve as a flight attendant on an airplane when required by paragraph (a) of this section, unless that person has demonstrated to the pilot in command that he is familiar with the necessary functions to be performed in an emergency or a situation requiring emergency evacuation and is capable of using the emergency equipment installed on that airplane for the performance of those functions.

[§ 91.217   [Reserved.]]

[§ 91.219   [Reserved.]]

Subpart E—Operating Noise Limits

§ 91.301   Applicability; relation to Part 36.

(a) This subpart prescribes operating noise limits and related requirements that apply, as follows, to the operation of civil aircraft in the United States:

(1) Sections 91.302, 91.303, 91.305, 91.306, and 91.307 apply to civil subsonic turbojet airplanes with maximum weights of more than 75,000 pounds and—

(i) If U.S. registered, that have standard airworthiness certificates; or

(ii) If foreign registered, that would be required by this chapter to have a U.S. standard airworthiness certificate in order to conduct the operations intended for the airplane were it registered in the United States.

Those sections apply to operations to or from airports in the United States under this Part and Parts 121, 123, 125, 129, and 135 of this chapter.

(2) Section 91.308 applies to U.S. operators of civil subsonic turbojet airplanes covered by this subpart. That section applies to operators operating to or from airports in the United States under this Part and Parts 121, 123, 125, and 135 but not to those operating under Part 129 of this chapter.

(3) Sections 91.302, 91.309 and 91.311 apply to U.S. registered civil supersonic airplanes having standard airworthiness cer-

tificates, and to foreign registered civil supersonic airplanes that, if registered in the United States, would be required by this chapter to have a U.S. standard airworthiness certificate in order to conduct the operations intended for the airplane. Those sections apply to operations under this Part and under Parts 121, 123, 125, 129, and 135 of this chapter.

(b) Unless otherwise specified, as used in this subpart "Part 36" refers to 14 CFR Part 36, including the noise levels under Appendix C of that Part, notwithstanding the provisions of that Part excepting certain airplanes from the specified noise requirements. For purposes of this subpart, the various stages of noise levels, the terms used to describe airplanes with respect to those levels, and the terms "subsonic airplane" and "supersonic airplane" have the meanings specified under Part 36 of this chapter. For purposes of this subpart, for subsonic airplanes operated in foreign air commerce in the United States, the Administrator may accept compliance with the noise requirements under Annex 16 of the International Civil Aviation Organization when those requirements have been shown to be substantially compatible with, and achieve results equivalent to those achievable under, Part 36 for that airplane. Determinations made under these provisions are subject to the limitations of § 36.5 of this chapter as if those noise levels were Part 36 noise levels.

(c) [Reserved]

§ 91.302   Part 125 operators: designation of applicable regulations.

For airplanes covered by this subpart and operated under Part 125, the following regulations apply as specified:

(a) For each airplane operation to which requirements prescribed under this subpart applied before November 29, 1980, those requirements of this subpart continue to apply.

(b) For each subsonic airplane operation to which requirements prescribed under this subpart did not apply before November 29, 1980, because the airplane was not operated in the United States under this Part or Part 121, 123, 129, or 135, the requirements prescribed under §§ 91.303, 91.306, 91.307, and 91.308 of this subpart apply.

(c) For each supersonic airplane operation to which requirements prescribed under this subpart did not apply before November 29, 1980, because the airplane was not operated in the United States under this Part or Part 121, 123, 129, or 135, the requirements of §§ 91.309 and 91.311 of this subpart apply.

(d) For each airplane required to operate under Part 125 for which a deviation under that Part is approved to operate, in whole or in part, under this Part or Parts 121, 123, 129, or 135, notwithstanding the approval, the requirements prescribed under paragraphs (a), (b), and (c) of this section continue to apply.

§ 91.303   Final compliance: subsonic airplanes.

Except as provided in §§ 91.306 and 91.307, on and after January 1, 1985, no person may operate to or from an airport in the United

States any subsonic airplane covered by this subpart, unless that airplane has been shown to comply with Stage 2 or Stage 3 noise levels under Part 36 of this chapter.

§ 91.305 Phased compliance under Parts 121, 125, and 135: subsonic airplanes.

(a) *General.* Each person operating airplanes under Parts 121 or 135 of this chapter, or under Part 125 of this chapter, as prescribed under § 91.302 of this subpart regardless of the State of registry of the airplane, shall comply with this section with respect to subsonic airplanes covered by this subpart.

(b) *Compliance schedule.* Except for airplanes shown to be operated in foreign air commerce under paragraph (c) of this section or covered by an exemption (including those issued under § 91.307), airplanes operated by U.S. operators in air commerce in the United States must be shown to comply with Stage 2 or Stage 3 noise levels under Part 36, in accordance with the following schedule, or they may not be operated to or from airports in the United States:

(1) By January 1, 1981:

(i) At least one quarter of the airplanes that have four engines with no bypass ratio or with a bypass ratio less than two.

(ii) At least half of the airplanes powered by engines with any other bypass ratio or by another number of engines.

(2) By January 1, 1983:

(i) At least one half of the airplanes that have four engines with no bypass ratio or with bypass ratio less than two.

(ii) All airplanes powered by engines with any other bypass ratio or by another number of engines.

(c) *Apportionment of airplanes.* For purposes of paragraph (b) of this section, a person operating airplanes engaged in domestic and foreign air commerce in the United States may elect not to comply with the phased schedule with respect to that portion of the airplanes operated by that person shown, under an approved method of apportionment, to be engaged in foreign air commerce in the United States.

§ 91.306 Replacement airplanes.

A Stage 1 airplane may be operated after the otherwise applicable compliance dates prescribed under §§ 91.303 and 91.305 if, under an approved plan, a replacement airplane has been ordered by the operator under a binding contract as follows:

(a) For replacement of an airplane powered by two engines, until January 1, 1986, but not after the date specified in the plan, if the contract is entered into by January 1, 1983, and specifies delivery before January 1, 1986, of a replacement airplane which has been shown to comply with Stage 3 noise levels under Part 36 of this chapter.

(b) For replacement of an airplane powered by three engines, until January 1, 1985, but not after the date specified in the plan, if the contract is entered into by January 1, 1983, and specifies delivery before January 1, 1985, of a replacement airplane which has been shown to comply with Stage 3 noise levels under Part 36 of this chapter.

(c) For replacement of any other airplane, until January 1, 1985, but not after the date specified in the plan, if the contract specifies delivery before January 1, 1985, of a replacement airplane which—

(1) Has been shown to comply with Stage 2 or Stage 3 noise levels under Part 36 of this chapter prior to issuance of an original standard airworthines certificate; or

(2) Has been shown to comply with Stage 3 noise levels under Part 36 of this chapter prior to issuance of a standard airworthiness certificate other than original issue.

(d) Each operator of a Stage 1 airplane for which approval of a replacement plan is requested under this section shall submit to the FAA Director of the Office of Environement and Energy an application constituting the proposed replacement plan (or revised plan) that contains the information specified under this paragraph and which is certified (under penalty of 18 U.S.C. § 1001) as true and correct. Each application for approval must provide information corresponding to that specified in the contract, upon which the FAA may rely in considering its approval, as follows:

(1) Name and address of the applicant.

(2) Aircraft type and model and registration number for each airplane to be replaced under the plan.

(3) Aircraft type and model of each replacement airplane.

(4) Scheduled dates of delivery and introduction into service of each replacement airplane.

(5) Name and address of the parties to the contract and any other persons who may effectively cancel the contract or otherwise control the performance of any party.

(6) Information specifying the anticipated disposition of the airplanes to be replaced.

(7) A statement that the contract represents a legally enforceable, mutual agreement for delivery of an eligible replacement airplane.

(8) Any other information or documentation requested by the Director, Office of Environment and Energy reasonably necessary to determine whether the plan should be approved.

§ 91.307 Service to small communities exemption: two-engine, subsonic airplanes.

(a) A Stage 1 airplane powered by two engines may be operated after the compliance dates prescribed under §§ 91.303, 91.305, and 91.306, when, with respect to that airplane, the Administrator issues an exemption to the operator from the noise level requirements under this subpart. Each exemption issued under this section terminates on the earlier of the following dates—

(1) For an exempted airplane sold, or otherwise disposed of, to another person on or after January 1, 1983—on the date of delivery to that person;

(2) For an exempted airplane with a seating configuration of 100 passenger seats

or less—on January 1, 1988; or

(3) For an exempted airplane with a seating configuration of more than 100 passenger seats—on January 1, 1985.

(b) For purposes of this section, the seating configuration of an airplane is governed by that shown to exist on December 1, 1979, or an earlier date established for that airplane by the Administrator.

§ 91.308 Compliance plans and status: U.S. operators of subsonic airplanes.

(a) Each U.S. operator of a civil subsonic airplane covered by this subpart (regardless of the State of registry) shall submit to the FAA, Director of the Office of Environment and Energy, in accordance with this section, the operator's current compliance status and plan for achieving and maintaining compliance with the applicable noise level requirements of this subpart. If appropriate, an operator may substitute for the required plan a notice, certified as true (under penalty of 18 U.S.C. § 1001) by that operator, that no change in the plan or status of any airplane affected by the plan has occurred since the date of the plan most recently submitted under this section.

(b) Each compliance plan, including any revised plans, must contain the information specified under paragraph (c) of this section for each airplane covered by this section that is operated by the operator. Unless otherwise approved by the Administrator, compliance plans must provide the required plan and status information as it exists on the date 30 days before the date specified for submission of the plan. Plans must be certified by the operator as true and complete (under penalty of 18 U.S.C. § 1001) and be submitted for each airplane covered by this section on or before the following dates—

(1) May 1, 1980 or 90 days after initially commencing operation of airplanes covered by this section, whichever is later, and thereafter—

(2) Thirty days after any change in the operator's fleet or compliance planning decisions that has a separate or cumulative effect on 10 percent or more of the airplanes in either class of airplanes covered by § 91.305(b); and

(3) Thirty days after each compliance date applicable to that airplane type under this subpart and annually thereafter through 1985 or until any later compliance date for that airplane prescribed under this subpart, on the anniversary of that submission date, to show continuous compliance with this subpart.

(c) Each compliance plan submitted under this section must identify the operator and include information regarding the compliance plan and status for each airplane covered by this section as follows:

(1) Name and address of the airplane operator.

(2) Name and telephone number of the person designated by the operator to be responsible for the preparation of the compliance plan and its submission.

(3) The total number of airplanes covered by this section and in each of the following classes and subclasses:

(i) Airplanes engaged in domestic air commerce.

(A) Airplanes powered by four turbojet engines with no bypass ratio or with a bypass ratio less than two.

(B) Airplanes powered by engines with any other bypass ratio or by another number of engines.

(C) Airplanes covered by an exemption issued under § 91.307 of this subpart.

(ii) Airplanes engaged in foreign air commerce under an approved apportionment plan.

(A) Airplanes powered by four turbojet engines with no bypass ratio or with a bypass ratio less than two.

(B) Airplanes powered by engines with any other bypass ratio or by another number of engines.

(C) Airplanes covered by an exemption issued under § 91.307 of this subpart.

(4) For each airplane covered by this section—

(i) Aircraft type and model;

(ii) Aircraft registration number;

(iii) Aircraft manufacturer serial number;

(iv) Aircraft power plant make and model;

(v) Aircraft year of manufacture;

(vi) Whether Part 36 noise level compliance has been shown: Yes/No;

(vii) [Reserved];

(viii) The appropriate code prescribed under paragraph (c)(5) of this section which indicates the acoustical technology installed, or to be installed, on the airplane;

(ix) For airplanes on which acoustical technology has been or will be applied, following the appropriate code entry, the actual or scheduled month and year of installation on the airplane;

(x) For DC-8 and B-707 airplanes operated in domestic U.S. air commerce which have been or will be retired from service in the United States without replacement between January 24, 1977, and January 1, 1985, the appropriate code prescribed under paragraph (c)(5) of this section followed by the actual or scheduled month and year of retirement of the airplane from service;

(xi) For DC-8 and B-707 airplanes operated in foreign air commerce in the United States, which have been or will be retired from service in the United States without replacement between April 14, 1980, and January 1, 1985, the appropriate code prescribed under paragraph (c)(5) of this section followed by the actual or scheduled month and year of retirement of the airplane from service;

(xii) For airplanes covered by an approved replacement plan under § 91.305(c) of this subpart, the appropriate code prescribed under paragraph (c)(5) of this

section followed by the scheduled month and year for replacement of the airplane;

(xiii) For airplanes designated as "engaged in foreign commerce" in accordance with an approved method of apportionment under § 91.305(c) of this subpart, the appropriate code prescribed under paragraph (c)(5) of this section;

(xiv) For airplanes covered by an exemption issued to the operator granting relief from noise level requirements of this subpart, the appropriate code prescribed under paragraph (c)(5) of this section followed by the actual or scheduled month and year of expiration of the exemption and the appropriate code and applicable dates which indicate the compliance strategy planned or implemented for the airplane;

(xv) For all airplanes covered by this section, the number of spare shipsets of acoustical components need for continuous compliance and the number available on demand to the operator in support of those airplanes; and

(xvi) For airplanes for which none of the other codes prescribed under paragraph (c)(5) of this section describes either the technology applied, or to be applied to the airplane in accordance with the certification requirements under Parts 21 and 36 of this chapter, or the compliance strategy or methodology, following the code "OTH" enter the date of any certificate action and attach an addendum to the plan explaining the nature and extent of the certificated technology, strategy, or methodology employed, with reference to the type certificate documentation.

(5) TABLE OF ACOUSTICAL TECHNOLOGY/STRATEGY CODES

Code	Airplane Type/Model	Certificated Technology
A	B-707-120B B-707-320B/C B-720B	Quiet Nacelles + 1-Ring
B	B-727-100	Double Wall Fan Duct Treatment
C	B-727-200	Double Wall Fan Duct Treatment (Pre-January 1977 Installations and Amended Type Certificate)
D	B-727-200 B-737-100 B-737-200	Quiet Nacelles + Double Wall Fan Duct Treatment
E	B-747-100 (pre-December 1971) B-747-200 (pre-December 1971)	Fixed Lip Inlets + Sound Absorbing Material Treatment
F	DC-8	New Extended Inlet and Bullet with Treatment + Fan Duct Treatment Areas
G	DC-9	P-36 Sound Absorbing Material Treatment Kit
H	BAC-111-200	Silencer Kit (BAC Acoustic Report 522)
I	BAC-111-400	(To be identified later if certificated)
J	B-707 DC-8	Reengined with High Bypass Ratio Turbojet Engines + Quiet Nacelles (if certificated under Stage 3 noise level requirements)

REP—For airplanes covered by an approved replacement under § 91.305(c) of this subpart.

EFC—For airplanes designated as "engaged in foreign commerce" in accordance with an approved method of apportionment under § 91.307 of this subpart.

RET—For DC-8 and B-707 airplanes operated in domestic U.S. air commerce and retired from service in the

United States without replacement between January 24, 1977, and January 1, 1985.

RFC—For DC-8 and B-707 airplanes operated by U.S. operators in foreign air commerce in the United States and retired from service in the United States without replacement between April 14, 1980, and January 1, 1985.

EXD—For airplanes exempted from showing compliance with the noise level requirements of this subpart.

OTH—For airplanes for which no other prescribed code describes either the certificated technology applied, or to be applied to the airplane, or the compliance strategy or methodology. (An addendum must explain the nature and extent of technology, strategy or methodology and reference the type certificate documentation.

§ 91.309 Civil supersonic airplanes that do not comply with Part 36.

(a) *Applicability.* This section applies to civil supersonic airplanes that have not been shown to comply with the Stage 2 noise limits of Part 36 in effect on October 13, 1977, using applicable tradeoff provisions, and that are operated in the United States after July 31, 1978.

(b) *Airport use.* Except in an emergency, the following apply to each person who operates a civil supersonic airplane to or from an airport in the United States:

(1) Regardless of whether a type design change approval is applied for under Part 21 of this chapter, no person may land or take off an airplane, covered by this section, for which the type design is changed, after July 31, 1978, in a manner constituting an "acoustical change" under § 21.93, unless the acoustical change requirements of Part 36 are complied with.

(2) No flight may be scheduled, or otherwise planned, for takeoff or landing after 10 p.m. and before 7 a.m. local time.

§ 91.311 Civil supersonic airplanes: noise limits.

Except for Concorde airplanes having flight time before January 1, 1980, no person may, after July 31, 1978, operate, in the United States, a civil supersonic airplane that does not comply with the Stage 2 noise limits of Part 36 in effect on October 13, 1977, using applicable trade-off provisions.

APPENDIX A
Category II Operations:
Manual, Instruments, Equipment and Maintenance

1. Category II Manual.

(a) *Application for approval.* An applicant for approval of a Category II manual or an amendment to an approved Category II manual must submit the proposed manual or amendment to the [Flight Standards District Office] having jurisdiction of the area in which the applicant is located. If the application requests an evaluation program, it must include the following:

(1) The location of the airplane and the place where the demonstrations are to be conducted; and

(2) The date the demonstrations are to commence (at least 10 days after filing the application.)

(b) *Contents.* Each Category II manual must contain—

(1) The registration number, make, and model of the [aircraft] to which it applies;

(2) A maintenance program as specified in § 4 of this Appendix; and

(3) The procedures and instructions related to recognition of decision height, use of runway visual range information, approach monitoring, the decision region (the region between the middle marker and the decision height), the maximum permissible deviations of the basic ILS indicator within the decision region, a missed approach, use of airborne low approach equipment, minimum altitude for the use of the autopilot, instrument and equipment failure warning systems, instrument failure, and other procedures, instructions, and limitations that may be found necessary by the Administrator.

2. Required Instruments and Equipment.

The instruments and equipment listed in this section must be installed in each [aircraft] operated in a Category II operation. This section does not require duplication of instruments and equipment required by § 91.33 or any other provisions of this chapter.

(a) *Group I.*

(1) Two localizer and glide slope receiving systems. Each system must provide a basic ILS display and each side of the instrument panel must have a basic ILS display. However, a single localizer antenna and a single glide slope antenna may be used.

(2) A communications system that does not affect the operation of at least one of the ILS systems.

(3) A marker beacon receiver that provides distinctive aural and visual indications of the outer and the middle marker.

(4) Two gyroscopic pitch and bank indicating systems.

(5) Two gyroscopic direction indicating systems.

(6) Two airspeed indicators.

(7) Two sensitive altimeters adjustable for barometric pressure, each having a placarded correction for altimeter scale error and for the wheel height of the [aircraft.] After June 26, 1979, two sensitive altimeters adjustable for barometric pressure, having markings at 20-foot intervals and each having a placarded correction for altimeter scale error and for the wheel height of the [aircraft.]

(8) Two vertical speed indicators.

(9) A flight control guidance system that consists of either an automatic approach coupler or a flight director system. A flight director system must display computed information as steering command in relation to an ILS localizer and, on the same instrument, either computed information as pitch command in relation to an ILS glide slope or basic ILS glide slope information. An automatic approach coupler must provide, at least automatic steering in relation to an ILS localizer. The flight control guidance system may be operated from one of the receiving systems required by subparagraph (1) of this paragraph.

(10) For Category II operations with decision heights below 150 feet, either a marker beacon receiver providing aural and visual indications of the inner marker or a radio altimeter.

(b) *Group II.*

(1) Warning systems for immediate detection by the pilot of system faults in items (1), (4), (5) , and (9) of Group I and, if installed, for use in Category II operations, the radio altimeter and auto throttle system.

(2) Dual controls.

(3) An externally vented static pressure system with an alternate static pressure source.

(4) A windshield wiper or equivalent means of providing adequate cockpit visibility for a safe visual transition by either pilot to touch down and roll out.

(5) A heat source for each airspeed system pitot tube installed or an equivalent means of preventing malfunctioning due to icing of the pitot system.

3. Instruments and Equipment Approval.

(a) *General.* The instruments and equipment required by § 2 of this Appendix must be approved as provided in this section before being used in Category II operations. Before presenting an [aircraft] for approval of the instruments and equipment, it must be shown that, since the beginning of the 12th calendar month before the date of submission—

(1) The ILS localizer and glide slope equipment were bench checked according to the manufacturer's instructions and found to meet those standards specified in RTCA Paper 23-63/DO-117, dated March 14, 1963, "Standard Adjustment Criteria for Airborne Localizer and Glide Slope Receivers", which may be obtained from the RTCA Secretariat, 2000 K St., N.W., Washington, D.C. 20006, at cost of 50 cents per copy, payment in cash or by check or money order payable to the RADIO TECHNICAL COMMISSION FOR AERONAUTICS;

(2) The altimeters and the static pressure systems were tested and inspected in accordance with Appendix E to Part 43 of this chapter; and

(3) All other instruments and items of equipment specified in § 2(a) of this Appendix that are listed in the proposed maintenance program were bench checked and found to meet the manufacturer's specifications.

(b) *Flight Control Guidance System.* All components of the flight control guidance system must be approved as installed by the evaluation program specified in paragraph (e) of this section if they have not been approved for Category II operations under applicable type or supplemental type certification procedures. In addition, subsequent changes to make, model or design of these components must be approved under this paragraph. Related systems or devices such as the auto throttle and computed missed approach guidance system must be approved in the same manner if they are to be used for Category II operations.

(c) *Radio Altimeter.* A radio altimeter must meet the performance criteria of this paragraph for original approval and after each subsequent alteration.

(1) It must display to the flight crew clearly and positively the wheel height of the main landing gear above the terrain.

(2) It must display wheel height above the terrain to an accuracy of plus or minus five feet or five percent, whichever is greater, under the following conditions:

(i) Pitch angles of zero to plus or minus five degrees about the mean approach attitude.

(ii) Roll angles of zero to 20 degrees in either direction.

(iii) Forward velocities from minimum approach speed up to 200 knots.

(iv) Sink rates from zero to 15 feet per second at altitudes from 100 to 200 feet.

(3) Over level ground, it must track the actual altitude of the [aircraft] without significant lag or oscillation.

(4) With the [aircraft] at an altitude of 200 feet or less, any abrupt change in terrain representing no more than 10 percent of the [aircraft's] altitude must not cause the altimeter to unlock, and indicator response to such changes must not exceed 0.1 seconds, and in addition, if the system unlocks for greater changes, it must reacquire the signal in less than 1 second.

(5) Systems that contain a push-to-test feature must test the entire system (with or without an antenna) at a simulated altitude of less than 500 feet.

(6) The system must provide to the flight crew a positive failure warning display any time there is a loss of power or an absence of ground return signals within the designed range of operating altitudes.

(d) *Other Instruments and Equipment.* All other instruments and items of equipment required by § 2 of this Appendix must be capable of performing as necessary for Category II operations. Approval is also required after each subsequent alteration to these instruments and items of equipment.

(e) *Evaluation program.*

(1) *Application.* Approval by evaluation is requested as a part of the application for approval of the Category II manual.

(2) *Demonstrations.* Unless otherwise authorized by the Administrator, the evaluation program for each [aircraft] requires the demonstrations specified in this subparagraph. At least 50 ILS approaches must be flown with at least five approaches on each of three different ILS facilities and no more than one half of the total approaches on any one ILS facility. All approaches shall be flown under simulated instrument conditions to a 100 foot decision height and ninety per cent of the total approaches made must be successful. A successful approach is one in which—

(i) At the 100 foot decision height, the indicated airspeed and heading are satisfactory for a normal flare and landing (speed must be plus or minus five knots of programmed airspeed but may not be less than computed threshold speed, if auto throttles are used);

(ii) The [aircraft], at the 100 foot decision height, is positioned so that the cockpit is within, and tracking so as to remain within, the lateral confines of the runway extended;

(iii) Deviation from glide slope after leaving the outer marker does not exceed 50 percent of full scale deflection as displayed on the ILS indicator;

(iv) No unusual roughness or excessive attitude changes occur after leaving the middle marker; and

(v) In the case of an [aircraft] equipped with an approach coupler, the [aircraft] is sufficiently in trim when the approach coupler is disconnected at the decision height to allow for the continuation of a normal approach and landing.

(3) *Records.* During the evaluation program the following information must be maintained by the applicant for the airplane with respect to each approach and made available to the Administrator upon request:

(i) Each deficiency in airborne instruments and equipment that prevented the initiation of an approach.

(ii) The reasons for discontinuing an approach including the altitude above the runway at which it was discontinued.

(iii) Speed control at the 100 foot decision height if auto throttles are used.

(iv) Trim condition of the [aircraft] upon disconnecting the auto coupler with respect to continuation to flare and landing.

(v) Position of the [airplane] at the middle marker and at the decision height indicated both on a diagram of the basic ILS display and a diagram of the runway extended to the middle marker. Estimated touch down point must be indicated on the runway diagram.

(iv) Compatibility of flight director with the auto coupler, if applicable.

(vii) Quality of overall system performance.

(4) *Evaluation.* A final evaluation of the flight control guidance system is made upon successful completion of the demonstrations. If no hazardous tendencies have been displayed or are otherwise known to exist, the system is approved.

4. Maintenance program.

(a) Each maintenance program must contain the following:

(1) A list of each instrument and item of equipment specified in § 2 of this Appendix that is installed in the [aircraft] and approved for Category II operations, including the make and model of those specified in § 2(a).

(2) A schedule that provides for the performance of inspections under subparagraph (5) of this paragraph within 3 calendar months after the date of the previous inspection. The inspection must be performed by a person authorized by Part 43 of this chapter, except that each alternate inspection may be replaced by a functional flight check. This functional flight check must be performed by a pilot holding a Category II pilot authorization for the type [aircraft] checked.

(3) A schedule that provides for the performance of bench checks for each listed instrument and item of equipment that is specified in § 2(a) within 12 calendar months after the date of the previous bench check.

(4) A schedule that provides for the performance of a test and inspection of each static pressure system in accordance with Appendix E to Part 43 of this chapter within 12 calendar months after the date of the previous test and inspection.

(5) The procedures for the performance of the periodic inspections and functional flight checks to determine the ability of each listed instrument and item of equipment specified in § 2(a) of this Appendix to perform as approved for Category II operations including a procedure for recording functional flight checks.

(6) A procedure for assuring that the pilot is informed of all defects in listed instruments and items of equipment.

(7) A procedure for assuring that the condition of each listed instrument and item of equipment upon which maintenance is performed is at least equal to its Category II approval condition before it is returned to service for Category II operations.

(8) A procedure for an entry in the maintenance records required by § 43.9 of this chapter that shows the date, airport, and reasons for each discontinued Category II operation because of a malfunction of a listed instrument or item of equipment.

(b) *Bench Check.* A bench check required by this section must comply with this paragraph.

(1) It must be performed by a certificated repair station holding one of the following ratings as appropriate to the equipment checked:

(i) An instrument rating.

(ii) A radio rating.

(iii) A rating issued under Subpart D of Part 145.

(2) It must consist of removal of an instrument or item of equipment and performance of the following:

(i) A visual inspection for cleanliness, impending failure, and the need for lubrication, repair, or replacement of parts;

(ii) Correction of items found by that visual inspection; and

(iii) Calibration to at least the manufacturer's specifications unless otherwise specified in the approved Category II manual for the [aircraft] in which the instrument or item of equipment is installed.

(c) *Extensions.* After the completion of one maintenance cycle of 12 calendar months a request to extend the period for checks, tests, and inspections is approved if it is shown that the performance of particular equipment justifies the requested extension.

APPENDIX B
Authorizations to Exceed Mach 1 (§ 91.55)

SECTION 1. *Application.*

(a) An applicant for an authorization to exceed Mach 1 must apply in a form and manner prescribed by the Administrator and must comply with this Appendix.

(b) In addition, each application for an authorization to exceed Mach 1 covered by section 2(a) of this Appendix must contain all information, requested by the Administrator, that he deems necessary to assist him in determining whether the designation of a particular test area, or issuance of a particular authorization, is a "major Federal action significantly affecting the quality of the human environment" within the meaning of the National Environmental Policy Act of 1969 (42 U.S.C. 4321 *et seq.*), and to assist him in complying with that Act, and with related Executive orders, guidelines, and orders, prior to such action.

(c) In addition, each application for an authorization to exceed Mach 1 covered by section 2(a) of this Appendix must contain—

(1) Information showing that operation at a speed greater than Mach 1 is necessary to accomplish one or more of the purposes specified in section 2(a) of this Appendix, including a showing that the purpose of the test cannot be safely or properly accomplished by overocean testing;

(2) A description of the test area proposed by the applicant, including an en-

vironmental analysis of that area meeting the requirements of paragraph (b) of this section; and

(3) Conditions and limitations that will ensure that no measurable sonic boom overpressure will reach the surface outside of the designated test area.

(d) An application is denied if the Administrator finds that such action is necessary to protect or enhance the environment.

SEC. 2. *Issuance.*

(a) For a flight in a designated test area, an authorization to exceed Mach 1 may be issued when the Administrator has taken the environmental protective actions specified in section 1(b) of this Appendix, and the applicant shows one or more of the following:

(1) The flight is necessary to show compliance with airworthiness requirements.

(2) The flight is necessary to determine the sonic boom characteristics of the airplane, or is necessary to establish means of reducing or eliminating the effects of sonic boom.

(3) The flight is necessary to demonstrate the conditions and limitations under which speeds greater than a true flight Mach number of 1 will not cause a measurable sonic boom overpressure to reach the surface.

(b) For a flight outside of a designated test area, an authorization to exceed Mach 1 may be issued if the applicant shows conservatively under paragraph (a)(3) of this section that—

(1) The flight will not cause a measurable sonic boom overpressure to reach the surface when the aircraft is operated under conditions and limitations demonstrated under paragraph (a)(3) of this section; and

(2) Those conditions and limitations represent all foreseeable operating conditions.

SEC. 3. *Duration.*

(a) An authorization to exceed Mach 1 is effective until it expires or is surrendered, or until it is suspended or terminated by the Administrator. Such an authorization may be amended or suspended by the Administrator at any time if he finds that such action is necessary to protect the environment. Within 30 days of notification of amendment, the holder of the authorization must request reconsideration or the amendment becomes final. Within 30 days of notification of suspension, the holder of the authorization must request reconsideration or the authorization is automatically terminated. If reconsideration is requested within the 30-day period, the amendment or suspension continues until the holder shows why, in his opinion, the authorization should not be amended or terminated. Upon such showing, the Administrator may terminate or amend the authorization if he finds that such action is necessary to protect the environment, or he may reinstate the authorization without amendment if he finds that termination or amendment is not necessary to protect the environment.

(b) Findings and actions by the Administrator under this section do not affect any certificate issued under Title VI of the Federal Aviation Act of 1958.

APPENDIX C

Operations in the North Atlantic (NAT) Minimum Navigation Performance Specifications (MNPS) Airspace

[Section 1. NAT MNPS airspace is that volume of airspace between flight level 275 and flight level 400 extending between latitude 27 degrees north and the North Pole, bounded in the east by the eastern boundaries of control areas Santa Maria Oceanic, Shanwick Oceanic and Reykjavik, Oceanic and in the west by the western boundary of Reykjavik Oceanic Control Area, the western boundary of Gander Oceanic Control Area, and the western boundary of New York Oceanic Control Area, excluding the areas west of 60 degrees west and south of 38 degrees 30 minutes north.]

Section 2. The navigation performance capability required for aircraft to be operated in the airspace defined in § 1 of this Appendix is as follows:

(a) The standard deviation of lateral track errors shall be less than 6.3 NM (11.7 Km). Standard deviation is a statistical measure of data about a mean value. The mean is zero nautical miles. The overall form of data is such that the plus and minus one standard deviation about the mean encompasses approximately 68 percent of the data and plus or minus two deviations encompasses approximately 95 percent.

(b) The proportion of the total flight time spent by aircraft 30 NM (55.6 Km) or more off the cleared track shall be less than 5.3×10^{-4} (less than one hour in 1,887 flight hours.)

(c) The proportion of the total flight time spent by aircraft between 50 NM and 70 NM (92.6 Km and 129.6 Km) off the cleared track shall be less than 13×10^{-5} (less than one hour in 7,693 flight hours).

Section 3. Air traffic control (ATC) may authorize an aircraft operator to deviate from the requirements of § 91.20 for a specific flight if, at the time of flight plan filing for that flight, ATC determines that the aircraft may be provided appropriate separation and that the flight will not interfere with, or impose a burden upon, the operations of other aircraft which meet the requirements of § 91.20.

APPENDIX D

Airports/Locations Where the Transponder Requirements of § 91.24(b)(5)(ii) Apply

[Section 1. The requirements of § 91.24(b)(5)(ii) apply to operations in the vicinity of each of the following airports:

Logan International Airport, Billings, MT.
Hector International Airport, Fargo, ND.]

APPENDIX E—AIRPLANE FLIGHT RECORDER SPECIFICATIONS

Parameters	Range	Installed system[1] minimum accuracy (to recovered data)	Sampling interval (per second)	Resolution[4] read out
Relative Time (From Recorded on Prior to Takeoff).	8 hr minimum	±0.125% per hour	1	1 sec.
Indicated Airspeed.	V_{so} to V_D (KIAS).	±5% or ±10 kts., whichever is greater. Resolution 2 kts. below 175 KIAS.	1	1%[3]
Altitude.	−1,000 ft. to max cert. alt. of A/C	±100 to ±700 ft. (see Table 1, TSO C51–a).	1	25 to 150 ft.
Magnetic Heading.	360°	±5°	1	1°.
Vertical Acceleration.	−3g to +6g.	±0.2g in addition to ±0.3g maximum datum.	4 (or 1 per second where peaks, ref. to 1g are recorded).	0.03g.
Longitudinal Acceleration.	±1.0g	±1.5% max. range excluding datum error of ±5%.	2	0.01g.
Pitch Altitude.	100% of usable	±2°	1	0.8°.
Roll Altitude.	±60° or 100% of useable range, whichever is greater.	±2°	1	0.8°.
Stabilizer Trim Position, or.	Full Range.	±3% unless higher uniquely required	1	1%[3]
Pitch Control Position.	Full Range.	±3% unless higher uniquely required	1	1%[3]
Engine Power, Each Engine: Fan or N_1 Speed or EPR or Cockpit indications Used for Aircraft Certification OR.	Maximum Range.	±5%	1	1%[3]
Prop. speed and Torque (Sample Once/Sec as Close together as Practicable).			1 (prop Speed). 1 (torque).	1%[3] 1%[3]
Altitude Rate[2] (need depends on altitude resolution).	±8,000 fpm	±10%. Resolution 250 fpm below 12,000 ft. indicated.	1	250 fpm. below 12,000. 0.8%[3]
Angle of Attack[2] (need depends on altitude resolution).	−20° to 40° or of usable range	±2°	1	
Radio Transmitter Keying (Discrete).	On/Off.		1	
TE Flaps (Discrete or Analog).	Each discrete position (U, D, T/O, AAP) OR Analog 0-100% range.		1	
LE Flaps (Discrete or Analog).	Each discrete position (U, D, T/O, AAP) OR Analog 0-100% range.	±3°	1	1%[3]
Thrust Reverser, Each Engine (Discrete).	Stowed or full reverse.		1	
Spoiler/Speedbrake (Discrete).	Stowed or out.	±3°	1	1%[3]
Autopilot Engaged (Discrete).	Engaged or Disengaged.		1	

[1] When data sources are aircraft instruments (except altimeters) of acceptable quality to fly the aircraft the recording system excluding these sensors (but including all other characteristics of the recording system) shall contribute no more than half of the values in this column.
[2] If data from the altitude encoding altimeter (100 ft. resolution) is used, then either one of these parameters should also be recorded. If however, altitude is recorded at a minimum resolution of 25 feet, then these two parameters can be omitted.
[3] Per cent of full range.
[4] This column applies to aircraft manufactured after October 11, 1991.

APPENDIX F—HELICOPTER FLIGHT RECORDER SPECIFICATION

Parameters	Range	Installed system [1] minimum accuracy (to recovered data)	Sampling interval (per second)	Resolution [3] read out
Relative Time (From Recorded on Prior to Takeoff).	4 hr minimum	±0.125% per hour	1	1 sec.
Indicated Airspeed	V_m in to V_D (KIAS) (minimum airspeed signal attainable with installed pilot-static system).	±5% or ±10 kts., whichever is greater.	1	1 kt.
Altitude	−1,000 ft. to 20,000 ft. pressure altitude	±100 to ±700 ft. (see Table 1, TSO C51–a).	1	25 to 150 ft.
Magnetic Heading	360°	±5°	4 (or 1 per second where peaks, ref. to 1g are recorded).	1°.
Vertical Acceleration.	−3g to +6g.	±0.2g in addition to ±0.3g maximum datum.		0.05g.
Longitudinal Acceleration:	±1.0g.	±1.5% max. range excluding datum error of ±5%.	2.	0.03g.
Pitch Attitude	100% of usable range.	±2°	1	0.8°.
Roll Attitude.	±60° or 100% of usable range, whichever is greater.	±2°	1	0.8°.
Altitude Rate.	±8,000 fpm.	±10% Resolution 250 fpm below 12,000 ft. indicated.	1	250 fpm below 12,000.
Engine Power, Each Engine				
Main Rotor Speed.	Maximum Range.	±5%	1	1% [2].
Free or Power Turbine.	Maximum Range.	±5%	1	1% [2].
Engine Torque.	Maximum Range.	±5%	1	1% [2].
Flight Control Hydraulic Pressure				
Primary (Discrete)	High/Low		1	
Secondary-if applicable (Discrete).	High/Low		1	
Radio Transmitter Keying (Discrete).	On/Off.		1	
Autopilot Engaged (Discrete)	Engaged or Disengaged		1	
SAS Status-Engaged (Discrete).	Engaged or Disengaged		1	
SAS Fault Status (Discrete).	Fault/OK.		1	
Flight Controls				
Collective	Full range	±3%	2	1% [2].
Pedal Position.	Full range	±3%	2	1% [2].
Lat. Cyclic.	Full range	±3%	2	1% [2].
Long. Cyclic.	Full range	±3%	2	1% [2].
Controllable Stabilator Position.	Full range	±3%	2	1% [2].

[1] When data sources are aircraft instruments (except altimeters) of acceptable quality to fly the aircraft the recording system excluding these sensors (but including all other characteristics of the recording system) shall contribute no more than half of the values in this column.
[2] Per cent of full range.
[3] This column applies to aircraft manufactured after October 11, 1991.

FEDERAL AVIATION REGULATIONS

Part 97—Standard Instrument Approach Procedures

Contents

Subpart A—General

§ 97.1 Applicability.

This Part prescribes standard instrument approach procedures for instrument letdown to airports in the United States and the weather minimums that apply to takeoffs and landings under IFR at those airports.

§ 97.3 Symbols and terms used in procedures.

As used in the standard terminal instrument procedures prescribed in this part—

(a) "A" means alternate airport weather minimum.

[(b) "Aircraft approach category" means a grouping of aircraft based on a speed of 1.3 V_{so} (at maximum certificated landing weight). V_{so} and the maximum certificated landing weight are those values as established for the aircraft by the certificating authority of the country of registry. The categories are as follows:

[(1) Category A: Speed less than 91 knots.

[(2) Category B: Speed 91 knots or more but less than 121 knots.

[(3) Category C: Speed 121 knots or more but less than 141 knots.

[(4) Category D: Speed 141 knots or more but less than 166 knots.

[(5) Category E: Speed 166 knots or more.]

(c) Approach procedure segments for which altitudes (all altitudes prescribed are minimum altitudes unless otherwise specified) or courses, or both, are prescribed in procedures, are as follows:

(1) "Initial approach" is the segment between the initial approach fix and the intermediate fix or the point where the aircraft is established on the intermediate course or final approach course.

(2) "Initial approach altitude" means the altitude (or altitudes, in High Altitude Procedures) prescribed for the initial approach segment of an instrument approach.

(3) "Intermediate approach" is the segment between the intermediate fix or point and the final approach fix.

(4) "Final approach" is the segment between the final approach fix or point and the runway, airport, or missed-approach point.

(5) "Missed approach" is the segment between the missed-approach point, or point of arrival at decision height, and the missed-approach fix at the prescribed altitude.

(d) "C" means circling landing minimum, a statement of ceiling and visibility values, or minimum descent altitude and visibility, required for the circle-to-land maneuver.

(d–1) "Copter procedures" means helicopter procedures, with applicable minimums as prescribed in § 97.35 of this Part. Helicopters may also use other procedures prescribed in Subpart C of this Part and may use the Category A minimum descent altitude (MDA) or decision height (DH). The required visibility minimum may be reduced to one-half the published visibility minimum for Category A aircraft, but in no case may it be reduced to less than one-quarter mile or 1,200 feet RVR.

(e) "Ceiling minimum" means the minimum ceiling, expressed in feet above the surface of the airport, required for takeoff or required for designating an airport as an alternate airport.

(f) "d" means day.

(g) "FAF" means final approach fix.

(h) "HAA" means height above airport.

(h–1) "HAL" means height above a designated helicopter landing area used for helicopter instrument approach procedures.

(i) "HAT" means height above touchdown.

(j) "MAP" means missed approach point.

(k) "More than 65 knots" means an aircraft that has a stalling speed of more than 65 knots (as established in an approved flight manual) at maximum certificated landing weight with full flaps, landing gear extended, and power off.

(l) "MSA" means minimum safe altitude, an emergency altitude expressed in feet above mean sea level, which provides 1,000 feet clearance over all obstructions in that sector within 25 miles of the facility on which the procedure is based (LOM in ILS procedures).

(m) "n" means night.

(n) "NA" means not authorized.

(o) "NOPT" means no procedure turn required (altitude prescribed applies only if procedure turn is not executed).

(o–1) "Point in space approach" means a helicopter instrument approach procedure to a missed approach point that is more than 2,600 feet from an associated helicopter landing area.

(p) "Procedure turn" means the maneuver prescribed when it is necessary to reverse direction to establish the aircraft on an intermediate or final approach course. The outbound course, direction of turn, distance within which the turn must be completed, and minimum altitude are specified in the procedure. However, the point at which the turn may be commenced, and the type and rate of turn, is left to the discretion of the pilot.

(q) "RA" means radio altimeter setting height.

(r) "RVV" means runway visibility value.

(s) "S" means straight-in landing minimum, a statement of ceiling and visibility, minimum descent altitude and visibility, or decision height and visibility, required for a straight-in landing on a specified runway.

The number appearing with the "S" indicates the runway to which the minimum applies. If a straight-in minimum is not prescribed in the procedure, the circling minimum specified applies to a straight-in landing.

(t) "Shuttle" means a shuttle, or race-track-type, pattern with 2-minute legs prescribed in lieu of a procedure turn.

(u) "65 knots or less" means an aircraft that has a stalling speed of 65 knots or less (as established in an approved flight manual) at maximum certificated landing weight with full flaps, landing gear extended, and power off.

(v) "T" means takeoff minimum.

(w) "TDZ" means touchdown zone.

(x) "Visibility minimum" means the minimum visibility specified for approach, or landing, or takeoff, expressed in statute miles, or in feet where RVR is reported.

§ 97.5 Bearings; courses; headings; radials; miles.

(a) All bearings, courses, headings, and radials in this Part are magnetic.

(b) RVR values are stated in feet. Other visibility values are stated in statute miles. All other mileages are stated in nautical miles.

Subpart B—Procedures

§ 97.10 General.

This subpart prescribes standard instrument approach procedures other than those based on the criteria contained in the U.S. Standard for Terminal Instrument Approach Procedures (TERPS). Standard instrument approach procedures adopted by the FAA and described on FAA Form 3139 are incorporated into this Part and made a part hereof as provided in 5 U.S.C. 522(a)(1) and pursuant to 1 CFR Part 20. The incorporated standard instrument approach procedures are available for examination at the Rules Docket and at the National Flight Data Center, Federal Aviation Administration, 800 Independence Avenue, S.W., Washington, D.C. 20590. Copies of SIAPs adopted in a particular FAA Region

are also available for examination at the Headquarters of that Region. Moreover, copies of SIAPs originating in a particular Flight Inspection District Office are available for examination at that Office. Based on the information contained on FAA Form 3139, standard instrument approach procedures are portrayed on charts prepared for the use of pilots by the United States Coast and Geodetic Survey and other publishers of aeronautical charts.

§ 97.11 Low or medium frequency range, automatic direction finding, and very high frequency omnirange procedures.

Section 609.100 of the Regulations of the Administrator is hereby designated as § 97.11.

§ 97.13 Terminal very high frequency omnirange procedures.

Section 609.200 of the Regulations of the Administrator is hereby designated as § 97.13.

§ 97.15 Very high frequency omnirange-distance measuring equipment procedures.

Section 609.300 of the Regulations of the Administrator is hereby designated as § 97.15.

§ 97.17 Instrument landing system procedures.

Section 609.400 of the Regulations of the Administrator is hereby designated as § 97.17.

§ 97.19 Radar procedures.

Section 609.500 of the Regulations of the Administrator is hereby designated as § 97.19.

Subpart C—TERPS Procedures

§ 97.20 General.

This subpart prescribes standard instrument approach procedures based on the criteria contained in the U.S. Standard for Terminal Instrument Approach Procedures (TERPS). The standard instrument approach procedures adopted by the FAA and described on FAA Forms 8260-3, 8260-4, or 8260-5 are incorporated into this Part and made a part hereof as provided in 5 U.S.C. 552(a)(1) and pursu-

ant to 1 CFR Part 20. The incorporated standard instrument approach procedures are available for examination at the Rules Docket and at the National Flight Data Center, Federal Aviation Administration, 800 Independence Avenue, S.W., Washington, D.C. 20590. Copies of SIAPs adopted in a particular FAA Region are also available for examination at the headquarters of that Region. Moreover, copies of SIAPs originating in a particular Flight Inspection District Office are available for examination at that Office. Based on the information contained on FAA Forms 8260-3, 8260-4, and 8260-5, standard instrument approach procedures are portrayed on charts prepared for the use of pilots by the U.S. Coast and Geodetic Survey and other publishers of aeronautical charts.

§ 97.21 Low or medium frequency range (L/MF) procedures.

§ 97.23 Very high frequency omnirange (VOR) and very high frequency distance measuring equipment (VOR/DME) procedures.

§ 97.25 Localizer (LOC) and localizer-type directional aid (LDA) procedures.

§ 97.27 Nondirectional beacon (automatic direction finder) (NDB(ADF)) procedures.

§ 97.29 Instrument landing system (ILS) procedures.

§ 97.31 Precision approach radar (PAR) and airport surveillance radar (ASR) procedures.

§ 97.33 Area navigation (RNAV) procedures.

§ 97.35 Helicopter procedures.

NOTE: The procedures set forth in § 97.35 are not carried in the Code of Federal Regulations. For Federal Register citations affecting these procedures see List of CFR Sections Affected.

FEDERAL AVIATION REGULATIONS

Part 99—Security Control of Air Traffic

Subpart A—General

§ 99.1 Applicability.

(a) This subpart prescribes rules for operating civil aircraft in a defense area, or into, within, or out of the United States through an Air Defense Identification Zone (ADIZ), designated in Subpart B.

(b) Except for §§ 99.7 and 99.12, this subpart does not apply to the operation of any aircraft—

(1) Within the 48 contiguous States and the District of Columbia, or within the State of Alaska, on a flight which remains within 10 nautical miles of the point of departure;

(2) Operating at true airspeed of less than 180 knots in the Hawaii ADIZ or over any island, or within 3 nautical miles of the coastline of any island, in the Hawaii ADIZ;

(3) Operating at true airspeed of less than 180 knots in the Alaska ADIZ while the pilot maintains a continuous listening watch on the appropriate frequency; or

(4) Operating at true airspeed of less than 180 knots in the Guam ADIZ.

(c) Except as provided in § 99.7, the radio and position reporting requirements of this subpart do not apply to the operation of an aircraft within the 48 contiguous States and the District of Columbia, or within the State of Alaska, if that aircraft does not have two-way radio and is operated in accordance with a filed DVFR flight plan containing the time and point of ADIZ penetration and that aircraft departs within 5 minutes of the estimated departure time contained in the flight plan.

(d) An FAA ATC center may exempt the following operations from this subpart (except Section 99.7), on a local basis only, with the concurrence of the military commanders concerned:

(1) Aircraft operations that are conducted wholly within the boundaries of an ADIZ and are not currently significant to the air defense system.

(2) Aircraft operations conducted in accordance with special procedures prescribed by the military authorities concerned.

§ 99.3 General.

(a) The Air Defense Identification Zone (ADIZ) is an area of airspace over land or water in which the ready identification, location, and control of civil aircraft is required in the interest of national security.

(b) Unless designated as an ADIZ, a Defense Area is any airspace of the United States in which the control of aircraft is required for reasons of national security.

(c) For the purpose of this Part, a Defense Visual Flight Rules (DVFR) flight is a flight within an ADIZ conducted under the visual flight rules in Part 91.

§ 99.5 Emergency situations.

In an emergency that requires immediate decision and action for the safety of the flight, the pilot in command of an aircraft may deviate from the rules in this Part to the extent required by that emergency. He shall report the reasons for the deviation to the communications facility where flight plans or position reports are normally filed (referred to in this Part as "an appropriate aeronautical facility") as soon as possible.

§ 99.7 Special security instructions.

Each person operating an aircraft in an ADIZ or Defense Area shall, in addition to the applicable rules of this Part, comply with special security instructions issued by the Administrator in the interest of national security and that are consistent with appropriate agreements between the FAA and the Department of Defense.

§ 99.9 Radio requirements.

Except as provided in § 99.1(c), no person may operate an aircraft in an ADIZ unless the aircraft has a functioning two-way radio.

§ 99.11 ADIZ flight plan requirements.

(a) Unless otherwise authorized by ATC, no person may operate an aircraft into, within, or across an ADIZ unless that person has filed a flight plan with an appropriate aeronautical facility.

(b) Unless ATC authorizes an abbreviated flight plan—

(1) A flight plan for IFR flight must contain the information specified in § 91.83; and

(2) A flight plan for VFR flight must contain the information specified in § 91.83(a) (1) through (7).

(3) If airport of departure is within the Alaskan ADIZ and there is no facility for filing a flight plan then:

(i) Immediately after takeoff or when within range of an appropriate aeronautical facility, comply with provisions of paragraph (b)(1) or (b)(2) as appropriate.

(ii) Proceed according to the instructions issued by the appropriate aeronautical facility.

(c) The pilot shall designate a flight plan for VFR flight as a DVFR flight plan.

§ 99.12 Transponder-on requirements.

Unless otherwise authorized by ATC, each person who operates a civil aircraft into or out of the United States into, within, or across an ADIZ designated in Subpart B, if that aircraft is equipped with an operable radar beacon transponder, shall operate that transponder, including the altitude encoder, if installed, and reply on the appropriate code or a code assigned by ATC.

§ 99.15   Arrival or completion notice.

The pilot in command of an aircraft for which a flight plan has been filed shall file an arrival or completion notice with an appropriate aeronautical facility, unless the flight plan states that no notice will be filed.

§ 99.17   Position reports; aircraft operating in or penetrating an ADIZ; IFR.

The pilot of an aircraft operating in or penetrating an ADIZ under IFR—

(a) In controlled airspace, shall make the position reports required in § 91.125; and

(b) In uncontrolled airspace, shall make the position reports required in § 99.19.

§ 99.19   Position reports; aircraft operating in or penetrating an ADIZ; DVFR.

No pilot may operate an aircraft penetrating an ADIZ under DVFR unless—

(a) That pilot reports to an appropriate aeronautical facility before penetration: The time, position, and altitude at which the aircraft passed the last reporting point before penetration and the estimated time of arrival over the next appropriate reporting point along the flight route;

(b) If there is no appropriate reporting point along the flight route, that pilot reports at least 15 minutes before penetration: The estimated time, position, and altitude at which he will penetrate; or

(c) If the airport departure is within an ADIZ or so close to the ADIZ boundary that it prevents his complying with paragraphs (a) or (b) of this section, that pilot has reported immediately after taking off: the time of departure, altitude, and estimated time of arrival over the first reporting point along the flight route.

§ 99.21   Position reports; aircraft entering the United States through an ADIZ; United States aircraft.

The pilot of an aircraft entering the United States through an ADIZ shall make the reports required in §§ 99.17 or 99.19 to an appropriate aeronautical facility.

§ 99.23   Position reports; aircraft entering the United States through an ADIZ; foreign aircraft.

In addition to such other reports as ATC may require, no pilot in command of a foreign civil aircraft may enter the U.S. through an ADIZ unless that pilot makes the reports required in §§ 99.17 or 99.19 or reports the position of the aircraft when it is not less than one hour and not more than 2 hours average cruising distance from the United States.

§ 99.27   Deviation from flight plans and ATC clearances and instructions.

(a) No pilot may deviate from the provisions

of an ATC clearance or ATC instruction except in accordance with § 91.75 of this chapter.

(b) No pilot may deviate from the filed IFR flight plan when operating an aircraft in uncontrolled airspace unless that pilot notifies an appropriate aeronautical facility before deviating.

(c) No pilot may deviate from the filed DVFR flight plan unless that pilot notifies an appropriate aeronautical facility before deviating.

§ 99.29   Radio failure; DVFR.

If the pilot operating an aircraft under DVFR in an ADIZ cannot maintain two-way radio communications, the pilot may proceed in accordance with original DVFR flight plan or land as soon as practicable. The pilot shall report the radio failure to an appropriate aeronautical facility as soon as possible.

§ 99.31   Radio failure; IFR.

If a pilot operating an aircraft under IFR in an ADIZ cannot maintain two-way radio communications, the pilot shall proceed in accordance with § 91.127 of this chapter.

Subpart B—Designated Air Defense Identification Zones

§ 99.41   General.

The airspace above the areas described in this subpart is established as an ADIZ Defense Area. The lines between points described in this subpart are great circles except that the lines joining adjacent points on the same parallel of latitude are rhumb lines.

§ 99.42   Conterminous U.S. ADIZ.

The area bounded by a line 26°00′N, 96°35′W;   26°00′N,   95°00′W;   26°30′N, 95°00′W;   then along 26°30′N to;   26°30′N, 84°00′W;   24°00′N,   83°00′W;   24°00′N, 80°00′W;   24°00′N,   79°25′W;   25°40′N, 79°25′W;   27°30′N,   78°50′N;   30°45′N, 74°00′W;   39°30′N,   63°45′W;   43°00′N, 65°48′W;   41°15′N,   69°30′W;   40°32′N, 72°15′W;   39°55′N,   73°00′W;   39°38′N, 73°00′W; 39°36′30″N, 73°40′30″W; 39°30′N, 73°45′W;   37°00′N,   75°30′W;   36°10′N, 75°10′W;   35°10′N,   75°10′W;   32°01′N, 80°32′W;   30°50′N,   80°54′W;   30°05′N, 81°07′W;   27°59′N,   79°23′W;   24°49′N, 80°00′W;   24°49′N,   80°55′W;   25°10′N, 81°12′W; then along a line 3 nautical miles from a shoreline to; 25°45′N, 81°27′W; 25°45′N, 82°07′W;   28°55′N,   83°30′W;   29°20′N, 85°00′W;   30°00′N,   86°00′W;   30°00′N, 88°30′W;   29°00′N,   89°00′W;   28°45′N, 90°00′W;   29°26′N,   94°00′W;   28°42′N, 95°17′W;   28°05′N,   96°30′W;   26°25′N, 96°30′W;   26°00′N,   96°35′W;   25°58′N, 97°07′W; then westward along the Mexican Border to 32°32′03″N, 117°07′25″W; 32°30′N, 117°20′W;   32°00′N,   118°24′W;   30°45′N,

120°50′W;   29°00′N,   124°00′W;   37°42′N, 130°40′W;   48°20′N,   132°00′W;   48°20′N, 128°00′W; 48°30′N, 125°00′W; 48°29′38″N, 124°43′35″W; 48°00′N, 125°15′W; 46°15′N, 124°30′W;   43°00′N,   124°40′W;   40°00′N, 124°35′W;   38°50′N,   124°00′W;   34°50′N, 121°10′W;   34°00′N,   120°30′W;   32°00′N, 118°24′W; 32°30′N, 117°20′W; 32°32′03″N, 117°07′25″W.

§ 99.43   Alaska ADIZ.

The area bounded by a line 54°00′N, 136°00′W;   56°57′N,   144°00′W;   57°00′N, 145°00′W;   53°00′N,   158°00′W;   50°00′N, 169°00′W;   50°00′N,   180°00′;   50°00′N, 170°00′E;   53°00′N,   170°00′E;   60°00′N, 180°00′;   65°00′N,   169°00′W;   then along 169°00′W to; 75°00′N, 169°00′W; then along the 75°00′N parallel to; 75°00′N; 141°00′W to; 69°50′N,   141°00′W;   71°18′N,   156°44′W; 69°52′N, 163°00′W; then south along 163°00′W to;   54°00′N, 163°00′W; 56°30′N, 154°00′W; 59°20′N,   146°00′W;   59°30′N,   140°00′W; 57°00′N,   136°00′W;   54°35′N,   133°00′W;   to point of beginning.

§ 99.45   Guam ADIZ.

(a) *Inner boundary.* From a point 13°52′07″N, 143°59′16″E, counterclockwise along the 50-nautical-mile radius arc of the NIMITZ VORTAC (located at 13°27′11″N, 144°43′51″E);   to   a   point   13°02′08″N, 145°28′17″E); then to a point 14°49′07″N, 146°13′58″E; counterclockwise along the 35-nautical-mile radius arc of the SAIPAN NDB (located at 15°06′46″N, 145°42′42″E); to a point 15°24′21″N, 145°11′21″E; then to the point of origin.

(b) *Outer boundary.* The area bounded by a circle with a radius of 250 NM centered at latitude 13°32′41″N, longitude 144°50′30″E.

§ 99.47   Hawaii ADIZ.

(a) *Outer boundary.* The area included in the irregular octagonal figure formed by a line connecting 26°30′N, 156°00′W;   26°30′N, 161°00′W;   24°00′N,   164°00′W;   20°00′N, 164°00′W;   17°00′N,   160°00′W;   17°00′N, 156°00′W;   20°00′N,   153°00′W;   22°00′N, 153°00′W; to point of beginning.

(b) *Inner boundary.* The inner boundary to follow a line connecting 22°30′N, 157°00′W; 22°30′N,   160°00′W;   22°00′N,   161°00′W; 21°00′N,   161°00′W;   20°00′N,   160°00′W; 20°00′N, 156°30′W; 21°00′N, 155°30′W; to point of beginning.

§ 99.49   Defense Area.

All airspace of the United States is designated as Defense Area except that airspace already designated as Air Defense Identification Zone.

FEDERAL AVIATION REGULATIONS

Part 103—Ultralight Vehicles

Subpart A—General

§ 103.1   Applicability.

This Part prescribes rules governing the operation of ultralight vehicles in the United States. For the purposes of this Part, an ultralight vehicle is a vehicle that:

(a) Is used or intended to be used for manned operation in the air by a single occupant;

(b) Is used or intended to be used for recreation or sport purposes only;

(c) Does not have any U.S. or foreign airworthiness certificate; and,

(d) If unpowered, weighs less than 155 pounds; or

(e) If powered:

(1) Weighs less than 254 pounds empty weight, excluding floats and safety devices which are intended for deployment in a potentially catastrophic situation;

(2) Has a fuel capacity not exceeding 5 U.S. gallons;

(3) Is not capable of more than 55 knots calibrated airspeed at full power in level flight; and

(4) Has a power-off stall speed which does not exceed 24 knots calibrated airspeed.

§ 103.3   Inspection requirements.

(a) Any person operating an ultralight vehicle under this Part shall, upon request, allow the Administrator, or his designee, to inspect the vehicle to determine the applicability of this Part.

(b) The pilot or operator of an ultralight vehicle must, upon request of the Administrator, furnish satisfactory evidence that the vehicle is subject only to the provisions of this Part.

§ 103.5   Waivers.

No person may conduct operations that require a deviation from this Part except under a written waiver issued by the Administrator.

§ 103.7   Certification and registration.

(a) Notwithstanding any other section pertaining to certification of aircraft or their parts or equipment, ultralight vehicles and their component parts and equipment are not required to meet the airworthiness certification standards specified for aircraft or to have certificates of airworthiness.

(b) Notwithstanding any other section pertaining to airman certification, operators of ultralight vehicles are not required to meet any aeronautical knowledge, age, or experience requirements to operate those vehicles or to have airman or medical certificates.

(c) Notwithstanding any other section pertaining to registration and marking of aircraft, ultralight vehicles are not required to be registered or to bear markings of any type.

Subpart B—Operating Rules

§ 103.9   Hazardous operations.

(a) No person may operate any ultralight vehicle in a manner that creates a hazard to other persons or property.

(b) No person may allow an object to be dropped from an ultralight vehicle if such action creates a hazard to other persons or property.

§ 103.11   Daylight operations.

(a) No person may operate an ultralight vehicle except between the hours of sunrise and sunset.

(b) Notwithstanding paragraph (a) of this section, ultralight vehicles may be operated during the twilight periods 30 minutes before official sunrise and 30 minutes after official sunset or, in Alaska, during the period of civil twilight as defined in the Air Almanac, if:

(1) The vehicle is equipped with an operating anticollision light visible for at least 3 statute miles; and

(2) All operations are conducted in uncontrolled airspace.

§ 103.13   Operation near aircraft; Right-of-way rules.

(a) Each person operating an ultralight vehicle shall maintain vigilance so as to see and avoid aircraft and shall yield the right-of-way to all aircraft.

(b) No person may operate an ultralight vehicle in a manner that creates a collision hazard with respect to any aircraft.

(c) Powered ultralights shall yield the right-of-way to unpowered ultralights.

§ 103.15   Operations over congested areas.

No person may operate an ultralight vehicle over any congested area of a city, town, or settlement, or over any open air assembly of persons.

§ 103.17   Operations in certain airspace.

[No person may operate an ultralight vehicle within an airport traffic area, control zone, terminal control area, or positive control area unless that person has prior authorization from the air traffic control facility having jurisdiction over that airspace.]

§ 103.19   Operations in prohibited or restricted areas.

No person may operate an ultralight vehicle in prohibited or restricted areas unless that person has permission from the using or controlling agency, as appropriate.

§ 103.20   Flight restrictions in the proximity of certain areas designated by Notice to Airmen.

No person may operate an ultralight vehicle in areas designated in a Notice to Airmen under § 91.102 or § 91.104 of this chapter, unless authorized by ATC.

§ 103.21   Visual reference with the surface.

No person may operate an ultralight vehicle except by visual reference with the surface.

§ 103.23   Flight visibility and cloud clearance requirements.

No person may operate an ultralight vehicle when the flight visibility or distance from clouds is less than that in the following table, as appropriate:

Flight Altitudes	Minimum Flight Visibility	Minimum Distance from Clouds
1,200 feet or less above the surface regardless of MSL altitude:		
(1) Within controlled airspace—	3 statute miles	500 feet below 1,000 feet above 2,000 feet horizontal.
(2) Outside controlled airspace:	1 statute mile	Clear of clouds.
More than 1,200 feet above the surface but less than 10,000 feet MSL:		
(1) Within controlled airspace—	3 statute miles	500 feet below 1,000 feet above 2,000 feet horizontal.
(2) Outside controlled airspace—	1 statute mile	500 feet below 1,000 feet above 2,000 feet horizontal.
More than 1,200 feet above the surface and at or above 10,000 feet MSL:	5 statute miles	1,000 feet below 1,000 feet above 1 statute mile horizontal.

FEDERAL AVIATION REGULATIONS

Part 105—Parachute Jumping

Contents

Subpart A—General

§ 105.1 Applicability.

(a) This Part prescribes rules governing parachute jumps made in the United States except parachute jumps necessary because of an in-flight emergency.

(b) For the purposes of this Part, a "parachute jump" means the descent of a person, to the surface from an aircraft in flight, when he intends to use, or uses, a parachute during all or part of that descent.

Subpart B—Operating Rules

§ 105.11 Applicability.

(a) Except as provided in paragraphs (b) and (c) of this section, this subpart prescribes operating rules governing parachute jumps to which this Part applies.

(b) This subpart does not apply to a parachute jump necessary to meet an emergency on the surface, when it is made at the direction, or with the approval, of an agency of the United States, or of a State, Puerto Rico, the District of Columbia, or a possession of the United States, or of a political subdivision of any of them.

(c) Sections 105.13 through 105.17 and §§105.27 through 105.37 of this subpart do not apply to a parachute jump made by a member of an Armed Force—

(1) Over or within a restricted area when that area is under the control of an Armed Force; or

(2) In military operations in uncontrolled airspace.

(d) Section 105.23 does not apply to a parachute jump made by a member of an Armed Force within a restricted area that extends upward from the surface when that area is under the control of an Armed Force.

§ 105.13 General.

No person may make a parachute jump, and no pilot in command of an aircraft may allow a parachute jump to be made from that aircraft, if that jump creates a hazard to air traffic or to persons or property on the surface.

§ 105.14 Radio equipment and use requirements.

(a) Except when otherwise authorized by ATC—

(1) No person may make a parachute jump, and no pilot in command of an aircraft may allow a parachute jump to be made from that aircraft, in or into controlled airspace unless, during that flight—

(i) The aircraft is equipped with a functioning two-way radio communications system appropriate to the ATC facilities to be used;

(ii) Radio communications have been established between the aircraft and the nearest FAA Air Traffic Control Facility or FAA Flight Service Station at least 5 minutes before the jumping activity is to begin, for the purpose of receiving information in the aircraft about known air traffic in the vicinity of the jumping activity; and

(iii) The information described in subdivision (ii) has been received by the pilot in command and the jumpers in that flight; and

(2) The pilot in command of an aircraft used for any jumping activity in or into controlled airspace shall, during each flight—

(i) Maintain or have maintained a continuous watch on the appropriate frequency of the aircraft's radio communications system from the time radio communications are first established between the aircraft and ATC, until he advises ATC that the jumping activity is ended for that flight; and

(ii) Advise ATC that the jumping activity is ended for that flight when the last parachute jumper from the aircraft reaches the ground.

(b) If, during any flight, the required radio communications system is or becomes inoperative, any jumping activity from the aircraft in or into controlled airspace shall be abandoned. However, if the communications system becomes inoperative in flight after receipt of a required ATC authorization, the jumping activity from that flight may be continued.

§ 105.15 Jumps over or into congested areas or open air assembly of persons.

(a) No person may make a parachute jump, and no pilot in command of an aircraft may allow a parachute jump to be made from that aircraft, over or into a congested area of a city, town, or settlement, or an open air assembly of persons unless a certificate of authorization for that jump has been issued under this section. However, a parachutist may drift over that congested area or open air assembly with a fully deployed and properly functioning parachute if he is at a sufficient altitude to avoid creating a hazard to persons and property on the ground.

(b) An application for a certificate of authorization issued under this section is made in a form and in a manner prescribed by the Administrator and must be submitted to the FAA Flight Standards District Office having jurisdiction over the area in which the parachute jump is to be made, at least four days before the day of that jump.

(c) Each holder of a certificate of authorization issued under this section shall present that certificate for inspection upon the request of the Administrator, or any Federal, State, or local official.

§ 105.17 Jumps over or onto airports.

Unless prior approval have been given by the airport management, no person may make a parachute jump, and no pilot in command of an aircraft may allow a parachute jump to be made from that aircraft—

(a) Over an airport that does not have a functioning control tower operated by the United States; or

(b) Onto any airport.

However, a parachutist may drift over that airport with a fully deployed and properly functioning parachute if he is at least 2,000 feet above that airport's traffic pattern, and avoids creating a hazard to air traffic or to persons and property on the ground.

§ 105.19 Jumps in or into control zones with functioning control towers operated by the United States.

(a) No person may make a parachute jump, and no pilot in command may allow a parachute jump to be made from that aircraft, in or into a control zone in which there is a functioning control tower operated by the United States without, or in violation of the terms of, an authorization issued under this section.

(b) Each request for an authorization under this section must be submitted to the control tower having jurisdiction over the control zone concerned and must include the information prescribed in § 105.25.

§ 105.20 Jumps in or into airport radar service areas.

(a) No person may make a parachute jump and no pilot in command may allow a parachute jump to be made from that aircraft in or into an airport radar service area without, or in violation of, the terms of an ATC authorization issued under this section.

(b) Each request for an authorization under this section must be submitted to the control tower at the airport for which the airport radar service area is designated.

[§ 105.21 Jumps into or within positive control areas and terminal control areas.

[(a) No person may make a parachute jump, and no pilot in command of an aircraft may allow a parachute jump to be made from that aircraft, in or into a positive control area or terminal control area without, or in violation of, an authorization issued under this section.]

(b) Each request for an authorization issued under this section must be submitted to the nearest FAA Air Traffic Control Facility or FAA Flight Service Station and must include the information prescribed by § 105.25(a).

§ 105.23 Jumps in or into other airspace.

(a) No person may make a parachute jump, and no pilot in command of an aircraft may allow a parachute jump to be made from that aircraft, in or into airspace unless the nearest FAA Air Traffic Control Facility or FAA Flight Service Station was notified of that jump at least 1 hour before the jump is to be made, but not more than 24 hours before the jumping is to be completed, and the notice contained the information prescribed in § 105.25(a).

(b) Notwithstanding paragraph (a) of this section, ATC may accept from a parachute jumping organization a written notification of a scheduled series of jumps to be made over a stated period of time not longer than 12 calendar months. The notification must contain the information prescribed by § 105.25(a), identify the responsible persons associated with that jumping activity, and be submitted at least 15 days, but before 30 days, before the jumping is

to begin. ATC may revoke the acceptance of the notification for any failure of the jumping organization to comply with its terms.

(c) This section does not apply to parachute jumps in or into any airspace or place described in § 105.15, 105.19, or 105.21.

§ 105.25 Information required, and notice of cancellation or postponement of jump.

(a) Each person requesting an authorization under § 105.19 or § 105.21, and each person submitting a notice under § 105.23, must include the following information (on an individual or group basis) in that request or notice:

(1) The date and time jumping will begin.

(2) The size of the jump zone expressed in nautical mile radius around the target.

(3) The location of the center of the jump zone in relation to—

(i) The nearest VOR facility in terms of the VOR radial on which it is located, and its distance in nautical miles from the VOR facility when that facility is 30 nautical miles or less from the drop zone target; or

(ii) The nearest airport, town or city depicted on the appropriate Coast and Geodetic Survey WAC or Sectional Aeronautical Chart, when the nearest VOR facility is more than 30 nautical miles from the drop zone target.

[(4) The altitudes above mean sea level at which jumping will take place.]

(5) The duration of the intended jump.

(6) The name, address, and telephone number of the person requesting the authorization or giving notice.

(7) The identification of the aircraft to be used.

(8) The radio frequencies, if any, available in the aircraft.

(b) Each person requesting an authorization under § 105.19 or § 105.21, and each person submitting a notice under § 105.23, must promptly notify the FAA Air Traffic Control Facility or FAA Flight Service Station from which it requested authorization or which it notified, if the proposed or scheduled jumping activity is cancelled or postponed.

§ 105.27 Jumps over or within restricted or prohibited areas.

No person may make a prachute jump, and no pilot in command may allow a parachute jump to be made from that aircraft, over or within a restricted area or prohibited area unless the controlling agency of the area concerned has authorized that jump.

§ 105.29 Flight visibility and clearance from clouds requirements.

No person may make a parachute jump, and no pilot in command of an aircraft may allow a parachute jump to be made from that aircraft—

(a) Into or through a cloud; or

(b) When the flight visibility is less, or

at a distance from clouds that is less, than that prescribed in the following table:

Altitude	Flight Visibility	Distance from Clouds
(1) 1,200 feet or less above the surface regardless of the MSL Altitude	3 statute miles	500 feet below 1,000 feet above 2,000 feet horizontal
(2) More than 1,200 feet above the surface but less than 10,000 feet MSL	3 statute miles	500 feet below 1,000 feet above 2,000 feet horizontal
(3) More than 1,200 feet above the surface and at or above 10,000 feet MSL	5 statute miles	1,000 feet below 1,000 feet above 1 mile horizontal

§ 105.31 [Deleted]

§ 105.33 Parachute jumps between sunset and sunrise.

(a) No person may make a parachute jump, and no pilot in command of an aircraft may allow any person to make a parachute jump from that aircraft, between sunset and sunrise, unless that person is equipped with a means of producing a light visible for at least three statute miles.

(b) Each person making a parachute jump between sunset and sunrise shall display the light required by paragraph (a) of this section from the time that person exists the aircraft until that person reaches the surface.

§ 105.35 Liquor and drugs.

No person may make a parachute jump while, and no pilot in command of an aircraft may allow a person to make a parachute jump from that aircraft if that person appears to be,—

(a) Under the influence of intoxicating liquor; or

(b) Using any drug that affects his faculties in any way contrary to safety.

§ 105.37 Inspections.

The Administrator may inspect (including inspections at the jump site) any parachute jump operation to which this Part applies, to determine compliance with the regulations of this Part.

Subpart C—Parachute Equipment

§ 105.41 Applicability.

(a) Except as provided in paragraph (b) of this section, this subpart prescribes rules governing parachute equipment used in parachute jumps to which this Part applies.

(b) This subpart does not apply to a parachute jump made by a member of an Armed Force using parachute equipment of an Armed Force.

§ 105.43 Parachute equipment and packing requirements.

(a) No person may make a parachute jump, and no pilot in command of an aircraft may allow any person to make a parachute jump from that aircraft, unless that person is wearing a single harness dual parachute pack, having at least one main parachute and one approved auxiliary parachute that are packed as follows:

(1) The main parachute must have been packed by a certificated parachute rigger, or by the person making the jump, within 120 days before the date of its use.

(2) The auxiliary must have been packed by a certificated and appropriately rated parachute rigger:

(i) Within 120 days before the date of use, if its canopy, shroud, and harness are composed exclusively of nylon, rayon or other similar synthetic fiber or material that is substantially resistant to damage from mold, mildew, or other fungi and other rotting agents propagated in a moist environment; or

(ii) Within 60 days before the date of use, if it is composed in any amount of silk, pongee, or other natural fiber, or material not specified in paragraph (a)(2)(i) of this section.

(b) No person may make a parachute jump using a static line attached to the aircraft and the main parachute unless an assist device, described and attached as follows, is used to aid the pilot chute in performing its function, or, if no pilot chute is used, to aid in the direct deployment of the main parachute canopy.

(1) The assist device must be long enough to allow the container to open before a load is placed on the device.

(2) The assist device must have a static load strength of—

(i) At least 28 pounds but not more than 160 pounds, if it is used to aid the pilot chute in performing its function; or

(ii) At least 56 pounds but not more than 320 pounds, if it is used to aid in the direct deployment of the main parachute canopy.

(3) The assist device must be attached—

(i) At one end, to the static line above the static line pins, or, if static pins are not used, above the static line ties to the parachute cone; and

(ii) At the other end, to the pilot chute apex, bridle cord or bridle loop, or, if no pilot chute is used, to the main parachute canopy.

(c) No person may attach an assist device required by paragraph (b) of this section to any main parachute unless he has a current parachute rigger certificate issued under Part 65 of this chapter or is the person who makes the jump with that parachute.

(d) For the purposes of this section, an "approved" parachute is—

(1) A parachute manufactured under a type certificate or a technical standard order (C-23 series) ; or

(2) A personnel-carrying military parachute (other than a high altitude, high-speed, or ejection kind) identified by an NAF, AAF, or AN drawing number, an AAF order number, or any other military designation or specification number.

FEDERAL AVIATION REGULATIONS

Part 121—Appendix I

Appendix I—Drug Testing Program

This appendix contains the standards and components that must be included in an anti-drug program required by this chapter.

I. *DOT Procedures.* Each employer shall ensure that drug testing programs conducted pursuant to this regulation comply with the requirements of this appendix and the "Procedures for Transportation Workplace Drug Testing Programs" published by the Department of Transportation (DOT) (49 CFR Part 40). An employer may not use or contract with any drug testing laboratory that is not certified by the Department of Health and Human Services (DHHS) pursuant to the DHHS "Mandatory Guidelines for Federal Workplace Drug Testing Programs" (53 FR 11970; April 11, 1988).

II. *Definitions.* For the purpose of this appendix, the following definitions apply:

"Accident" means an occurrence associated with the operation of an aircraft which takes place between the time any person boards the aircraft with the intention of flight and all such persons have disembarked, and in which any person suffers death or serious injury, or in which the aircraft receives substantial damage (49 CFR 830.2).

"Annualized rate" for the purposes of unannounced testing of employees based on random selection means the percentage of specimen collection and testing of employees performing a function listed in section III of this appendix during a calendar year. The employer shall determine the annualized percentage rate by referring to the total number of employees performing a sensitive safety- or security-related function for the employer at the beginning of a calendar year or by an alternative method specified in the employer's drug testing plan approved by the FAA.

"Employee" is a person who performs, either directly or by contract, a function listed in section III of this appendix for a Part 121 certificate holder, a Part 135 certificate holder, an operator as defined in § 135.1(c) of this chapter (except operations of foreign civil aircraft navigated within the United States pursuant to Part 375 or emergency mail service operations pursuant to section 405(h) of the Federal Aviation Act of 1958), or an air traffic control facility not operated by, or under contract with, the FAA or the U.S. military. Provided however that an employee who works for an employer who holds a Part 135 certificate and who also holds a Part 121 certificate is considered to be an employee of the Part 121 certificate holder for the purposes of this appendix.

"Employer" is a Part 121 certificate holder, a Part 135 certificate holder, an operator as defined in § 135.1(c) of this chapter (except operations of foreign civil aircraft navigated within the United States pursuant to Part 375 or emergency mail service operations pursuant to Section 405(h) of the Federal Aviation Act of 1958, or an air traffic control facility not operated by, or under contract with, the FAA or the U.S. military. Provided, however, that an employer may use a person to perform a function listed in section III of this appendix, who is not included under that employer's drug program, if that person is subject to the requirements of another employer's FAA-approved anti-drug program.

"Failing a drug test" means that the test result shows positive evidence of the presence of a prohibited drug or drug metabolite in an employee's system.

"Passing a drug test" means that the test result does not show positive evidence of the presence of a prohibited drug or drug metabolite in an employee's system.

"Positive evidence" means the presence of a drug or drug metabolite in a urine sample at or above the test levels listed in the DOT "Procedures for Transportation Workplace Drug Testing Programs" (49 CFR Part 40).

"Prohibited drug" means marijuana, cocaine, opiates, phencyclidine (PCP), amphetamines, or a substance specified in

Schedule I or Schedule II of the Controlled Substances Act, 21 U.S.C. 811, 812 (1981 & 1987 Cum.P.P.), unless the drug is being used as authorized by a legal prescription or other exemption under Federal, state, or local law.

"Refusal to submit" means refusal by an individual to provide a urine sample after he or she has received notice of the requirement to be tested in accordance with this appendix.

III. *Employees Who Must Be Tested.* Each person who performs a function listed in this section must be tested pursuant to an FAA-approved anti-drug program conducted in accordance with this appendix:

a. Flight crewmember duties.

b. Flight attendant duties.

c. Flight instruction or ground instruction duties.

d. Flight testing duties.

e. Aircraft dispatcher or ground dispatcher duties.

f. Aircraft maintenance or preventive maintenance duties.

g. Aviation security or screening duties.

h. Air traffic control duties.

IV. *Substances For Which Testing Must Be Conducted.* Each employer shall test each employee who performs a function listed in section III of this appendix for evidence of marijuana, cocaine, opiates, phencyclidine (PCP), and amphetamines during each test required by section V of this appendix. As part of reasonable cause drug testing program established pursuant to this part, employers may test for drugs in addition to those specified in this part only with approval granted by the FAA under 49 CFR Part 40 and for substances for which the Department of Health and Human Services has established an approved testing protocol and positive threshhold.

V. *Types of Drug Testing Required.* Each employer shall conduct the following types of testing in accordance with the procedures set forth in this appendix and the DOT "Procedures for Transportation Workplace Drug Testing Programs" (49 CFR Part 40):

A. *Preemployment testing.* No employer may hire any person to perform a function listed in section III of this appendix unless the applicant passes a drug test for that employer. The employer shall advise an applicant at the time of application that preemployment testing will be conducted to determine the presence of marijuana, cocaine, opiates, phencyclidine (PCP), and amphetamines or a metabolite of those drugs in the applicant's system.

B. *Periodic testing.* Each employee who performs a function listed in section III of this appendix for an employer and who is required to undergo a medical examination under Part 67 of this chapter, shall submit to a periodic drug test. The employee shall be tested for the presence of marijuana, cocaine, opiates, phencyclidine (PCP), and amphetamines or a metabolite of those drugs as part of the first medical evaluation of the employee during the first calendar year of implementation of the employer's anti-drug program. An employer may discontinue periodic testing of its employees after the first calendar year of implementation of the employer's anti-drug program when the employer has implemented an unannounced testing program based on random selection of employees.

C. *Random testing.* Each employer shall randomly select employees who perform a function listed in section III of this appendix for the employer for unannounced drug testing. The employer shall randomly select employees for unannounced testing for the presence of marijuana, cocaine, opiates, phencyclidine (PCP), and amphetamines or a metabolite of those drugs in an employee's system using a random number table or a computer-based, number generator that is matched with an employee's social security number, payroll identification number, or any other alternative method approved by the FAA.

(1) During the first 12 months following implementation of unannounced testing based on random selection pursuant to this appendix, an employer shall meet the following conditions:

(a) The unannounced testing based on random selection of employees shall be spread reasonably throughout the 12-month period.

(b) The last collection of specimens for random testing during the year shall be conducted at an annualized rate equal to not less than 50 percent of employees performing a function listed in section III of this appendix.

(c) The total number of unannounced tests based on random selection during the 12-months shall be equal to not less than 25 percent of the employees performing a function listed in section III of this appendix.

(2) Following the first 12 months, an employer shall achieve and maintain an annualized rate equal to not less than 50 percent of employees performing a function listed in section III of this appendix.

D. *Postaccident testing.* Each employer shall test each employee who performs a function listed in section III of this appendix for the presence of marijuana, cocaine, opiates, phencyclidine (PCP), and amphetamines or a metabolite of those drugs in the employee's system if that employee's performance either contributed to an accident or cannot be completely discounted as a contributing factor to the accident. The employee shall be tested as soon as possible but not later than 32 hours after the accident. The decision not to administer a test under this section must be based on a determination, using the best information available at the time of the accident, that the employee's performance could not have contributed to the accident. The employee shall submit to postaccident testing under this section.

E. *Testing based on reasonable cause.* Each employer shall test each employee who performs a function listed in section III of this appendix and who is reasonably suspected of using a prohibited drug. Each employer shall test an employee's specimen for the presence of marijuana, cocaine, opiates, phencyclidine (PCP), and amphetamines or a metabolite of those drugs. An employer may test an employee's specimen for the presence of other prohibited drugs or drug metabolites only in accordance with this appendix and the DOT "Procedures for Transportation Workplace Drug Testing Programs" (49 CFR Part 40). At least two of the employee's supervisors, one of whom is trained in detection of the possible symptoms of drug use, shall substantiate and concur in the decision to test an employee who is reasonably suspected of drug use. In the case of an employer holding a Part 135 certificate who employs 50 or fewer employees who perform a function listed in section III of this appendix or an operator as defined in § 135.1(c) of this chapter, one supervisor, who is trained in detection of possible symptoms of drug use, shall substantiate the decision to test an employee who is reasonably suspected of drug use. The decision to test must be based on a reasonable and articulable belief that the employee is using a prohibited drug on the basis of specific, contemporaneous physical, behavioral, or performance indicators of probable drug use.

F. *Testing after return to duty.* Each employer shall implement a reasonable program of unannounced testing of each individual who has been hired and each employee who has returned to duty to perform a function listed in section III of this appendix after failing a drug test conducted in accordance with this appendix or after refusing to submit to a drug test required by this appendix. The individual or employee shall be subject to unannounced testing for not more than 60 months after the individual has been hired or the employee has returned to duty to perform a function listed in section III of this appendix.

VI. *Administrative Matters.*—A. *Collection, testing, and rehabilitation records.* Each employer shall maintain all records related to the collection process, including all logbooks and certification statements, for two years. Each employer shall maintain records of employee confirmed positive drug test results and employee rehabilitation for five years. The employer shall maintain records of negative test results for 12 months. The employer shall permit the Administrator or the Administrator's representative to examine these records.

B. *Laboratory inspections.* The employer shall contract only with a laboratory that permits pre-award inspections by the employer before the laboratory is awarded a testing contract and unannounced inspections, including examination of any and all records at any time by the employer, the Administrator, or the Administrator's representative.

C. *Employee request to retest a specimen.* Not later than 60 days after receipt of a confirmed positive test result, an employee may submit a written request to the MRO for retesting of the specimen producing the positive test result. Each employee may make one written request that a sample of the specimen be provided to the original or another DHHS-certified laboratory for testing. The laboratories shall follow chain-of-custody procedures. The employee shall pay the costs of the additional test and all handling and shipping costs associated with the transfer of the specimen to the laboratory.

D. *Release of Drug Testing Information.* An employer may release information regarding an employee's drug testing results or rehabilitation to a third party only with the specific, written consent of the employee authorizing release of the information to an identified person. Information regarding an employee's drug testing results or rehabilitation may be released to the National Transportation Safety Board as part of an accident investigation, to the FAA upon request, or as required by section VII.C.5 of this appendix.

VII. *Review of Drug Testing Results.* The employer shall designate or appoint a medical review officer (MRO). If the employer does not have a qualified individual on staff to serve as MRO, the employer may contract for the provision of MRO services as part of its drug testing program.

A. *MRO qualifications.* The MRO must be a licensed physician with knowledge of drug abuse disorders.

B. *MRO duties.* The MRO shall perform the following functions for the employer:

1. Review the results of the employer's drug testing program before the results are reported to the employer and summarized for the FAA.

2. Within a reasonable time, notify an employee of a confirmed positive test result.

3. Review and interpret each confirmed positive test result in order to determine if there is an alternative medical explanation for the confirmed positive test result. The MRO shall perform the following functions as part of the review of a confirmed positive test result:

a. Provide an opportunity for the employee to discuss a positive test result with the MRO.

b. Review the employee's medical history and any relevant biomedical factors.

c. Review all medical records made available by the employee to determine if a confirmed positive test resulted from legally prescribed medication.

d. Verify that the laboratory report and assessment are correct. The MRO shall be authorized to request that the original specimen be reanalyzed to determine the accuracy of the reported test result.

4. Process employee requests to retest a specimen in accordance with section VI.C of this appendix.

5. Determine whether and when, consistent with an employer's anti-drug program, a return-to-duty recommendation for a current employee or a decision to hire an individual to perform a function listed in section III of this appendix after failing a test conducted in accordance with this appendix or after refusing to submit to a test required by this appendix, including review of any rehabilitation program in which the individual or employee participated, may be made.

6. Ensure that an individual or employee has been tested in accordance with the procedures of this appendix and the DOT "Procedures for Transportation Workplace Drug Testing Programs" (49 CFR Part 40) before the individual is hired or the employee returns to duty.

7. Determine a schedule of unannounced testing for an individual who has been hired or an employee who has returned to duty to perform a function listed in section III of this appendix after the individual or employee has failed a drug test conducted in accordance with this appendix or has refused to submit to a drug test required by this appendix.

C. *MRO determinations.* 1. If the MRO determines, after appropriate review, that there is a legitimate medical explanation for the confirmed positive test result that is consistent with legal drug use, the MRO shall conclude that the test result is negative and shall report the test as a negative test result.

2. If the MRO determines, after appropriate review, that there is no legitimate medical explanation for the confirmed positive test result that is consistent with legal drug use, the MRO shall refer the employee to an employer's rehabilitation program is available or to a personnel or administrative officer for further proceedings in accordance with the employer's anti-drug program.

3. Based on a review of laboratory inspection reports, quality assurance and quality control data, and other drug test results, the MRO may conclude that a particular drug test result is scientifically insufficient for futher action. Under these circumstances, the MRO shall conclude that the test is negative for the presence of drugs or drug metabolites in an employee's system.

4. In order to make a recommendation to hire an individual to perform a function listed in section III of this appendix or to return an employee to duty to perform a function listed in section III of this appendix after the individual or employee has failed a drug test conducted in accordance with this appendix or refused to submit to a drug test required by this appendix, the MRO shall—

a. Ensure that the individual or employee is drug free based on a drug test that shows no positive evidence of the presence of a drug or a drug metabolite in the person's system;

b. Ensure that the individual or employee has been evaluated by a rehabilitation program counselor for drug use or abuse; and

c. Ensure that the individual or employee demonstrates compliance with any conditions or requirements of a rehabilitation program in which the person participated.

5. Notwithstanding any other section in this appendix, the MRO shall make the following determinations in the case of an employee or applicant who holds, or is required to hold, a medical certificate issued pursuant to Part 67 of this chapter in order to perform a function listed in section III of this appendix for an employer:

a. The MRO shall make a determination of probable drug dependence or nondependence as specified in Part 67 of this chapter. If the MRO makes a determination of nondependence, the MRO has authority to recommend that the employee return to duty in a position that requires the employee to hold a certificate issued under Part 67 of this chapter. The MRO shall forward the determination of nondependence, the return-to-duty decision, and any supporting documentation to the Federal Air Surgeon for review.

b. If the MRO makes a determination of probable drug dependence at any time, the MRO shall report the name of the individual and identifying information, the determination of probable drug dependence, and any supporting documentation to the Federal Air Surgeon. The MRO does not have the authority to recommend that the employee return to duty in a position that requires the employee to hold a certificate issued under Part 67 of this chapter. The Federal Air Surgeon shall determine if the individual may retain or may be issued a medical certificate consistent with the requirements of Part 67 of this chapter.

c. The MRO shall report to the Federal Air Surgeon the name of any employee who is required to hold a medical certificate issued pursuant to Part 67 of this chapter and who fails a drug test. The MRO shall report to the Federal Air Surgeon the name of any person who applies for a position that requires the person to hold a medical certificate issued pursuant to Part 67 of this chapter and who fails a preemployment drug test.

d. The MRO shall forward the information specified in paragraphs (a), (b), and (c) of this section to the Federal Air Surgeon, Federal Aviation Administration, Drug Abatement Branch (AAM-220), 800 Independence Avenue, SW., Washington, DC 20591.

VIII. *Employee Assistance Program (EAP).* The employer shall provide an EAP for employees. The employer may establish the EAP as a part of its internal personnel services or the employer may contract with an entity that will provide EAP services to an employee. Each EAP must include education and training on drug use for employees and training for supervisors making determinations for testing of employees based on reasonable cause.

A. *EAP education program.* Each EAP education program must include at least the following elements: display and distribution of informational material; display and distribution of a community service hot-line telephone number for employee assistance; and display and distribution of the employer's policy regarding drug use in the workplace.

B. *EAP training program.* Each employer shall implement a reasonable program of initial training for employees. The employee training program must include at least the following elements: The effects and consequences of drug use on personal health, safety, and work environment; the manifestations and behavioral cues that may indicate drug use and abuse; and documentation of training given to employees and employer's supervisory personnel. The employer's supervisory personnel who will determine when an employee is subject to testing based on reasonable cause shall receive specific training on the specific, contemporaneous physical, behavioral, and performance indicators of probable drug use in addition to the training specified above. The employer shall ensure that supervisors who will make reasonable cause determinations receive at least 60 minutes of initial training. The employer shall implement a reasonable recurrent training program for supervisory personnel making reasonable cause determinations during subsequent years. The employer shall identify the employee and supervisor EAP training in the employer's drug testing plan submitted to the FAA for approval.

IX. *Employer's Drug Testing Plan.*— A. *Schedule for submission of plans and implementation.* (1) Each employer shall submit a drug testing plan to the Federal Aviation Administration, Office of Aviation Medicine, Drug Abatement Branch (AAM-220), 800 Independence Avenue, SW., Washington, DC 20591.

(2) Each employer who holds a Part 121 certificate and each employer who holds a Part 135 certificate and employs more than 50 employees who perform a function listed in section III of this appendix shall submit an anti-drug program to the FAA (specifying the procedures for all testing required by this appendix) not later than 120 days after December 21, 1988. Each employer shall implement preemployment testing of applicants for a position to perform a function listed in section III of this appendix not later than 10 days after approval of the plan by the FAA. Each employer shall implement the remainder of the employer's anti-drug program no later than 180 days after approval of the plan by the FAA.

(3) Each employer who holds a Part 135 certificate and employs from 11 to 50 employees who perform a function listed in section III of this appendix shall submit an interim anti-drug program to the FAA (specifying the procedures for preemployment testing, periodic testing, postaccident testing, testing based on reasonable cause, and testing after return to duty) not later than 180 days after December 21, 1988. Each employer shall implement the interim anti-drug program not later than 180 days after approval of the plan by the FAA. Each employer shall submit an amendment to its approved anti-drug program to the FAA (specifying the procedures for unannounced testing based on random selection) not later than 120 days after approval of the interim anti-drug program by the FAA. Each employer shall implement the random testing provision of its amended anti-drug program not later than 180 days after approval of the amendment.

(4) Each employer who holds a Part 135 certificate and employs 10 or fewer employees who perform a function listed in section III of this appendix, each operator as defined in § 135.1(c) of this chapter, and each air traffic control facility not operated by, or under contract with the FAA or the U.S. military, shall submit an anti-drug program to the FAA (specifying the procedures for all testing required by this appendix) not later than 360 days after December 21, 1988. Each employer shall implement the employer's anti-drug program not later than 180 days after approval of the plan by the FAA.

(5) Each employer or operator, who becomes subject to the rule as a result of the FAA's issuance of a Part 121 or Part 135 certificate or as a result of beginning operations listed in § 135.1(b) for compensation or hire (except operations of foreign civil aircraft navigated within the United States pursuant to Part 375 or emergency mail service operations pursuant to section 405(h) of the Federal Aviation Act of 1958) shall submit an anti-drug plan to the FAA for approval, within the timeframes of paragraphs (2), (3), or (4) of this section, according to the type and size of the category of operations. For purposes of applicability of the timeframes, the date that an employer becomes subject to the requirements of this appendix is substituted for [the effective date of the rule].

B. An employer's anti-drug plan must specify the methods by which the employer will comply with the testing requirements of this appendix. The plan must provide the name and address of the laboratory which has been selected by the employer for analysis of the specimens collected during the employer's anti-drug testing program.

C. An employer's anti-drug plan must specify the procedures and personnel the employer will use to ensure that a determination is made as to the veracity of test results and possible legitimate explanations for an employee failing a test.

D. The employer shall consider its anti-drug program to be approved by the Administrator, unless notified to the contrary by the FAA, within 60 days after submission of the plan to the FAA.

X. *Reporting Results of Drug Testing Program.* A. Each employer shall submit a semiannual report to the FAA summarizing the results of its drug testing program and covering the period from January 1–June 30. Each employer shall submit a annual report to the FAA summarizing the results of its drug testing program and covering the period from January 1–December 31. Each employer shall submit these reports no later than 45 days after the last day of the report period.

B. Each report shall contain:

1. The total number of tests performed and the total number of tests performed for each category of test.

2. The total number of positive test results by category of test; the total number of positive test results by each function listed in section III of this appendix; and the total number of positive test results by the type of drug shown in a positive test result.

3. The disposition of an individual who failed a drug test conducted in accordance with this appendix or who refused to submit to a drug test required by this appendix by each category of test.

XI. *Preemption.* A. The issuance of these regulations by the FAA preempts any State or local law, rule, regulation, order, or standard covering the subject matter of this rule, including but not limited to, drug testing of aviation personnel performing sensitive safety- or security-related functions.

B. The issuance of these regulations does not preempt provisions of State criminal law that impose sanctions for reckless conduct of an individual that leads to actual loss of life, injury, or damage to property whether such provisions apply specifically to aviation employees or generally to the public.

XII. *Conflict with foreign laws or international law.* A. This appendix shall not apply to any person for whom compliance with this appendix would violate the domestic laws or policies of another country.

B. This appendix is not effective until January 1, 1990, with respect to any person for whom a foreign government contends that application of this appendix raises questions of compatability with that country's domestic laws or policies. On or before December 1, 1989, the Administrator shall issue any necessary amendment resolving the applicability of this appendix to such person on or after January 1, 1990.

FEDERAL AVIATION REGULATIONS

Part 135—Air Taxi Operators and Commercial Operators

Contents

Subpart A—General

§ 135.1 **Applicability.**

(a) Except as provided in paragraph (b) of this section, this Part prescribes rules governing—

(1) Air taxi operations conducted under the exemption authority of Part 298 of this title;

(2) The transportation of mail by aircraft conducted under a postal service contract awarded under section 5402c of Title 39, United States Code;

(3) The carriage in air commerce of persons or property for compensation or hire as a commercial operator (not an air carrier) in

aircraft having a maximum seating capacity of less than 20 passengers or a maximum payload capacity of less than 6,000 pounds, or the carriage in air commerce of persons or property in common carriage operations solely between points entirely within any state of the United States in aircraft having a maximum seating capacity of 30 seats or less or a maximum payload capacity of 7,500 pounds or less; and

(4) Each person who is on board an aircraft being operated under this Part.

(b) Except as provided in paragraph (c) of this section, this part does not apply to—

(1) Student instruction;

(2) Nonstop sightseeing flights that begin and end at the same airport, and are conducted within a 25 statute mile radius of that airport;

(3) Ferry or training flights;

(4) Aerial work operations, including—

(i) Crop dusting, seeding, spraying, and bird chasing;

(ii) Banner towing;

(iii) Aerial photography or survey;

(iv) Fire fighting;

(v) Helicopter operations in construction or repair work (but not including transportation to and from the site of operations); and

(vi) [Powerline or pipeline patrol, or similar types of patrol approved by the Administrator;]

(5) Sightseeing flights conducted in hot air balloons;

(6) Nonstop flights conducted within a 25 statute mile radius of the airport of takeoff carrying persons for the purpose of intentional parachute jumps;

(7) Helicopter flights conducted within a 25 statute mile radius of the airport of takeoff, if—

(i) Not more than two passengers are carried in the helicopter in addition to the required flight crew;

(ii) Each flight is made under VFR during the day;

(iii) The helicopter used is certificated in the standard category and complies with the 100-hour inspection requirements of Part 91 of this chapter;

(iv) The operator notifies the FAA Flight Standards District Office responsible for the geographic area concerned at least 72 hours before each flight and furnishes any essential information that the office requests;

(v) The number of flights does not exceed a total of six in any calendar year;

(vi) Each flight has been approved by the Administrator; and

(vii) Cargo is not carried in or on the helicopter;

(8) Operations conducted under Part 133 or 375 of this title;

(9) Emergency mail service conducted under section 405(h) of the Federal Aviation Act of 1958; or

[(10) This Part does not apply to operations conducted under the provisions of § 91.59.]

(c) For the purpose of §§ 135.249, 135.251, and 135.353, "operator" means any person or entity conducting an operation listed in paragraph (b) of this section for compensation or hire except operation of foreign civil aircraft navigated within the United States pursuant to Part 375 described in paragraph (b)(8) and emergency mail service operation pursuant to section 405(h) of the Federal Aviation Act of 1958 described in paragraph (b)(9). Each operator and each employee of an operator shall comply with the requirements of §§ 135.249, 135.251, and 135.353 of this part.

(d) Notwithstanding the provisions of paragraph (c) of this section, an operator who does not hold a Part 121 certificate or a Part 135 certificate is permitted to use a person, who is otherwise authorized to perform aircraft maintenance or preventive maintenance duties and who is not subject to the requirements of an FAA-approved anti-drug program, to perform—

(1) Aircraft maintenance or preventive maintenance on the operator's aircraft if the operator would be required to transport the aircraft more than 50 nautical miles further than the closest available repair point from the operator's principal place of operations to obtain these services; or

(2) Emergency repairs on the operator's aircraft if the aircraft cannot be safely operated to a location where an employee subject to the requirements of this appendix can perform the emergency repairs.

§ 135.2   Air taxi operations with large aircraft.

(a) Except as provided in paragraph (d) of this section, no person may conduct air taxi operations in large aircraft under an individual exemption and authorization issued by the Civil Aeronautics Board or under the exemption authority of Part 298 of this title, unless that person—

(1) Complies with the certification requirements for supplemental air carriers in Part 121 of this chapter, except that the person need not obtain, and that person is not eligible for, a certificate under that Part; and

(2) Conducts those operations under the rules of Part 121 of this chapter that apply to supplemental air carriers.

However, the Administrator may issue operations specifications which require an operator to comply with the rules of Part 121 of this chapter that apply to domestic or flag air carriers, as appropriate, in place of the rules required by paragraph (a)(2) of this section, if the Administrator determines compliance with those rules is necessary to provide an appropriate level of safety for the operation.

(b) The holder of an operating certificate issued under this Part who is required to comply with Subpart L of Part 121 of this chapter, under paragraph (a) of this section, may perform and approve maintenance, preventive maintenance, and alterations on aircraft having a maximum passenger seating configuration, excluding any pilot seat, of 30 seats or less and a maximum payload capacity of 7,500 pounds or less as provided in that subpart. The aircraft so maintained shall be identified by registration number in the operations specifications of the certificate holder using the aircraft.

(c) Operations that are subject to paragraph (a) of this section are not subject to §§ 135.21 through 135.43 of Subpart A and Subparts B through J of this Part. Seaplanes used in operations that are subject to paragraph (a) of this section are not subject to § 121.291(a) of this chapter.

(d) Operations conducted with aircraft having a maximum passenger seating configuration, excluding any pilot seat, of 30 seats or less and a maximum payload capacity of 7,500 pounds or less shall be conducted under the rules of this Part. However, a certificate holder who is conducting operations on December 1, 1978, in aircraft described in this paragraph may continue to operate under paragraph (a) of this section.

(e) For the purposes of this Part—

(1) "Maximum payload capacity" means:

(i) For an aircraft for which a maximum zero fuel weight is prescribed in FAA technical specifications, the maximum zero fuel weight, less empty weight, less all justifiable aircraft equipment, and less the operating load (consisting of minimum flight crew, foods and beverages and supplies and equipment related to foods and beverages, but not including disposable fuel or oil);

(ii) For all other aircraft, the maximum certificated takeoff weight of an aircraft, less the empty weight, less all justifiable aircraft equipment, and less the operating load (consisting of minimum fuel load, oil, and flight crew). The allowance for the weight of the crew, oil, and fuel is as follows:

(A) Crew—200 pounds for each crewmember required under this chapter.

(B) Oil—350 pounds.

(C) Fuel—the minimum weight of fuel required under this chapter for a flight between domestic points 174 nautical miles apart under VFR weather conditions that does not involve extended overwater operations.

(2) "Empty weight" means the weight of the airframe, engines, propellers, rotors, and fixed equipment. Empty weight excludes the weight of the crew and payload, but includes the weight of all fixed ballast, unusable fuel supply, undrainable oil, total quantity of engine coolant, and total quantity of hydraulic fluid.

(3) "Maximum zero fuel weight" means the maximum permissible weight of an aircraft with no disposable fuel or oil. The zero fuel weight figure may be found in either the aircraft type certificate data sheet or the approved Aircraft Flight Manual, or both.

(4) For the purposes of this paragraph, "justifiable aircraft equipment" means any equipment necessary for the operation of the aircraft. It does not include equipment or ballast specifically installed, permanently or otherwise, for the purpose of altering the empty weight of an aircraft to meet the maximum payload capacity specified in paragraph (d) of this section.

§ 135.3 Rules applicable to operations subject to this Part.

Each person operating an aircraft in operations under this Part shall—

(a) While operating inside the United States, comply with the applicable rules of this chapter; and

(b) While operating outside the United States, comply with Annex 2, Rules of the Air, to the Convention on International Civil Aviation or the regulations of any foreign country, whichever applies, and with any rules of Part 61 and 91 of this chapter and this Part that are more restrictive than that Annex or those regulations and that can be complied with without violating that Annex or those regulations. Annex 2 is incorporated by reference in § 91.1(c) of this chapter.

§ 135.5 Certificate and operations specifications required.

No person may operate an aircraft under this Part without, or in violation of, an air taxi/commercial operator (ATCO) operating certificate and appropriate operations specifications issued under this Part, or for operations with large aircraft having a maximum passenger seating configuration, excluding any pilot seat, of more than 30 seats, or a maximum payload capacity of more than 7,500 pounds, without, or in violation of appropriate operations specifications issued under Part 121 of this chapter.

§ 135.7 Applicability of rules to unauthorized operators.

The rules in this Part which apply to a person certificated under § 135.5 also apply to a person who engages in any operation governed by this Part without an appropriate certificate and operations specifications required by § 135.5.

§ 135.9 Duration of certificate.

(a) An ATCO operating certificate is effective until surrendered, suspended or revoked. The holder of an ATCO operative certificate that is suspended or revoked shall return it to the Administrator.

(b) Except as provided in paragraphs (c) and (d) of this section, an ATCO operating certificate in effect on December 1, 1978, expires on February 1, 1979. The certificate holder must continue to conduct operations under Part 135 and the operations specifications in effect on November 30, 1978, until the certificate expires.

(c) If the certificate holder applies before February 1, 1979, for new operations specifications under this Part, the operating certificate held continues in effect and the certificate holder must continue operations under Part 135 and operations specifications in effect on November 30, 1978, until the earliest of the following—

(1) The date on which new operations specifications are issued; or

(2) The date on which the Administrator notifies the certificate holder that the application is denied; or

(3) August 1, 1979.

If new operations specifications are issued under paragraph (c)(1) of this paragraph, the ATCO operating certificate continues in effect until surrendered, suspended or revoked under paragraph (a) of this section.

(d) A certificate holder may obtain an extension of the expiration date in paragraph (c) of this section, but not beyond December 1, 1979, from the Director, Flight Standards Service, if before July 1, 1979, the certificate holder—

(1) Shows that due to the circumstances beyond its control it cannot comply by the expiration date; and

(2) Submits a schedule for compliance, acceptable to the Director, indicating that compliance will be achieved at the earliest practicable date.

(e) The holder of an ATCO operating certificate that expires, under paragraphs (b), (c), or (d) of this section, shall return it to the Administrator.

§ 135.10 Compliance dates for certain rules.

After January 2, 1991, no certificate holder may use a person as a flight crewmember unless that person has completed the windshear ground training required by §§ 135.345(b)(6) and 135.351(b)(2) of this part.

§ 135.11 Application and issue of certificate and operations specifications.

(a) An application for an ATCO operating certificate and appropriate operations specifications is made on a form and in a manner prescribed by the Administrator and filed with the FAA Flight Standards District Office that has jurisdiction over the area in which the applicant's principal business office is located.

(b) An applicant who meets the requirements of this Part is entitled to—

(1) An ATCO operating certificate containing all business names under which the certificate holder may conduct operations and the address of each business office used by the certificate holder; and

(2) Separate operations specifications, issued to the certificate holder, containing:

(i) The type and area of operations authorized.

(ii) The category and class of aircraft that may be used in those operations.

(iii) Registration numbers and types of aircraft that are subject to an airworthiness maintenance program required by § 135.411(a)(2), including time limitations or standards for determining time limitations, for overhauls, inspections, and checks for airframes, aircraft engines, propellers, rotors, appliances, and emergency equipment.

(iv) Registration numbers of aircraft that are to be inspected under an approved aircraft inspection program under § 135.419.

(v) Additional maintenance items required by the Administrator under § 135.421.

(vi) Any authorized deviation from this Part.

(vii) Any other items the Administrator may require or allow to meet any particular situation.

[(c) No person holding operations specifications issued under this part may list on its operations specifications or on the current list of aircraft required by § 135.63(a)(3) any airplane listed on operations specifications issued under Part 125.]

§ 135.13 Eligibility for certificate and operations specifications.

(a) To be eligible for an ATCO operating certificate and appropriate operations specifications, a person must—

(1) Be a citizen of the United States, a partnership of which each member is a citizen of the United States, or a corporation or association created or organized under the laws of the United States or any state, territory, or possession of the United States, of which the president and two-thirds or more of the board of directors and other managing officers are citizens of the United States and in which at least 75 percent of the voting interest is owned or controlled by citizens of the United States or one of its possessions; and

(2) Show, to the satisfaction of the Administrator, that the person is able to conduct each kind of operation for which the person seeks authorization in compliance with applicable regulations; and

(3) Hold any economic authority that may be required by the Civil Aeronautics Board.

However, no person holding a commercial operator operating certificate issued under Part 121 of this chapter is eligible for an ATCO operating certificate unless the person shows to the satisfaction of the Administrator that the person's contract carriage business in large aircraft, having a maximum passenger seating configuration, excluding any pilot seat, of more than 30 seats or a maximum payload capacity of more than 7,500 pounds, will not result directly or indirectly from the person's air taxi business.

(b) The Administrator may deny any applicant a certificate under this Part if the Administrator finds—

(1) That an air carrier or commercial operator operating certificate under Part 121 or an ATCO operating certificate previously issued to the applicant was revoked; or

(2) That a person who was employed in a position similar to general manager, director of operations, director of maintenance, chief pilot, or chief inspector, or who has exercised control with respect to any ATCO operating certificate holder, air carrier, or commercial operator, whose operating certificate has been revoked, will be employed in any of those positions or a similar position, or will be in control of or have a substantial ownership interest in the applicant, and that the person's employment or control contributed materially to the reasons for revoking that certificate.

§ 135.15 Amendment of certificate.

(a) The Administrator may amend an ATCO operating certificate—

(1) On the Administrator's own initiative, under section 609 of the Federal Aviation Act of 1958 (49 U.S.C. 1492) and Part 13 of this chapter; or

(2) Upon application by the holder of that certificate.

(b) The certificate holder must file an application to amend an ATCO operating certificate at least 15 days before the date proposed by the applicant for the amendment to become effective, unless a shorter filing period is approved. The application must be on a form and in a manner prescribed by the Administrator and must be submitted to the FAA Flight Standards District Office charged with the overall inspection of the certificate holder.

(c) The FAA Flight Standards District Office charged with the overall inspection of the certificate holder grants an amendment to the ATCO operating certificate if it is determined that safety in air commerce and the public interest allow that amendment.

(d) Within 30 days after receiving a refusal to amend the operating certificate, the certificate holder may petition the Director, Flight Standards Service, to reconsider the request.

§ 135.17 Amendment of operations specifications

(a) The FAA Flight Standards District Office charged with the overall inspection of the certificate holder may amend any operations specifications issued under this Part if—

(1) It determines that safety in air commerce requires that amendment; or

(2) Upon application by the holder, that District Office determines that safety in air commerce allows that amendment.

(b) The certificate holder must file an application to amend operations specifications at least 15 days before the date proposed by the applicant for the amendment to become effective, unless a shorter filing period is approved. The application must be on a form and in a manner prescribed by the Administrator and be submitted to the FAA Flight Standards District Office charged with the overall inspection of the certificate holder.

(c) Within 30 days after a notice of refusal to approve a holder's application for amendment is received, the holder may petition the Director of Airworthiness for amendments pertaining to airworthiness or the Director of Flight Operations for amendments pertaining to flight operations, to reconsider the refusal to amend.

(d) When the FAA Flight Standards District Office charged with the overall inspection of the certificate holder amends operations specifications, that District Office gives notice in writing to the holder of a proposed amendment to the operations specifications, fixing a period of not less than 7 days within which the holder may submit written information, views, and arguments concerning the proposed amendment. After consideration of all relevant matter presented, that District Office notifies the holder of any amendment adopted, or a recission of the notice. The amendment becomes effective not less than 30 days after the holder receives notice of the adoption of the amendment, unless the holder petitions the Director of Airworthiness for amendments pertaining to airworthiness or the Director of Flight Operations for amendments pertaining to flight operations, for reconsideration of the amendment. In that case, the effective date of the amendment is stayed pending a decision by the Director. If the Director finds there is an emergency requiring immediate action as to safety in air commerce that makes the provisions of this paragraph impracticable or contrary to the public interest, the Director notifies the certificate holder that the amendment is effective on the date of receipt, without previous notice.

§ 135.19 Emergency operations.

(a) In an emergency involving the safety of persons or property, the certificate holder may deviate from the rules of this Part relating to aircraft and equipment and weather minimums to the extent required to meet that emergency.

(b) In an emergency involving the safety of persons or property, the pilot in command may deviate from the rules of this Part to the extent required to meet that emergency.

(c) Each person who, under the authority of this section, deviates from a rule of this Part shall, within 10 days, excluding Saturdays, Sundays, and Federal holidays, after the deviation, send to the FAA Flight Standards District Office charged with the overall inspection of the aircraft operation involved a complete report of the aircraft operation involved, including a description of the deviation and reasons for it.

§ 135.21 Manual requirements.

[(a) Each certificate holder, other than one who uses only one pilot in the certificate holder's operations,] shall prepare and keep current a manual setting forth the certificate holder's procedures and policies acceptable to the Administrator. This manual must be used by the certificate holder's flight, ground, and maintenance personnel in conducting its operations. However, the Administrator may authorize a deviation from this paragraph if the Administrator finds that, because of the limited size of the operation, all or part of the manual is not necessary for guidance of flight, ground, or maintenance personnel.

(b) Each certificate holder shall maintain at least one copy of the manual at its principal operations base.

(c) The manual must not be contrary to any applicable Federal regulations, foreign regulation applicable to the certificate holder's operations in foreign countries, or the certificate holder's operating certificate or operations specifications.

(d) A copy of the manual, or appropriate portions of the manual (and changes and additions) shall be made available to maintenance and ground operations personnel by the certificate holder and furnished to—

(1) Its flight crewmembers; and

(2) Representatives of the Administrator assigned to the certificate holder.

(e) Each employee of the certificate holder to whom a manual or appropriate portions of it are furnished under paragraph (d)(1) of this section shall keep it up to date with the changes and additions furnished to them.

(f) Except as provided in paragraph (g) of this section, each certificate holder shall carry appropriate parts of the manual on each aircraft when away from the principal operations base. The appropriate parts must be available for use by ground or flight personnel.

(g) If a certificate holder conducts aircraft inspections or maintenance at specified stations where it keeps the approved inspection program manual, it is not required to carry the manual aboard the aircraft en route to those stations.

§ 135.23 Manual contents.

Each manual shall have the date of the last revision on each revised page. The manual must include—

[(a) The name of each management person required under § 135.37(a) who is authorized to act for the certificate holder, the person's assigned area of responsibility, the person's duties, responsibilities, and authority, and the name and title of each person authorized to exercise operational control under § 135.77;]

(b) Procedures for ensuring compliance with aircraft weight and balance limitations and, for multiengine aircraft, for determining compliance with § 135.185;

(c) Copies of the certificate holder's operations specifications or appropriate extracted information, including area of operations authorized, category and class of aircraft authorized, crew complements, and types of operations authorized;

(d) Procedures for complying with accident notification requirements.

(e) Procedures for ensuring that the pilot in command knows that required airworthiness inspections have been made and that the aircraft has been approved for return to service in compliance with applicable maintenance requirements;

(f) Procedures for reporting and recording mechanical irregularities that come to the attention of the pilot in command before, during, and after completion of a flight;

(g) Procedures to be followed by the pilot in command for determining that mechanical irregularities or defects reported for previous flights have been corrected or that correction has been deferred;

(h) Procedures to be followed by the pilot in command to obtain maintenance, preventive maintenance, and servicing of the aircraft at a place where previous arrangements have not been made by the operator, when the pilot is authorized to so act for the operator;

(i) Procedures under § 135.179 for the release for, or continuation of, flight if any item of equipment required for the particular type of operation becomes inoperative or unserviceable en route;

(j) Procedures for refueling aircraft, eliminating fuel contamination, protecting from fire (including electrostatic protection), and supervising and protecting passengers during refueling;

(k) Procedures to be followed by the pilot in command in the briefing under § 135.117;

(l) Flight locating procedures, when applicable;

(m) Procedures for ensuring compliance with emergency procedures, including a list of the functions assigned each category of required crewmembers in connection with an emergency and emergency evacuation duties under § 135.123;

(n) En route qualification procedures for pilots, when applicable;

(o) The approved aircraft inspection program, when applicable;

(p) Procedures and instructions to enable personnel to recognize hazardous materials, as defined in Title 49 CFR, and if these materials are to be carried, stored, or handled, procedures and instructions for—

(1) Accepting shipment of hazardous material required by Title 49 CFR, to assure proper packaging, marking, labeling, shipping documents, compatibility of articles, and instructions on their loading, storage, and handling;

(2) Notification and reporting hazardous material incidents as required by Title 49 CFR; and

(3) Notification of the pilot in command when there are hazardous materials aboard, as required by Title 49 CFR;

(q) Procedures for the evacuation of persons who may need the assistance of another person to move expeditiously to an exit if an emergency occurs; and

(r) Other procedures and policy instructions regarding the certificate holder's operations, that are issued by the certificate holder.

§ 135.25 Aircraft requirements.

[(a) Except as provided in paragraph (d) of this section, no certificate holder may operate an aircraft under this Part unless that aircraft—]

(1) Is registered as a civil aircraft of the United States and carries an appropriate and current airworthiness certificate issued under this chapter; and

(2) Is in an airworthy condition and meets the applicable airworthiness requirements of this chapter, including those relating to identification and equipment.

(b) Each certificate holder must have the exclusive use of a least one aircraft that meets the requirements for at least one kind of operation authorized in the certificate holder's operations specifications. In addition, for each kind of operation for which the certificate holder does not have the exclusive use of an aircraft, the certificate holder must have available for use under a written agreement (including arrangements for performing required maintenance) at least one aircraft that meets the requirements for that kind of operation. However, this paragraph does not prohibit the operator from using or authorizing the use of the aircraft for other than air taxi or commercial operations and does not require the certificate holder to have exclusive use of all aircraft that the certificate holder uses.

(c) For the purposes of paragraph (b) of this section, a person has exclusive use of an aircraft if that person has the sole possession, control, and use of it for flight, as owner, or has a written agreement (including arrangements for performing required maintenance), in effect when the aircraft is operated, giving the person that possession, control, and use for at least 6 consecutive months.

[(d) A certificate holder may operate in common carriage, and for the carriage of mail, a civil aircraft which is leased or chartered to it without crew and is registered in a country which is a party to the Convention on International Civil Aviation if—

[(1) The aircraft carries an appropriate airworthiness certificate issued by the country of registration and meets the registration and identification requirements of that country;

[(2) The aircraft is of a type design which is approved under a U.S. type certificate and complies with all of the requirements of this chapter (14 CFR Chapter 1) that would be applicable to that aircraft were it registered in the United States, including the requirements which must be met for issuance of a U.S. standard airworthiness certificate (including type design conformity, condition for safe operation, and the noise, fuel venting, and engine emission requirements of this

chapter), except that a U.S. registration certificate and a U.S. standard airworthiness certificate will not be issued for the aircraft;

[(3) The aircraft is operated by U.S.-certificated airmen employed by the certificate holder; and

[(4) The certificate holder files a copy of the aircraft lease or charter agreement with the FAA Aircraft Registry, Department of Transportation, 6400 South MacArthur Boulevard, Oklahoma City, Oklahoma (Mailing address: P.O. Box 25504, Oklahoma City, Oklahoma 73125).]

§ 135.27 Business office and operations base.

(a) Each certificate holder shall maintain a principal business office.

(b) Each certificate holder shall, before establishing or changing the location of any business office or operations base, except a temporary operations base, notify in writing the FAA Flight Standards District Office charged with the overall inspection of the certificate holder.

(c) No certificate holder who establishes or changes the location of any business office or operations base, except a temporary operations base, may operate an aircraft under this Part unless the certificate holder complies with paragraph (b) of this section.

§ 135.29 Use of business names.

No certificate holder may obtain an aircraft under this Part under a business name that is not on the certificate holder's operating certificate.

§ 135.31 Advertising.

No certificate holder may advertise or otherwise offer to perform operations subject to this Part that are not authorized by the certificate holder's operating certificate and operations specifications.

§ 135.33 Area limitations on operations.

(a) No person may operate an aircraft in a geographical area that is not specifically authorized by appropriate operations specifications issued under this Part.

(b) No person may operate an aircraft in a foreign country unless that person is authorized to do so by that country.

§ 135.35 Termination of operations.

Within 30 days after a certificate holder terminates operations under this Part, the operating certificate and operations specifications must be surrendered by the certificate holder to the FAA Flight Standards District Office charged with the overall inspection of the certificate holder.

§ 135.37 Management personnel required.

[(a) Each certificate holder, other than one who uses only one pilot in a certificate holder's operations,] must have enough qualified management personnel in the following or equivalent positions to ensure safety in its operations:

(1) Director of operations.

(2) Chief pilot.

(3) Director of maintenance.

(b) Upon application by the certificate holder, the Administrator may approve different positions or numbers of positions than those listed in paragraph (a) of this section for a particular operation if the certificate holder shows that it can perform its operations safely under the direction of fewer or different categories of management personnel.

(c) Each certificate holder shall—

(1) Set forth the duties, responsibilities, and authority of the personnel required by this section in the manual required by § 135.21;

(2) List in the manual required by § 135.21 the name of the person or persons assigned to those positions; and

(3) Within 10 working days, notify the FAA Flight Standards District Office charged with the overall inspection of the certificate holder of any change made in the assignment of persons to the listed positions.

§ 135.39 Management personnel qualifications.

(a) *Director of operations.* No person may serve as director of operations under § 135.37(a) unless that person knows the contents of the manual required by § 135.21, the operations specifications, the provisions of this Part and other applicable regulations necessary for the proper performance of the person's duties and responsibilities and:

(1) The director of operations for a certificate holder conducting any operations for which the pilot in command is required to hold an airline transport pilot certificate must—

(i) Hold or have held an airline transport pilot certificate; and

(ii) Have at least 3 years of experience as pilot in command of an aircraft operated under this Part, Part 121 or Part 127 of this chapter; or

(iii) Have at least 3 years of experience as director of operations with a certificate holder operating under this Part, Part 121 or Part 127 of this chapter.

(2) The director of operations for a certificate holder who is not conducting any operation for which the pilot in command is required to hold an airline transport pilot certificate must—

(i) Hold or have held a commercial pilot certificate; and

(ii) Have at least 3 years of experience as a pilot in command of an aircraft operated under this Part, Part 121 or Part 127 of this chapter; or

(iii) Have at least 3 years of experience as director of operations with a certificate holder operating under this Part, Part 121 or Part 127 of this chapter.

(b) *Chief pilot.* No person may serve as chief pilot under § 135.37(a) unless that person knows the contents of the manual required by § 135.21, the operations specifications, the provisions of this Part and other applicable regulations necessary for the proper performance of the person's duties, and:

(1) The chief pilot of a certificate holder conducting any operation for which the pilot in command is required to hold an airline transport pilot certificate must—

(i) Hold a current airline transport pilot certificate with appropriate ratings for at least one of the types of aircraft used; and

(ii) Have at least 3 years of experience as a pilot in command of an aircraft under this Part, Part 121 or Part 127 of this chapter.

(2) The chief pilot of a certificate holder who is not conducting any operation for which the pilot in command is required to hold an airline transport pilot certificate must—

[(i) Hold a current, commercial pilot certificate with an instrument rating. If an instrument rating is not required for the pilot in command under this part, the chief pilot must hold a current, commercial pilot certificate; and]

(ii) Have at least 3 years of experience as a pilot in command of an aircraft under this Part, Part 121 or Part 127 of this chapter.

(c) *Director of maintenance.* No person may serve as director of maintenance under § 135.37(a) unless that person knows the maintenance sections of the certificate holder's manual, the operations specifications, the provisions of this Part and other applicable regulations necessary for the proper performance of the person's duties, and—

(1) Holds a mechanic certificate with both airframe and powerplant ratings; and

(2) Has at least 3 years of maintenance experience as a certificated mechanic on aircraft, including, at the time of appointment as director of maintenance, the recent experience requirements of § 65.83 of this chapter in the same category and class of aircraft as used by the certificate holder, or at least 3 years of experience with a certificated airframe repair station including 1 year in the capacity of approving aircraft for return to service.

(d) Deviations from this section may be authorized if the person has had equivalent aeronautical experience. The Chief of the Flight Standards Division in the region of the certificate holding district office may authorize a deviation for the director of operations, chief pilot, and the director of maintenance.

§ 135.41 Carriage of narcotic drugs, marihuana, and depressant or stimulant drugs or substances.

If the holder of a certificate issued under this Part allows any aircraft owned or leased by that holder to be engaged in any operation that the certificate holder knows to be in violation of § 91.12(a) of this chapter, that operation is a basis for suspending or revoking the certificate.

§ 135.43 Crewmember certificate: international operations: application and issue.

(a) This section provides for the issuance of a crewmember certificate to United States citizens who are employed by certificate holders as crewmembers on United States registered aircraft engaged in international air commerce. The purpose of the certificate is to facilitate the entry and clearance of those crewmembers into ICAO contracting states. They are issued under Annex 9, as amended, to the Convention on International Civil Aviation.

(b) An application for a crewmember certificate is made on FAA Form 8060-6 "Application for Crewmember Certificate," to the FAA Flight Standards District Office charged with the overall inspection of the certificate holder by whom the applicant is employed. The certificate is issued on FAA Form 8060-42, "Crewmember Certificate."

(c) The holder of a certificate issued under this section, or the certificate holder by whom the holder is employed, shall surrender the certificate for cancellation at the nearest FAA Flight Standards District Office or submit it for cancellation to the Airmen Certification Branch, AAC-260, P.O. Box 25082, Oklahoma City, Oklahoma 73125, at the termination of the holder's employment with that certificate holder.

Subpart B—Flight Operations

§ 135.61 General.

This subpart prescribes rules, in addition to those in Part 91 of this chapter, that apply to operations under this Part.

§ 135.63 Recordkeeping requirements.

(a) Each certificate holder shall keep at its principal business office or at other places approved by the Administrator, and shall make available for inspection by the Administrator the following—

(1) The certificate holder's operating certificate;

(2) The certificate holder's operations specifications;

(3) A current list of the aircraft used or available for use in operations under this Part and the operations for which each is equipped; and

(4) An individual record of each pilot used in operations under this Part, including the following information:

(i) The full name of the pilot.

(ii) The pilot certificate (by type and number) and ratings that the pilot holds.

(iii) The pilot's aeronautical experience in sufficient detail to determine the pilot's qualifications to pilot aircraft in operation under this Part.

(iv) The pilot's current duties and the date of the pilot's assignment to those duties.

(v) The effective date and class of the medical certificate that the pilot holds.

(vi) The date and result of each of the initial and recurrent competency tests and proficiency and route checks required by this Part and the type of aircraft flown during that test or check.

(vii) The pilot's flight time in sufficient detail to determine compliance with the flight time limitations of this Part.

(viii) The pilot's check pilot authorization, if any.

(ix) Any reaction taken concerning the pilot's release from employment for physical or professional disqualification.

(x) The date of the completion of the initial phase and each recurrent phase of the training required by this Part.

(b) Each certificate holder shall keep each record required by paragraph (a)(3) of this section for at least 6 months, and each record required by paragraph (a)(4) of this section for at least 12 months, after it is made.

(c) For multiengine aircraft, each certificate holder is responsible for the preparation and accuracy of a load manifest in duplicate containing information concerning the loading of the aircraft. The manifest must be prepared before each takeoff and must include—

(1) The number of passengers;

(2) The total weight of the loaded aircraft;

(3) The maximum allowable takeoff weight for that flight;

(4) The center of gravity limits;

(5) The center of gravity of the loaded aircraft, except that the actual center of gravity need not be computed if the aircraft is loaded according to a loading schedule or other approved method that ensures that the center of gravity of the loaded aircraft is within approved limits. In those cases, an entry shall be made on the manifest indicating that the center of gravity is within limits according to a loading schedule or other approved method;

(6) The registration number of the aircraft or flight number;

(7) The origin and destination; and

(8) Identification of crewmembers and their crew position assignments.

(d) The pilot in command of the aircraft for which a load manifest must be prepared shall carry a copy of the completed load manifest in the aircraft to its destination. The certificate holder shall keep copies of completed load manifest for at least 30 days at its principal operations base, or at another location used by it and approved by the Administrator.

§ 135.65   Reporting mechanical irregularities.

(a) Each certificate holder shall provide an aircraft maintenance log to be carried on board each aircraft for recording or deferring mechanical irregularities and their correction.

(b) The pilot in command shall enter or have entered in the aircraft maintenance log each mechanical irregularity that comes to the pilot's attention during flight time. Before each flight, the pilot in command shall, if the pilot does not already know, determine the status of each irregularity entered in the maintenance log at the end of the preceding flight.

(c) Each person who takes corrective action or defers action concerning a reported or observed failure or malfunction of an airframe, powerplant, propeller, rotor, or appliance, shall record the action taken in the aircraft maintenance log under the applicable maintenance requirements of this chapter.

(d) Each certificate holder shall establish a procedure for keeping copies of the aircraft maintenance log required by this section in the aircraft for access by appropriate personnel and shall include that procedure in the manual required by § 135.21.

§ 135.67   Reporting potentially hazardous meteorological conditons and irregularities of communications or navigation facilities.

[Whenever a pilot encounters a potentially hazardous meteorological condition or an irregularity in a ground communications or navigational facility in flight, the knowledge of which the pilot considers essential to the safety of other flights, the pilot shall notify an appropriate ground radio station as soon as practicable.]

§ 135.69   Restriction or suspension of operations: continuation of flight in an emergency.

(a) During operations under this Part, if a certificate holder or pilot in command knows of conditions, including airport and runway conditions, that are a hazard to safe operations, the certificate holder or pilot in command, as the case may be, shall restrict or suspend operations as necessary until those conditions are corrected.

(b) No pilot in command may allow a flight to continue toward any airport of intended landing under the conditions set forth in paragraph (a) of this section, unless in the opinion of the pilot in command, the conditions that are a hazard to safe operations may reasonably be expected to be corrected by the estimated time of arrival or, unless there is no safer procedure. In the latter event, the continuation toward that airport is an emergency situation under § 135.19.

§ 135.71   Airworthiness check.

The pilot in command may not begin a flight unless the pilot determines that the airworthiness inspections required by § 91.169 of this chapter, or § 135.419, whichever is applicable, have been made.

§ 135.73   Inspections and tests.

Each certificate holder and each person employed by the certificate holder shall allow the Administrator, at any time or place, to make inspections or tests (including en route inspections) to determine the holder's compliance with the Federal Aviation Act of 1958, applicable regulations, and the certificate holder's operating certificate, and operations specifications.

§ 135.75   Inspectors credentials: admission to pilots' compartment: forward observer's seat.

(a) Whenever, in performing the duties of conducting an inspection, an FAA inspector presents an Aviation Safety Inspector credential, FAA Form 110A, to the pilot in command of an aircraft operated by the certificate holder, the inspector must be given free and uninterrupted access to the pilot compartment of that aircraft. However, this paragraph does not limit the emergency authority of the pilot in command to exclude any person from the pilot compartment in the interest of safety.

(b) A forward observer's seat on the flight deck, or forward passenger seat with headset or speaker must be provided for use by the Administrator while conducting en route inspections. The suitability of the location of the seat and the headset or speaker for use in conducting en route inspections is determined by the Administrator.

§ 135.77   Responsibility for operational control.

Each certificate holder is responsible for operational control and shall list, in the manual required by § 135.21, the name and title of each person authorized by it to exercise operational control.

§ 135.79   Flight locating requirements.

(a) Each certificate holder must have procedures established for locating each flight, for which an FAA flight plan is not filed, that—

(1) Provide the certificate holder with at least the information required to be included in a VFR flight plan;

(2) Provide for timely notification of an FAA facility or search and rescue facility, if an aircraft is overdue or missing; and

(3) Provide the certificate holder with the location, date, and estimated time for reestablishing radio or telephone communications, if the flight will operate in an area where communications cannot be maintained.

(b) Flight locating information shall be retained at the certificate holder's principal place of business, or at other places designated by the certificate holder in the flight locating procedures, until the completion of the flight.

(c) Each certificate holder shall furnish the representative of the Administrator assigned to it with a copy of its flight locating procedures and any changes or additions, unless those procedures are included in a manual required under this Part.

§ 135.81 Informing personnel of operational information and appropriate changes.

Each certificate holder shall inform each person in its employment of the operations specifications that apply to that person's duties and responsibilities and shall make available to each pilot in the certificate holder's employ the following materials in current form:

(a) Airman's Information Manual (Alaska Supplement in Alaska and Pacific Chart Supplement in Pacific-Asia Regions) or a commercial publication that contains the same information.

(b) This Part and Part 91 of this chapter.

(c) Aircraft Equipment Manuals, and Aircraft Flight Manual or equivalent.

(d) For foreign operations, the International Flight Information Manual or a commercial publication that contains the same information concerning the pertinent operational and entry requirements of the foreign country or countries involved.

§ 135.83 Operating information required.

(a) The operator of an aircraft must provide the following materials, in current and appropriate form, accessible to the pilot at the pilot station, and the pilot shall use them:

(1) A cockpit checklist.

(2) For multiengine aircraft or for aircraft with retractable landing gear, an emergency cockpit checklist containing the procedures required by paragraph (c) of this section, as appropriate.

(3) Pertinent aeronautical charts.

(4) For IFR operations, each pertinent navigational en route, terminal area, and approach and letdown chart.

(5) For multiengine aircraft, one-engine-inoperative climb performance data and if the aircraft is approved for use in IFR or over-the-top operations, that data must be sufficient to enable the pilot to determine compliance with § 135.181(a)(2).

(b) Each cockpit checklist required by paragraph (a)(1) of this section must contain the following procedures: (1) Before starting engines; (2) Before takeoff; (3) Cruise; (4) Before landing; (5) After landing; (6) Stopping engines.

(c) Each emergency cockpit checklist required by paragraph (a)(2) of this section must contain the following procedures as appropriate:

(1) Emergency operation of fuel, hydraulic, electrical, and mechanical systems.

(2) Emergency operation of instruments and controls.

(3) Engine inoperative procedures.

(4) Any other emergency procedures necessary for safety.

§ 135.85 Carriage of persons without compliance with the passenger-carrying provisions of this Part.

The following persons may be carried aboard an aircraft without complying with the passenger-carrying requirements of this Part:

(a) A crewmember or other employee of the certificate holder.

(b) A person necessary for the safe handling of animals on the aircraft.

(c) A person necessary for the safe handling of hazardous materials (as defined in Subchapter C of Title 49 CFR).

(d) A person performing duty as a security or honor guard accompanying a shipment made by or under the authority of the U.S. Government.

(e) A military courier or a military route supervisor carried by a military cargo contract air carrier or commercial operator in operations under a military cargo contract, if that carriage is specifically authorized by the appropriate military service.

(f) An authorized representative of the Administrator conducting an en route inspection.

(g) A person, authorized by the Administrator, who is performing a duty connected with a cargo operation of the certificate holder.

§ 135.87 Carriage of cargo including carry-on baggage.

No person may carry cargo, including carry-on baggage, in or on any aircraft unless—

(a) It is carried in an approved cargo rack, bin, or compartment installed in or on the aircraft;

(b) It is secured by an approved means; or

(c) It is carried in accordance with each of the following:

(1) For cargo, it is properly secured by a safety belt or other tie-down having enough strength to eliminate the possibility of shifting under all normally anticipated flight and ground conditions, or for carry-on baggage, it is restrained so as to prevent its movement during air turbulence.

(2) It is packaged or covered to avoid possibly injuring occupants.

(3) It does not impose any load on seats or on the floor structure that exceeds the load limitation for those components.

(4) It is not located in a position that obstructs the access to, or use of, any required emergency or regular exit, or the use of the aisle between the crew and the passenger compartment, or located in a position that obscures any passenger's view of the "seat belt" sign, "no smoking" sign, or any

required exit sign, unless an auxiliary sign or other approved means for proper notification of the passengers is provided.

(5) It is not carried directly above seated occupants.

(6) It is stowed in compliance with this section for takeoff and landing.

(7) For cargo only operations, paragraph (c)(4) of this section does not apply if the cargo is loaded so that at least one emergency or regular exit is available to provide all occupants of the aircraft a means of unobstructed exit from the aircraft if an emergency occurs.

(d) Each passenger seat under which baggage is stowed shall be fitted with a means to prevent articles of baggage stowed under it from sliding under crash impacts severe enough to induce the ultimate inertia forces specified in the emergency landing condition regulations under which the aircraft was type certificated.

(e) When cargo is carried in cargo compartments that are designed to require the physical entry of a crewmember to extinguish any fire that may occur during flight, the cargo must be loaded so as to allow a crewmember to effectively reach all parts of the compartment with the contents of a hand fire extinguisher.

§ 135.89 Pilot requirements: use of oxygen.

☞ F121

(a) *Unpressurized aircraft.* Each pilot of an unpressurized aircraft shall use oxygen continuously when flying—

(1) At altitudes above 10,000 feet through 12,000 feet MSL for that part of the flight at those altitudes that is of more than 30 minutes duration; and

(2) Above 12,000 feet MSL.

(b) *Pressurized aircraft.*

(1) Whenever a pressurized aircraft is operated with the cabin pressure altitude more than 10,000 feet MSL, each pilot shall comply with paragraph (a) of this section.

(2) Whenever a pressurized aircraft is operated at altitudes above 25,000 feet through 35,000 feet MSL unless each pilot has an approved quick-donning type oxygen mask—

(i) At least one pilot at the controls shall wear, secured and sealed, an oxygen mask that either supplies oxygen at all times or automatically supplies oxygen whenever the cabin pressure altitude exceeds 12,000 feet MSL; and

(ii) During that flight, each other pilot on flight deck duty shall have an oxygen mask, connected to an oxygen supply, located so as to allow immediate placing of the mask on the pilot's face sealed and secured for use.

(3) Whenever a pressurized aircraft is operated at altitudes above 35,000 feet MSL, at least one pilot at the controls shall wear,

secured and sealed, an oxygen mask required by paragraph (2)(i) of this paragraph.

(4) If one pilot leaves a pilot duty station of an aircraft when operating at altitudes above 25,000 feet MSL, the remaining pilot at the controls shall put on and use an approved oxygen mask until the other pilot returns to the pilot duty station of the aircraft.

§ 135.91 Oxygen for medical use by passengers.

(a) Except as provided in paragraphs (d) and (e) of this section, no certificate holder may allow the carriage or operation of equipment for the storage, generation or dispensing of medical oxygen unless the unit to be carried is constructed so that all valves, fittings, and gauges are protected from damage during that carriage or operation and unless the following conditions are met—

(1) The equipment must be—

(i) Of an approved type or in conformity with the manufacturing, packaging, marking, labeling and maintenance requirements of Title 49 CFR Parts 171, 172, and 173, except § 173.24(a)(1);

(ii) When owned by the certificate holder, maintained under the certificate holder's approved maintenance program;

(iii) Free of flammable contaminants on all exterior surfaces; and ☞ **F121**

(iv) Appropriately secured.

(2) When the oxygen is stored in the form of a liquid, the equipment must have been under the certificate holder's approved maintenance program since its purchase new or since the storage container was last purged.

(3) When the oxygen is stored in the form of a compressed gas as defined in Title 49 CFR § 173.300(a) —

(i) When owned by the certificate holder, it must be maintained under its approved maintenance program; and

(ii) The pressure in any oxygen cylinder must not exceed the rated cylinder pressure.

(4) The pilot in command must be advised when the equipment is on board, and when it is intended to be used.

(5) The equipment must be stowed, and each person using the equipment must be seated, so as not to restrict access to or use of any required emergency or regular exit, or of the aisle in the passenger compartment.

(b) No person may smoke and no certificate holder may allow any person to smoke within 10 feet of oxygen storage and dispensing equipment carried under paragraph (a) of this section.

(c) No certificate holder may allow any person other than a person trained in the use of medical oxygen equipment to connect or disconnect oxygen bottles or any other ancillary

component while any passenger is aboard the aircraft.

(d) Paragraph (a)(1)(i) of this section does not apply when that equipment is furnished by a professional or medical emergency service for use on board an aircraft in a medical emergency when no other practical means of transportation (including any other properly equipped certificate holder) is reasonably available and the person carried under the medical emergency is accompanied by a person trained in the use of medical oxygen.

(e) Each certificate holder who, under the authority of paragraph (a)(1)(i) of this section deviates from paragraph (a)(1)(i) of this section under a medical emergency shall, within 10 days, excluding Saturdays, Sundays, and Federal holidays, after the deviation, send to the FAA Flight Standards District Office charged with the overall inspection of the certificate holder a complete report of the operation involved, including a description of the deviation and the reasons for it.

§ 135.93 Autopilot: minimum altitudes for use.

(a) Except as provided in paragraphs (b), (c), and (d) of this section, no person may use an autopilot at an altitude above the terrain which is less than 500 feet or less than twice the maximum altitude loss specified in the approved Aircraft Flight Manual or equivalent for a malfunction of the autopilot, whichever is higher.

(b) When using an instrument approach facility other than ILS, no person may use an autopilot at an altitude above the terrain that is less than 50 feet below the approved minimum descent altitude for that procedure, or less than twice the maximum loss specified in the approved Airplane Flight Manual or equivalent for a malfunction of the autopilot under approach conditions, whichever is higher.

(c) For ILS approaches, when reported weather conditions are less than the basic weather conditions in § 91.105 of this chapter, no person may use an autopilot with an approach coupler at an altitude above the terrain that is less than 50 feet above the terrain, or the maximum altitude loss specified in the approved Airplane Flight Manual or equivalent for the malfunction of the autopilot with approach coupler, whichever is higher.

(d) Without regard to paragraphs (a), (b), or (c) of this section, the Administrator may issue operations specifications to allow the use to touchdown, of an approved flight control guidance system with automatic capability, if—

(1) The system does not contain any altitude loss (above zero) specified in the approved Aircraft Flight Manual or equivalent for malfunction of the autopilot with approach coupler; and

(2) The Administrator finds that the use of the system to touchdown will not otherwise adversely affect the safety standards of this section.

(e) This section does not apply to the operations conducted in rotorcraft.

§ 135.95 Airmen: limitations on use of services.

No certificate holder may use the services of any person as a airman unless the person performing those services—

(a) Holds an appropriate and current airman certificate; and

(b) Is qualified, under this chapter, for the operation for which the person is to be used.

§ 135.97 Aircraft and facilities for recent flight experience.

Each certificate holder shall provide aircraft and facilities to enable each of its pilots to maintain and demonstrate the pilot's ability to conduct all operations for which the pilot is authorized.

§ 135.99 Composition of flight crew.

(a) No certificate holder may operate an aircraft with less than the minimum flight crew specified in the aircraft operating limitations or the Aircraft Flight Manual for that aircraft and required by this Part for the kind of operation being conducted.

(b) No certificate holder may operate an aircraft without a second in command if that aircraft has a passenger seating configuration, excluding any pilot seat, of ten seats or more.

[§ 135.100 Flight crewmember duties.

[(a) No certificate holder shall require, nor may any flight crewmember perform, any duties during a critical phase of flight except those duties required for the safe operation of the aircraft. Duties such as company required calls made for such nonsafety related purposes as ordering galley supplies and confirming passenger connections, announcements made to passengers promoting the air carrier or pointing out sights of interest, and filling out company payroll and related records are not required for the safe operation of the aircraft.

[(b) No flight crewmember may engage in, nor may any pilot in command permit, any activity during a critical phase of flight which could distract any flight crewmember from the performance of his or her duties or which could interfere in any way with the proper conduct of those duties. Activities such as eating meals, engaging in nonessential conversations within the cockpit and nonessential communications between the cabin and cockpit crews, and reading publications not related to the proper conduct of the flight are not required for the safe operation of the aircraft.

[(c) For the purposes of this section, critical phases of flight includes all ground operations involving taxi, takeoff and landing, and all other flight operations conducted below 10,000 feet, except cruise flight.

[NOTE—Taxi is defined as "movement of an airplane under its own power on the surface of an airport."]

§ 135.101 Second in command required in IFR conditions.

Except as provided in §§ 135.103 and 135.105, no person may operate an aircraft carrying passengers in IFR conditions, unless there is a second in command in the aircraft.

§ 135.103 Exception to second in command requirement: IFR operations.

The pilot in command of an aircraft carrying passengers may conduct IFR operations without a second in command under the following conditions:

(a) A takeoff may be conducted under IFR conditions if the weather reports or forecasts, or any combination of them, indicate that the weather along the planned route of flight allows flight under VFR within 15 minutes flying time, at normal cruise speed, from the takeoff airport.

(b) En route IFR may be conducted if unforecast weather conditions below the VFR minimums of this chapter are encountered on a flight that was planned to be conducted under VFR.

(c) An IFR approach may be conducted if, upon arrival at the destination airport, unforecast weather conditions do not allow an approach to be completed under VFR.

(d) When IFR operations are conducted under this section:

(1) The aircraft must be properly equipped for IFR operations under this Part.

(2) The pilot must be authorized to conduct IFR operations under this Part.

(3) The flight must be conducted in accordance with an ATC IFR clearance.

IFR operations without a second in command may not be conducted under this section in an aircraft requiring a second in command under § 135.99.

§ 135.105 Exception to second in command requirement: approval for use of autopilot system.

(a) Except as provided in §§ 135.99 and 135.111, unless two pilots are required by this chapter for operations under VFR, a person may operate an aircraft without a second in command, if it is equipped with an operative approved autopilot system and the use of that system is authorized by appropriate operations specifications. No certificate holder may use any person, nor may any person serve, as a pilot in command under this section of an aircraft operated by a Commuter Air Carrier (as defined in § 298.2 of this title) in passenger-carrying operations unless that person has at least 100 hours pilot in command flight time in the make and model of aircraft to be flown and has met all other applicable requirements of this Part.

(b) The certificate holder may apply for an amendment of its operations specifications to authorize the use of an autopilot system in place of a second in command.

(c) The Administrator issues an amendment to the operations specifications authorizing the use of an autopilot system, in place of a second in command, if—

(1) The autopilot is capable of operating the aircraft controls to maintain flight and maneuver it about the three axes; and

(2) The certificate holder shows, to the satisfaction of the Administrator, that opera-

tions using the autopilot system can be conducted safely and in compliance with this Part.

The amendment contains any conditions or limitations on the use of the autopilot system that the Administrator determines are needed in the interest of safety.

§ 135.107 Flight attendant crewmember requirement.

No certificate holder may operate an aircraft that has a passenger seating configuration, excluding any pilot seat, of more than 19 unless there is a flight attendant crewmember on board the aircraft.

§ 135.109 Pilot in command or second in command: designation required.

(a) Each certificate holder shall designate a—

(1) Pilot in command for each flight; and

(2) Second in command for each flight requiring two pilots.

(b) The pilot in command, as designated by the certificate holder, shall remain the pilot in command at all times during that flight.

§ 135.111 Second in command required in Category II operations.

No person may operate an aircraft in a Category II operation unless there is a second in command of the aircraft.

§ 135.113 Passenger occupancy of pilot seat.

No certificate holder may operate an aircraft type certificate after October 15, 1971, that has a passenger seating configuration, excluding any pilot seat, of more than eight seats if any person other than the pilot in command, a second in command, a company check airman, or an authorized representative of the Administrator, the National Transportation Safety Board, or the United States Postal Service occupies a pilot seat.

§ 135.115 Manipulation of controls.

No pilot in command may allow any person to manipulate the flight controls of an aircraft during flight conducted under this Part, nor may any person manipulate the controls during such flight unless that person is—

(a) A pilot employed by the certificate holder and qualified in the aircraft; or

(b) An authorized safety representative of the Administrator who has the permission of the pilot in command, is qualified in the aircraft, and is checking flight operations.

§ 135.117 Briefing of passengers before flight.

(a) Before each takeoff each pilot in command of an aircraft carrying passengers shall ensure that all passengers have been orally briefed on—

(1) *Smoking.* Each passenger shall be briefed on when, where, and under what conditions smoking is prohibited (including, but not limited to, the pertinent requirements of Part 252 of this title). This briefing shall include a statement that the Federal Aviation Regulations require passenger compliance with the lighted passenger information signs (if such signs are required) and posted placards. The briefing shall

also include a statement (if the aircraft is equipped with a lavatory) that Federal law prohibits tampering with, disabling, or destroying any smoke detector installed in an aircraft lavatory.

(2) Use of seat belts;

(3) The placement of seat backs in an upright position before takeoff and landing;

(4) Location and means for opening the passenger entry door and emergency exits;

(5) Location of survival equipment;

(6) If the flight involves extended overwater operation, ditching procedures and the use of required flotation equipment;

(7) If the flight involves operations above 12,000 feet MSL, the normal and emergency use of oxygen; and

(8) Location and operation of fire extinguishers.

(b) Before each takeoff the pilot in command shall ensure that each person who may need the assistance of another person to move expeditiously to an exit if an emergency occurs and that person's attendant, if any, has received a briefing as to the procedures to be followed if an evacuation occurs. This paragraph does not apply to a person who has been given a briefing before a previous leg of a flight in the same aircraft.

[(c) The oral briefing required by paragraph (a) of this section shall be given by the pilot in command or a crewmember.

[(d) Notwithstanding the provisions of paragraph (c) of this section, for aircraft certificated to carry 19 passengers or less, the oral briefing required by paragraph (a) of this section shall be given by the pilot in command, a crewmember, or other qualified person designated by the certificate holder and approved by the Administrator.

[(e) The oral briefing required by paragraph (a) shall be supplemented by printed cards which must be carried in the aircraft in locations convenient for the use of each passenger.

The cards must—

[(1) Be appropriate for the aircraft on which they are to be used;

[(2) Contain a diagram of, and method of operating, the emergency exits; and

[(3) Contain other instructions necessary for the use of emergency equipment on board the aircraft.

[(f) The briefing required by paragraph (a) may be delivered by means of an approved recording playback device that is audible to each passenger under normal noise levels.]

§ 135.119 Prohibition against carriage of weapons.

No person may, while on board an aircraft being operated by a certificate holder, carry on or about that person a deadly or dangerous weapon, either concealed or unconcealed. This section does not apply to—

(a) Officials or employees of a municipality or a State, or of the United States, who are authorized to carry arms; or

(b) Crewmembers and other persons authorized by the certificate holder to carry arms.

§ 135.121 Alcoholic beverages.

(a) No person may drink any alcoholic beverage aboard an aircraft unless the certificate holder operating the aircraft has served that beverage.

(b) No certificate holder may serve any alcoholic beverage to any person aboard its aircraft if that person appears to be intoxicated.

(c) No certificate holder may allow any person to board any of its aircraft if that person appears to be intoxicated.

§ 135.123 Emergency and emergency evacuation duties.

(a) Each certificate holder shall assign to each required crewmember for each type of aircraft as appropriate, the necessary functions to be performed in an emergency or in a situation requiring emergency evacuation. The certificate holder shall ensure that those functions can be practicably accomplished, and will meet any reasonably anticipated emergency including incapacitation of individual crewmembers or their inability to reach the passenger cabin because of shifting cargo in combination cargo-passenger aircraft.

(b) The certificate holder shall describe in the manual required under § 135.21 the functions of each category of required crewmembers assigned under paragraph (a) of this section.

§ 135.125 Airplane security.

Certificate holders conducting operations under this Part shall comply with the applicable security requirements in Part 108 of this chapter.

§ 135.127 Passenger information.

(a) The no smoking signs required by § 135.177(a)(3) of this part must be turned on:

(1) During flight time on flight segments which are scheduled in the current North American Edition of the Official Airline Guide to be 2 hours or less in duration, except those flight segments between a point in the United States and a point in another country; or

(2) On flight segments other than those described in paragraph (a)(1) of this section, for each takeoff and landing, and at any other time considered necessary by the pilot in command.

(b) No person may smoke while a no smoking sign is lighted, except that the pilot in command may authorize smoking on the flight deck (if it is physically separated from the passenger cabin) except during takeoff and landing.

(c) No person may smoke in any aircraft lavatory.

(d) After December 31, 1988, no person may operate an aircraft with a lavatory equipped with a smoke detector unless there is in that lavatory a sign or placard which reads: "Federal law provides for a penalty of up to $2,000 for tampering with the smoke detector installed in this lavatory."

(e) The provisions of paragraph (a)(1) of this section shall cease to be effective on April 24, 1990.

Subpart C—Aircraft and Equipment

§ 135.141 Applicability.

This subpart prescribes aircraft and equipment requirements for operations under this Part. The requirements of this subpart are in addition to the aircraft and equipment requirements of Part 91 of this chapter. However, this Part does not require the duplication of any equipment required by this chapter.

§ 135.143 General requirements.

(a) No person may operate an aircraft under this Part unless that aircraft and its equipment meet the applicable regulations of this chapter.

(b) Except as provided in § 135.179, no person may operate an aircraft under this Part unless the required instruments and equipment in it have been approved and are in an operable condition.

(c) ATC transponder equipment installed within the time periods indicated below must meet the performance and environmental requirements of the following TSO's.

(1) Through January 1, 1992:
(i) Any class of TSO-C74b or any class of TSO-C74c as appropriate, provided that the equipment was manufactured before January 1, 1990; or
(ii) The appropriate class of TSO-C112 (Mode S).

(2) After January 1, 1992: The appropriate class of TSO-C112 (Mode S). For purposes of paragraph (c)(2) of this section, "installation" does not include—
(i) Temporary installation of TSO-C74b or TSO-C74c substitute equipment, as appropriate, during maintenance of the permanent equipment;
(ii) Reinstallation of equipment after temporary removal for maintenance; or
(iii) For fleet operations, installation of equipment in a fleet aircraft after removal of the equipment for maintenance from another aircraft in the same operator's fleet.

§ 135.145 Aircraft proving tests.

(a) No certificate holder may operate a turbojet airplane, or an aircraft for which two pilots are required by this chapter for operations under VFR, if it has not previously proved that aircraft or an aircraft of the same make and similar design in any operation under this Part unless, in addition to the aircraft certification tests, at least 25 hours of proving tests acceptable to the Administrator have been flown by that certificate holder including—

(1) Five hours of night time, if night flights are to be authorized;

(2) Five instrument approach procedures under simulated or actual instrument weather conditions, if IFR flights are to be authorized; and

(3) Entry into a representative number of en route airports as determined by the Administrator.

(b) No certificate holder may carry passengers in an aircraft during proving tests, except those needed to make the tests and those designated by the Administrator to observe the tests. However, pilot flight training may be conducted during the proving tests.

(c) For the purposes of paragraph (a) of this section, an aircraft is not considered to be of similar design if an alteration includes—

(1) The installation of powerplants other than those of a type similar to those with which it is certificated; or

(2) Alterations to the aircraft or its components that materially affect flight characteristics.

(d) The Administrator may authorize deviations from this section if the Administrator finds that special circumstances make full compliance with this section necessary.

§ 135.147 Dual controls required.

No person may operate an aircraft in operations requiring two pilots unless it is equipped with functioning dual controls. However, if the aircraft type certification operating limitations do not require two pilots, a throwover control wheel may be used in place of two control wheels.

§ 135.149 Equipment requirements: general.

No person may operate an aircraft unless it is equipped with—

(a) A sensitive altimeter that is adjustable for barometric pressure;

(b) Heating or deicing equipment for each carburetor or, for a pressure carburetor, an alternate air source;

(c) For turbojet airplanes, in addition to two gyroscopic bank-and-pitch indicators (artificial horizons) for use at the pilot stations, a third indicator that—

(1) Is powered from a source independent of the aircraft's electrical generating system;

(2) Continues reliable operation for at least 30 minutes after total failure of the aircraft's electrical generating systems;

(3) Operates independently of any other attitude indicating system;

(4) Is operative without selection after total failure of the aircraft's electrical generating system;

(5) Is located on the instrument panel in a position that will make it plainly visible to, and useable by, any pilot at the pilot's station; and

(6) Is appropriately lighted during all phases of operation;

(d) For aircraft having a passenger seating configuration, excluding any pilots seat, of more than 19, a public address system and a crewmember interphone system, approved under § 21.305 of this chapter, which meet §§ 121.318 and 121.319, respectively, of this chapter; and

(e) For turbine powered aircraft, any other equipment as the Administrator may require.

§ 135.151 Cockpit voice recorders.

(a) After October 11, 1991, no person may operate a multiengine, turbine-powered airplane or rotorcraft having a passenger seating configuration of six or more and for which two pilots are required by certification or operating rules unless it is equipped with an approved cockpit voice recorder that:

(1) Is installed in compliance with § 23.1457(a)(1) and (2), (b), (c), (d), (e), (f), and (g); § 25.1457(a) (1) and (2), (b), (c), (d), (e), (f), and (g); § 27.1457(a) (1) and (2), (b), (c), (d), (e), (f), and (g); or § 29.1457(a) (1) and (2), (b), (c), (d), (e), (f), and (g) of this chapter, as applicable; and

(2) Is operated continuously from the use of the check list before the flight to completion of the final check list at the end of the flight.

(b) After October 11, 1991, no person may operate a multiengine, turbine-powered airplane or rotorcraft having a passenger seating configuration of 20 or more seats unless it is equipped with an approved cockpit voice recorder that—

(1) Is installed in compliance with § 23.1457, § 25.1457, § 27.1457 or § 29.1457 of this chapter, as applicable; and

(2) Is operated continuously from the use of the check list before the flight to completion of the final check list at the end of the flight.

(c) In the event of an accident, or occurrence requiring immediate notification of the National Transportation Safety Board which results in termination of the flight, the certificate holder shall keep the recorded information for at least 60 days or, if requested by the Administrator or the Board, for a longer period. Information obtained from the record may be used to assist in determining the cause of accidents or occurrences in connection with investigations. The Administrator does not use the record in any civil penalty or certificate action.

(d) For those aircraft equipped to record the uninterrupted audio signals received by a boom or a mask microphone the flight crewmembers are required to use the boom microphone below 18,000 feet mean sea level. No person may operate a large turbine engine powered airplane manufactured after October 11, 1991, or on which a cockpit voice recorder has been installed after October 11, 1991, unless it is equipped to record the uninterrupted audio signal received by a boom or mask microphone in accordance with § 25.1457(c)(5) of this chapter.

(e) In complying with this section, an approved cockpit voice recorder having an erasure feature may be used, so that during the operation of the recorder, information:

(1) Recorded in accordance with paragraph (a) of this section and recorded more than 15 minutes earlier; or

(2) Recorded in accordance with paragraph (b) of this section and recorded more than 30 minutes earlier; may be erased or otherwise obliterated.

§ 135.152 Flight recorders.

(a) No person may operate a multiengine, turbine-powered airplane or rotorcraft having a passenger seating configuration, excluding any pilot seat, of 10 to 19 seats, that is brought onto the U.S. register after October 11, 1991, unless it is equipped with one or more approved flight recorders that utilize a digital method of recording and storing data, and a method of readily retrieving that data from the storage medium. The parameters specified in Appendix B or C, as applicable, of this part must be recorded within the range accuracy, resolution, and recording intervals as specified. The recorder shall retain no less than 8 hours of aircraft operation.

(b) After October 11, 1991, no person may operate a multiengine, turbine-powered airplane having a passenger seating configuration of 20 to 30 seats or a multiengine, turbine-powered rotorcraft having a passenger seating configuration of 20 or more seats unless it is equipped with one or more approved flight recorders that utilize a digital method of recording and storing data, and a method of readily retrieving that data from the storage medium. The parameters in Appendix D or E of this part, as applicable, that are set forth below, must be recorded within the ranges, accuracies, resolutions, and sampling intervals as specified.

(1) Except as provided in paragraph (b)(3) of this section for aircraft type certificated before October 1, 1969, the following parameters must be recorded:

(i) Time;

(ii) Altitude;

(iii) Airspeed;

(iv) Vertical acceleration;

(v) Heading;

(vi) Time of each radio transmission to or from air traffic control;

(vii) Pitch attitude;

(viii) Roll attitude;

(ix) Longitudinal acceleration;

(x) Control column or pitch control surface position; and

(xi) Thrust of each engine.

(2) Except as provided in paragraph (b)(3) of this section for aircraft type certificated after September 30, 1969, the following parameters must be recorded:

(i) Time;

(ii) Altitude;

(iii) Airspeed;

(iv) Vertical acceleration;

(v) Heading;

(vi) Time of each radio transmission either to or from air traffic control;

(vii) Pitch attitude;

(viii) Roll attitude;

(ix) Longitudinal acceleration;

(x) Pitch trim position;

(xi) Control column or pitch control surface position;

(xii) Control wheel or lateral control surface position;

(xiii) Rudder pedal or yaw control surface position;

(xiv) Thrust of each engine;

(xv) Position of each thrust reverser;

(xvi) Trailing edge flap or cockpit flap control position; and

(xvii) Leading edge flap or cockpit flap control position.

(3) For aircraft manufactured after October 11, 1991, all of the parameters listed in Appendix D or E of this part, as applicable, must be recorded.

(c) Whenever a flight recorder required by this section is installed, it must be operated continuously from the instant the airplane begins the takeoff roll or the rotorcraft begins the lift-off until the airplane has completed the landing roll or the rotorcraft has landed at its destination.

(d) Except as provided in paragraph (c) of this section, and except for recorded data erased as authorized in this paragraph, each certificate holder shall keep the recorded data prescribed in paragraph (a) of this section until the aircraft has been operating for at least 8 hours of the operating time specified in paragraph (c) of this section. In addition, each certificate holder shall keep the recorded data prescribed in paragraph (b) of this section for an airplane until the airplane has been operating for at least 25 hours, and for a rotorcraft until the rotorcraft has been operating for at least 10 hours, of the operating time specified in paragraph (c) of this section. A total of 1 hour of recorded data may be erased for the purpose of testing the flight recorder or the flight recorder system. Any erasure made in accordance with this paragraph must be of the oldest recorded data accumulated at the time of testing. Except as provided in paragraph (c) of this section, no record need be kept more than 60 days.

(e) In the event of an accident or occurrence that requires the immediate notification of the National Transportation Safety Board under 49 CFR Part 830 of its regulations and that results in termination of the flight, the certificate holder shall remove the recording media from the aircraft and keep the recorded data required by paragraphs (a) and (b) of this section for at least 60 days or for a longer period upon request of the Board or the Administrator.

(f) Each flight recorder required by this section must be installed in accordance with the requirements of §§ 23.1459, 25.1459, 27.1459, or 29.1459, as appropriate, of this chapter. The correlation required by paragraph (c) of §§ 23.1459, 25.1459, 27.1459, or 29.1459, as appropriate, of this chapter need be established only on one aircraft of a group of aircraft:

(1) That are of the same type;

(2) On which the flight recorder models and their installations are the same; and

(3) On which there are no differences in the type design with respect to the installation of the first pilot's instruments associated with the flight recorder. The most recent instrument calibration, including the recording medium from which this calibration is derived, and the recorder correlation must be retained by the certificate holder.

(g) Each flight recorder required by this section that records the data specified in paragraphs (a) and (b) of this section must have an approved device to assist in locating that recorder under water.

§ 135.153 Ground proximity warning system.

No person may operate a turbojet airplane

having a passenger seating configuration, excluding any pilot seat, of 10 seats or more, unless it is equipped with—

(a) A ground proximity warning system that meets § 37.201 of this chapter; or

(b) A system that conveys warnings of excessive closure rates with the terrain and any deviations below glide slope by visual and audible means. This system must—

(1) Be approved by the Director of Flight Operations; and

(2) Have a means of alerting the pilot when a malfunction occurs in the system.

(c) For the system required by this section, the Airplane Flight Manual shall contain—

(1) Appropriate procedures for—

(i) The use of the equipment;

(ii) Proper flight crew action with respect to the equipment; and

(iii) Deactivation for planned abnormal and emergency conditions; and

(2) An outline of all input sources that must be operating.

(d) No person may deactivate a system required by this section except under procedures in the Airplane Flight Manual.

(e) Whenever a system required by this section is deactivated, an entry shall be made in the airplane maintenance record that includes the date and time of deactivation.

(f) For a system required by paragraph (b) of this section, procedures acceptable to the FAA Flight Standards District Office charged with the overall inspection of the certificate holder shall be established by the certificate holder to ensure that the performance of the system can be appropriately monitored.

§ 135.155 Fire extinguishers: passenger-carrying aircraft.

No person may operate an aircraft carrying passengers unless it is equipped with hand fire extinguishers of an approved type for use in crew and passenger compartments as follows—

(a) The type and quantity of extinguishing agent must be suitable for all the kinds of fires likely to occur;

(b) At least one hand fire extinguisher must be provided and conveniently located on the flight deck for use by the flight crew; and

(c) At least one hand fire extinguisher must be conveniently located in the passenger compartment of each aircraft having a passenger seating configuration, excluding any pilot seat, of at least 10 seats but less than 31 seats.

§ 135.157 Oxygen equipment requirements. ☞ **F121**

(a) *Unpressurized aircraft.* No person may operate an unpressurized aircraft at altitudes prescribed in this section unless it is equipped with enough oxygen dispensers and oxygen to supply the pilots under § 135.89(a) and to supply, when flying—

(1) At altitudes above 10,000 feet through 15,000 feet MSL, oxygen to at least 10 percent of the occupants of the aircraft, other than the pilots, for that part of the flight at those altitudes that is of more than 30 minutes duration; and

(2) Above 15,000 feet MSL oxygen to each occupant of the aircraft other than the pilots.

(b) *Pressurized aircraft.* No person may operate a pressurized aircraft—

(1) At altitudes above 25,000 feet MSL, unless at least a 10-minute supply of supplemental oxygen is available for each occupant of the aircraft, other than the pilots, for use when a descent is necessary due to loss of cabin pressurization; and

(2) Unless it is equipped with enough oxygen dispensers and oxygen to comply with paragraph (a) of this section whenever the cabin pressure altitude exceeds 10,000 feet MSL and, if the cabin pressurization fails, to comply with § 135.89(a) or to provide a 2-hour supply for each pilot, whichever is greater, and to supply when flying—

(i) At altitudes above 10,000 feet through 15,000 feet MSL, oxygen to at least 10 percent of the occupants of the aircraft, other than the pilots, for that part of the flight at those altitudes that is of more than 30 minutes duration; and

(ii) Above 15,000 feet MSL, oxygen to each occupant of the aircraft, other than the pilots, for one hour unless, at all times during flight above that altitude, the aircraft can safely descend to 15,000 feet MSL within four minutes, in which case only a 30-minute supply is required.

(c) The equipment required by this section must have a means—

(1) To enable the pilots to readily determine, in flight, the amount of oxygen available in each source of supply and whether the oxygen is being delivered to the dispensing units; or

(2) In the case of individual dispensing units, to enable each user to make those determinations with respect to that person's oxygen supply and delivery; and

(3) To allow the pilots to use undiluted oxygen at their discretion at altitudes above 25,000 feet MSL.

§ 135.158 Pitot heat indication systems.

(a) Except as provided in paragraph (b) of this section, after April 12, 1981, no person may operate a transport category airplane equipped with a flight instrument pitot heating system unless the airplane is also equipped with an operable pitot heat indication system that complies with § 25.1326 of this chapter in effect on April 12, 1978.

(b) A certificate holder may obtain an extension of the April 12, 1981, compliance date specified in paragraph (a) of this section, but not beyond April 12, 1983, from the Director of Flight Operations if the certificate holder—

(1) Shows that due to circumstances beyond its control it cannot comply by the specified compliance date; and

(2) Submits by the specified compliance date a schedule for compliance, acceptable to the Director, indicating that compliance will be achieved at the earliest practicable date.

§ 135.159 Equipment requirements: carrying passengers under VFR at night or under VFR over-the-top conditions.

No person may operate an aircraft carrying passengers under VFR at night or under VFR over-the-top unless it is equipped with—

[(a) A gyroscopic rate-of-turn indicator except on the following aircraft:

[(1) Helicopters with a third attitude instrument system usable through flight attitudes of ±80 degrees of pitch and ±120 degrees of roll and installed in accordance with §29.1303(g) of this chapter.

[(2) Helicopters with a maximum certificated takeoff weight of 6,000 pounds or less.

[(b) A slip skid indicator.

[(c) A gyroscopic bank-and-pitch indicator.

[(d) A gyroscopic direction indicator.

[(e) A generator or generators able to supply all probable combinations of continuous in-flight electrical loads for required equipment and for recharging the battery.

[(f) For night flights—

[(1) An anticollision light system;

[(2) Instrument lights to make all instruments, switches, and gauges easily readable, the direct rays of which are shielded from the pilot's eyes; and

[(3) A flashlight having at least two size "D" cells or equivalent.

[(g) For the purpose of paragraph (e) of this section, a continuous in-flight electrical load includes one that draws current continuously during flight, such as radio equipment, electrically driven instruments and lights, but does not include occasional intermittent loads.

[(h) Notwithstanding provisions of paragraphs (b), (c), and (d), helicopters having a maximum certificated takeoff weight of 6,000 pounds or less may be operated until January 6, 1988 under visual flight rules at night without a slip skid indicator, a gyroscopic bank-and-pitch indicator, or a gyroscopic direction indicator.]

§ 135.161 Radio and navigational equipment: carrying passengers under VFR at night or under VFR over-the-top.

(a) No person may operate an aircraft carrying passengers under VFR at night, or under VFR over-the-top, unless it has two-way communications equipment able, at least in flight, to transmit to, and receive from, ground facilities 25 miles away.

(b) No person may operate an aircraft carrying passengers under VFR over-the-top unless it has radio navigational equipment able to receive radio signals from the ground facilities to be used.

(c) No person may operate an airplane carrying passengers under VFR at night unless it has radio navigational equipment able to receive radio signals from the ground facilities to be used.

§ 135.163 Equipment requirements: aircraft carrying passengers under IFR.

No person may operate an aircraft under IFR, carrying passengers, unless it has—

(a) A vertical speed indicator;

(b) A free-air temperature indicator;

(c) A heated pitot tube for each airspeed indicator;

(d) A power failure warning device or vacuum indicator to show the power available for gyroscopic instruments from each power source;

(e) An alternate source of static pressure for the altimeter and the airspeed and vertical speed indicators;

(f) For a single-engine aircraft, a generator or generators able to supply all probable combinations of continuous inflight electrical loads for required equipment and for recharging the battery;

(g) For multiengine aircraft, at least two generators each of which is on a separate engine, of which any combination of one-half of the total number are rated sufficiently to supply the electrical loads of all required instruments and equipment necessary for safe emergency operation of the aircraft except that for multiengine helicopters, the two required generators may be mounted on the main rotor drive train; and

(h) Two independent sources of energy (with means of selecting either), of which at least one is an engine-drive pump or generator, each of which is able to drive all gyroscopic instruments and installed so that failure of one instrument or source does not interfere with the energy supply to the remaining instruments or the other energy source, unless, for single-engine aircraft, the rate-of-turn indicator has a source of energy separate from the bank and pitch and direction indicators. For the purpose of this paragraph, for multiengine aircraft, each engine-driven source of energy must be on a different engine.

(i) For the purpose of paragraph (f) of this section, a continuous inflight electrical load includes one that draws current continuously during flight, such as radio equipment, electrically driven instruments, and lights, but does not include occasional intermittent loads.

§ 135.165 Radio and navigational equipment: extended overwater or IFR operations.

(a) No person may operate a turbojet airplane having a passenger seating configuration, excluding any pilot seat, of 10 seats or more, or a multiengine airplane carrying passengers as a "Commuter Air Carrier" as defined in Part 298 of this title, under IFR or in extended overwater operations unless it has at least the following radio communication and navigational equipment appropriate to the facilities to be used which are capable of transmitting to and receiving from, at any place on the route to be flown, at least one ground facility·

(1) Two transmitters, (2) two microphones, (3) two headsets or one headset and one speaker, (4) a marker beacon receiver, (5) two independent receivers for navigation, and (6) two independent receivers for communications.

(b) No person may operate an aircraft other than that specified in paragraph (a) of this section, under IFR or in extended overwater operations unless it has at least the following radio communication and navigational equipment appropriate to the facilities to be used and which are capable of transmitting to, and receiving from, at any place on the route, at least one ground facility:

(1) A transmitter, (2) two microphones, (3) two headsets or one headset and one speaker, (4) a marker beacon receiver, (5) two independent receivers for navigation, (6) two independent receivers for communications, and

(7) for extended overwater operations only, an additional transmitter.

(c) For the purpose of paragraphs (a) (5), (a) (6), (b) (5), and (b) (6) of this section, a receiver is independent if the function of any part of it does not depend on the functioning of any part of another receiver. However, a receiver that can receive both communications and navigational signals may be used in place of a separate communications receiver and a separate navigational signal receiver.

§ 135.167 Emergency equipment: extended overwater operations.

(a) No person may operate an aircraft in extended overwater operations unless it carries, installed in conspicuously marked locations easily accessible to the occupants if a ditching occurs, the following equipment:

(1) An approved life preserver equipped with an approved survivor locator light for each occupant of the aircraft. The life preserver must be easily accessible to each seated occupant.

(2) Enough approved life rafts of a rated capacity and buoyancy to accommodate the occupants of the aircraft.

(b) Each life raft required by paragraph (a) of this section must be equipped with or contain at least the following:

(1) One approved survivor locator light.

(2) One approved pyrotechnic signaling device.

(3) Either—

(i) One survival kit, appropriately equipped for the route to be flown; or

(ii) One canopy (for sail, sunshade, or rain catcher);

(iii) One radar reflector;

(iv) One life raft repair kit;

(v) One bailing bucket;

(vi) One signaling mirror;

(vii) One police whistle;

(viii) One raft knife;

(ix) One CO_2 bottle for emergency inflation;

(x) One inflation pump;

(xi) Two oars;

(xii) One 75-foot retaining line;

(xiii) One magnetic compass;

(xiv) One dye marker;

(xv) One flashlight having at least two size "D" cells or equivalent;

(xvi) A two-day supply of emergency food rations supplying at least 1,000 calories a day for each person;

(xvii) For each two persons the raft is rated to carry, two pints of water or one sea water desalting kit;

(xviii) One fishing kit; and

(xix) One book on survival appropriate for the area in which the aircraft is operated.

(c) No person may operate an aircraft in extended overwater operations unless there is attached to one of the life rafts required by paragraph (a) of this section, a survival type emergency locator transmitter that meets the applicable requirements of TSO–C91. Batteries used in this transmitter must be replaced (or recharged, if the battery is rechargeable) when the transmitter has been in use for more than 1 cumulative hour, and also when 50 percent of their useful life (or for rechargeable batteries, 50 percent of their useful life of charge), as established by the transmitter manufacturer under TSO–C91, paragraph (g) (2), has expired. The new expiration date for the replacement or recharged battery must be legibly marked on the outside of the transmitter. The battery useful life or useful life of charge requirements of this paragraph do not apply to batteries (such as water-activated batteries) that are essentially unaffected during probable storage intervals.

§ 135.169 Additional airworthiness requirements.

[(a) Except for commuter category airplanes, no person may operate a large airplane unless it meets the additional airworthiness requirements of §§ 121.213 through 121.283, 121.307, and 121.312 of this chapter.

[(b) No person may operate a reciprocating-engine or turbopropeller-powered small airplane that has a passenger seating configuration, excluding pilot seats, of 10 seats or more unless it is type certificated—]

(1) In the transport category;

(2) Before July 1, 1970, in the normal category and meets special conditions issued by the Administrator for airplanes intended for use in operations under this Part;

(3) Before July 19, 1970, in the normal category and meets the additional airworthiness standards in Special Federal Aviation Regulation No. 23;

(4) In the normal category and meets the additional airworthiness standards in Appendix A;

(5) In the normal category and complies with section 1.(a) of Special Federal Aviation Regulation No. 41;

(6) In the normal category and complies with section 1.(b) of Special Federal Aviation Regulation No. 41 [; or

[(7) In the commuter category.]

(c) No person may operate a small airplane with a passenger seating configuration, excluding any pilot seat, of 10 seats or more, with a seating configuration greater than the maximum seating configuration used in that type airplane in operations under this Part before August 19, 1977. This paragraph does not apply to

(1) An airplane that is type certificated in the transport category; or

(2) An airplane that complies with—

(i) Appendix A of this Part provided that its passenger seating configuration, excluding pilot seats, does not exceed 19 seats; or

(ii) Special Federal Aviation Regulation No. 41.

§ 135.170 Materials for compartment interiors.

No person may operate an airplane that conforms to an amended or supplemental type certificate issued in accordance with SFAR No. 41 for a maximum certificated takeoff weight in excess of 12,500 pounds, unless within one year after issuance of the initial airworthiness certificate under that SFAR, the airplane meets the compartment interior requirements set forth in § 25.853(a), (b), (b–1), (b–2), and (b–3) of this chapter in effect on September 26, 1978.

§ 135.171 Shoulder harness installation at flight crewmember stations.

(a) No person may operate a turbojet aircraft or an aircraft having a passenger seating configuration, excluding any pilot seat, of 10 seats or more unless it is equipped with an approved shoulder harness installed for each flight crewmember station.

(b) Each flight crewmember occupying a station equipped with a shoulder harness must fasten the shoulder harness during takeoff and landing, except that the shoulder harness may be unfastened if the crewmember cannot perform the required duties with the shoulder harness fastened.

§ 135.173 Airborne thunderstorm detection equipment requirements.

(a) No person may operate an aircraft that has a passenger seating configuration, excluding any pilot seat, of 10 seats or more in passenger-carrying operations, except a helicopter operating under day VFR conditions, unless the aircraft is equipped with either approved thunderstorm detection equipment or approved airborne weather radar equipment.

(b) After January 6, 1988, no person may operate a helicopter that has a passenger seating configuration, excluding any pilot seat, of 10 seats or more in passenger-carry operations, under night VFR when current weather reports indicate that thunderstorms or other potentially hazardous weather conditions that can be detected with airborne thunderstorm detection equipment may reasonably be expected along the route to be flown, unless the helicopter is equipped with either approved thunderstorm detection equipment or approved airborne weather radar equipment.

(c) No person may begin a flight under IFR or night VFR conditions when current weather reports indicate that thunderstorms or other potentially hazardous weather conditions that can be detected with airborne thunderstorm detection equipment, required by paragraph (a) or (b) of this section, may reasonably be expected along the route to be flown, unless the airborne thunderstorm detection equipment is in satisfactory operating condition.

(d) If the airborne thunderstorm detection equipment becomes inoperative en route, the aircraft must be operated under the instructions and procedures specified for that event in the manual required by § 135.21.

(e) This section does not apply to aircraft used solely within the State of Hawaii, within the State of Alaska, within the part of Canada west of longitude 130 degrees W, between latitude 70 degrees N, and latitude 53 degrees N, or during any training, test, or ferry flight.

(f) Without regard to any other provision of this Part, an alternate electrical power supply is not required for airborne thunderstorm detection equipment.

§ 135.175 Airborne weather radar equipment requirements.

(a) No person may operate a large, transport category aircraft in passenger-carrying operations unless approved airborne weather radar equipment is installed in the aircraft.

(b) No person may begin a flight under IFR or night VFR conditions when current weather

reports indicate that thunderstorms, or other potentially hazardous weather conditions that can be detected with airborne weather radar equipment, may reasonably be expected along the route to be flown, unless the airborne weather radar equipment required by paragraph (a) of this section is in satisfactory operating condition.

(c) If the airborne weather radar equipment becomes inoperative en route, the aircraft must be operated under the instructions and procedures specified for that event in the manual required by § 135.21.

(d) This section does not apply to aircraft used solely within State of Hawaii, within the State of Alaska, within the part of Canada west of longitude 130 degrees W, between latitude 70 degrees N, and latitude 53 degrees N, or during any training, test, or ferry flight.

(e) Without regard to any other provision of this Part, an alternate electrical power supply is not required for airborne weather radar equipment.

§ 135.177 Emergency equipment requirements for aircraft having a passenger seating configuration of more than 19 passengers.

(a) No person may operate an aircraft having a passenger seating configuration, excluding any pilot seat, of more than 19 seats unless it is equipped with the following emergency equipment:

(1) One approved first aid kit for treatment of injuries likely to occur in flight or in a minor accident, which meets the following specifications and requirements:

(i) Each first aid kit must be dust and moisture proof, and contain only materials that either meet Federal Specifications GGK–319a, as revised, or as approved by the Administrator.

(ii) Required first aid kits must be readily accessible to the cabin flight attendants.

(iii) At time of takeoff, each first aid kit must contain at least the following or other contents approved by the Administrator:

Contents	Quantity
Adhesive bandage compressors, 1 in	16
Antiseptic swabs	20
Ammonia inhalents	10
Bandage compressors, 4 in	8
Triangular bandage compressors, 40 in	5
Burn compound, ⅛ oz or an equivalent or other burn remedy	6
Arm splint, noninflatable	1
Leg splint, noninflatable	1
Roller bandage, 4 in	4
Adhesive tape, 1-in standard roll	2
Bandage scissors	1

(2) A crash axe carried so as to be accessible to the crew but inaccessible to passengers during normal operations.

(3) Signs that are visible to all occupants to notify them when smoking is prohibited and when safety belts should be fastened. The signs must be so constructed that they

can be turned on and off by a crewmember. Seat belt signs must be turned on for each takeoff and landing, and at other times considered necessary by the pilot in command. No smoking signs shall be turned on when required by § 135. 127 of this part.

(4) For airplanes, has the additional emergency equipment specified in § 121.310 of this chapter.

(b) Each item of equipment must be inspected regularly under inspection periods established in the operations specifications to ensure its condition for continued serviceability and immediate readiness to perform its intended emergency purposes.

§ 135.179 Inoperable instruments and equipment for multiengine aircraft.

(a) No person may take off a multiengine aircraft unless the following instruments and equipment are in an operable condition:

(1) Instruments and equipment that are either specifically or otherwise required by the airworthiness requirements under which the aircraft is type certificated and which are essential for safe operations under all operating conditions.

(2) Instruments and equipment required by an airworthiness directive to be in operable condition unless the airworthiness directive provides otherwise.

(b) No person may take off any multiengine aircraft with inoperable instruments or equipment installed, other than those described in paragraph (a) of this section, unless the following conditions are met:

(1) An approved Minimum Equipment List exists for the aircraft type.

(2) The aircraft has within it a letter of authorization, issued by the FAA Flight Standards District Office having certification responsibility for the certificate holder, authorizing operation of the aircraft under the Minimum Equipment List. The letter of authorization may be obtained by written request of the certificate holder. The Minimum Equipment List and the letter of authorization constitute a supplemental type certificate for the aircraft.

(3) The approved Minimum Equipment List must provide for the operation of the aircraft with the instruments and equipment in an inoperable condition.

(4) The aircraft records available to the pilot must include an entry describing the inoperable instruments and equipment.

(5) The aircraft is operated under all applicable conditions and limitations contained in the Minimum Equipment List and the letter authorizing the use of the list.

(c) Without regard to the requirements of paragraph (a) (1) of this section, an aircraft with inoperable instruments or equipment may be operated under a special flight permit under §§ 21.197 and 21.199 of this chapter.

§ 135.181 Performance requirements: aircraft operated over-the-top or in IFR conditions.

(a) Except as provided in paragraphs (b) and (c) of this section, no person may—

(1) Operate a single-engine aircraft carrying passengers over-the-top or in IFR conditions; or

(2) Operate a multiengine aircraft carrying passengers over-the-top or in IFR conditions at a weight that will not allow it to climb, with the critical engine inoperative, at least 50 feet a minute when operating at the MEAs of the route to be flown or 5,000 feet MSL, whichever is higher.

[(b) Notwithstanding the restrictions in paragraph (a)(2) of this section, multiengine helicopters carrying passengers may conduct such operations in over-the-top or in IFR conditions at a weight that will allow the helicopter to climb at least 50 feet per minute with the critical engine inoperative when operating at the MEA of the route to be flown or 1,500 feet MSL, whichever is higher.]

[(c)] Without regard to paragraph (a) of this section,

(1) If the latest weather reports or forecasts, or any combination of them, indicate that the weather along the planned route (including takeoff and landing) allows flight under VFR under the ceiling (if a ceiling exists) and that the weather is forecast to remain so until at least 1 hour after the estimated time of arrival at the destination, a person may operate an aircraft over-the-top; or

(2) If the latest weather reports or forecasts, or any combination of them, indicate that the weather along the planned route allows flight under VFR under the ceiling (if a ceiling exists) beginning at a point no more than 15 minutes flying time at normal cruise speed from the departure airport, a person may—

(i) Take off from the departure airport in IFR conditions and fly in IFR conditions to a point no more than 15 minutes flying time at normal cruise speed from that airport;

(ii) Operate an aircraft in IFR conditions if unforecast weather conditions are encountered while en route on a flight planned to be conducted under VFR; and

(iii) Make an IFR approach at the destination airport if unforecast weather conditions are encountered at the airport that do not allow an approach to be completed under VFR.

[(d)] Without regard to paragraph (a) of this section, a person may operate an aircraft over-the-top under conditions allowing—

(1) For multiengine aircraft, descent or continuance of the flight under VFR if its critical engine fails; or

(2) For single-engine aircraft, descent under VFR if its engine fails.

§ 135.183 Performance requirements: land aircraft operated over water.

No person may operate a land aircraft carrying passengers over the water unless—

(a) It is operated at an altitude that allows it to reach land in the case of engine failure;

(b) It is necessary for takeoff or landing;

(c) It is a multiengine aircraft operated at a weight that will allow it to climb, with the critical engine inoperative, at least 50 feet a minute, at an altitude of 1,000 feet above the surface; or

(d) It is a helicopter equipped with helicopter flotation devices.

§ 135.185 Empty weight and center of gravity: currency requirement.

(a) No person may operate a multiengine aircraft unless the current empty weight and center of gravity are calculated from values established by actual weighing of the aircraft within the preceding 36 calendar months.

(b) Paragraph (a) of this section does not apply to—

(1) Aircraft issued an original airworthiness certificate within the preceding 36 calendar months; and

(2) Aircraft operated under a weight and balance system approved in the operations specifications of the certificate holder.

Subpart D—VFR/IFR Operating Limitations and Weather Requirements

§ 135.201 Applicability.

This subpart prescribes the operating limitations for VFR/IFR flight operations and associated weather requirements for operations under this Part.

§ 135.203 VFR: minimum altitudes.

Except when necessary for takeoff and landing, no person may operate under VFR—

(a) An airplane—

(1) During the day, below 500 feet above the surface or less than 500 horizontally from any obstacle; or

(2) At night, at an altitude less than 1,000 feet above the highest obstacle within a horizontal distance of 5 miles from the course intended to be flown or, in designated mountainous terrain, less than 2,000 feet above the highest obstacle within a horizontal distance of 5 miles from the course intended to be flown; or

(b) A helicopter over a congested area at an altitude less than 300 feet above the surface.

§ 135.205 VFR: visibility requirements.

(a) No person may operate an airplane under VFR in uncontrolled airspace when the ceiling is less than 1,000 feet unless flight visibility is at least 2 miles.

(b) No person may operate a helicopter under VFR in uncontrolled airspace at an altitude of 1,200 feet or less above the surface or in control zones unless the visibility is at least—

(1) During the day—$\frac{1}{2}$ miles; or

(2) At night—1 mile.

§ 135.207 VFR: helicopter surface reference requirements.

No person may operate a helicopter under VFR unless that person has visual surface reference or, at night, visual surface light reference, sufficient to safely control the helicopter.

§ 135.209 VFR: fuel supply.

(a) No person may begin a flight operation in an airplane under VFR unless, considering wind and forecast weather conditions, it has enough fuel to fly to the first point of intended landing and, assuming normal cruising fuel consumption—

(1) During the day, to fly after that for at least 30 minutes; or

(2) At night, to fly after that for at least 45 minutes.

(b) No person may begin a flight operation in a helicopter under VFR unless, considering wind and forecast weather conditions, it has enough fuel to fly to the first point of intended landing and, assuming normal cruising fuel consumption, to fly after that for at least 20 minutes.

§ 135.211 VFR: over-the-top carrying passengers: operating limitations.

Subject to any additional limitations in § 135.181, no person may operate an aircraft under VFR over-the-top carrying passengers, unless—

(a) Weather reports or forecasts, or any combination of them, indicate that the weather at the intended point of termination of over-the-top flight—

(1) Allows descent to beneath the ceiling under VFR and is forecast to remain so until at least 1 hour after the estimated time of arrival at that point; or

(2)) Allows an IFR approach and landing with flight clear of the clouds until reaching the prescribed initial approach altitude over the final approach facility, unless the approach is made with the use of radar under § 91.116(f) of this chapter; or

(b) It is operated under conditions allowing

(1) For multiengine aircraft, descent or continuation of the flight under VFR if its critical engine fails; or

(2) For single-engine aircraft, descent under VFR if its engine fails.

§ 135.213 Weather reports and forecasts.

(a) Whenever a person operating an aircraft under this Part is required to use a weather report or forecast, that person shall use that of the U.S. National Weather Service, a source approved by the U.S. National Weather Service, or a source approved by the Administrator. However, for operations under VFR, the pilot in command may, if such a report is not available, use weather information based on that pilot's own observations or on those of other persons competent to supply appropriate observations.

(b) For the purposes of paragraph (a) of this section, weather observations made and furnished to pilots to conduct IFR operations at an airport must be taken at the airport where those IFR operations are conducted, unless the Administrator issues operations specifications allowing the use of weather observations taken at a location not at the airport where the IFR operations are conducted. The Administrator issues such operations specifications when, after investigation by the U.S. National Weather Service and the FAA Flight Standards District Office charged with the overall inspection of the certificate holder, it is found that the standards of safety for that operation would allow the deviation from this paragraph for a particular operation for which an ATCO operating certificate has been issued.

§ 135.215 IFR: operating limitations.

(a) Except as provided in paragraphs (b), (c) and (d) of this section, no person may operate an aircraft under IFR outside of controlled airspace or at any airport that does not have an approved standard instrument approach procedure.

(b) The Administrator may issue operations specifications to the certificate holder to allow it to operate under IFR over routes outside controlled airspace if—

(1) The certificate holder shows the Administrator that the flight crew is able to navigate, without visual reference to the ground, over an intended track without deviating more than 5 degrees or 5 miles, whichever is less, from that track; and

(2) The Administrator determines that the proposed operations can be conducted safely.

(c) A person may operate an aircraft under IFR outside of controlled airspace if the certificate holder has been approved for the operations and that operation is necessary to—

(1) Conduct an instrument approach to an airport for which there is in use a current approved standard or special instrument approach procedure; or

(2) Climb into controlled airspace during an approved missed approach procedure; or

(3) Make an IFR departure from an airport having an approved instrument approach procedure.

(d) The Administrator may issue operations specifications to the certificate holder to allow it to depart at an airport that does not have an approved standard instrument approach procedure when the Administrator

determines that it is necessary to make an IFR departure from that airport and that the proposed operations can be conducted safely. The approval to operate at that airport does not include an approval to make an IFR approach to that airport.

§ 135.217 IFR: takeoff limitations.

No person may takeoff an aircraft under IFR from an airport where weather conditions are at or above take off minimums but are below authorized IFR landing minimums unless there is an alternate airport within 1 hour's flying time (at normal cruising speed, in still air) of the airport of departure.

§ 135.219 IFR: destination airport weather minimums.

No person may take off an aircraft under IFR or being an IFR or over-the-top operation unless the latest weather reports or forecasts, or any combination of them, indicate that weather conditions at the estimated time of arrival at the next airport of intended landing will be at or above authorized IFR landing minimums.

§ 135.221 IFR: alternate airport weather minimums.

No person may designate an alternate airport unless the weather reports or forecasts, or any combination of them, indicate that the weather conditions will be at or above authorized alternate airport landing minimums for that airport at the estimated time of arrival.

§ 135.223 IFR: alternate airport requirements.

(a) Except as provided in paragraph (b) of this section, no person may operate an aircraft in IFR conditions unless it carries enough fuel (considering weather reports or forecasts or any combination of them) to—

(1) Complete the flight to the first airport of intended landing;

(2) Fly from that airport to the alternate airport; and

[(3) Fly after that for 45 minutes at normal cruising speed, or helicopters, fly after that for 30 minutes at normal cruising speed.]

(b) Paragraph (a) (2) of this section does not apply if Part 97 of this chapter prescribes a standard instrument approach procedure for the first airport of intended landing and, for at least one hour before and after the estimated time of arrival, the appropriate weather reports or forecasts, or any combination of them, indicate that—

(1) The ceiling will be at least 1,500 feet above the lowest circling approach MDA; or

(2) If a circling instrument approach is not authorized for the airport, the ceiling will be at least 1,500 feet above the lowest published minimum or 2,000 feet above the airport elevation, whichever is higher; and

(3) Visibility for that airport is forecast to be at least three miles, or two miles more than the lowest applicable visibility minimums, whichever is the greater, for the instrument approach procedure to be used at the destination airport.

§ 135.225 IFR: takeoff, approach and landing minimums.

(a) No pilot may begin instrument approach procedure to an airport unless—

(1) That airport has a weather reporting facility operated by the U.S. National Weather Service, a source approved by U.S. National Weather Service, or a source approved by the Administrator; and

(2) The latest weather report issued by that weather reporting facility indicates that weather conditions are at or above the authorized IFR landing minimums for that airport.

(b) No pilot may begin the final approach segment of an instrument approach procedure to an airport unless the latest weather reported by the facility described in paragraph (a) (1) of this section indicates that weather conditions are at or above the authorized IFR landing minimums for that procedure.

(c) If a pilot has begun the final approach segment of an instrument approach to an airport under paragraph (b) of this section and a later weather report indicating below minimum conditions is received after the aircraft is—

(1) On an ILS final approach and has passed the final approach fix; or ☞F112

(2) On an ASR or PAR final approach and has been turned over to the final approach controller; or

(3) On a final approach using a VOR, NDB, or comparable approach procedure; and the aircraft—

(i) Has passed the appropriate facility or final approach fix; or

(ii) Where a final approach fix is not specified, has completed the procedure turn and is established inbound toward the airport on the final approach course within the distance prescribed in the procedure; the approach may be continued and a landing made if the pilot finds, upon reaching the authorized MDA or DH, that actual weather conditions are at least equal to the minimums prescribed for the procedure.

(d) The MDA or DH and visibility landing minimums prescribed in Part 97 of this chapter or in the operator's operations specifications are increased by 100 feet and ½ mile respectively, but not to exceed the ceiling and visibility minimums for that airport when used as an alternate airport, for each pilot in command of a turbine-powered airplane who has not served at least 100 hours as pilot in command in that type of airplane.

(e) Each pilot making an IFR takeoff or approach and landing at a military or foreign airport shall comply with applicable instrument approach procedures and weather minimums prescribed by the authority having jurisdiction over the airport. In addition, no pilot may, at that airport—

(1) Take off under IFR when the visibility is less than 1 mile; or

(2) Make an instrument approach when the visibility is less than ½ mile.

(f) If takeoff minimums are specified in Part 97 of this chapter for the takeoff airport, no pilot may take off an aircraft under IFR when the weather conditions reported by the facility described in paragraph (a)(1) of this section are less than the takeoff minimums specified for the takeoff airport in Part 97 or in the certificate holder's operations specifications.

(g) Except as provided in paragraph (h) of this section, if takeoff minimums are not prescribed in Part 97 of this chapter for the takeoff airport, no pilot may take off an aircraft under IFR when the weather conditions reported by the facility described in paragraph (a)(1) of this section are less than that prescribed in Part 91 of this chapter or in the certificate holder's operations specifications.

(h) At airports where straight-in instrument approach procedures are authorized, a pilot may take off an aircraft under IFR when the weather conditions reported by the facility described in paragraph (a)(1) of this section are equal to or better than the lowest straight-in landing minimums, unless otherwise restricted, if—

(1) The wind direction and velocity at the time of takeoff are such that a straight-in instrument approach can be made to the runway served by the instrument approach;

(2) The associated ground facilities upon which the landing minimums are predicated and the related airborne equipment are in the normal operation; and

(3) The certificate holder has been approved for such operations.

§ 135.227 Icing conditions: operating limitations. ☞ F83

(a) No pilot may take off an aircraft that has—

(1) Frost, snow, or ice adhering to any rotor blade, propeller, windshield, or powerplant installation, or to an airspeed, altimeter, rate of climb, or flight attitude instrument system;

(2) Snow or ice adhering to the wings or stabilizing or control surfaces; or

(3) Any frost adhering to the wings, or stabilizing or control surfaces, unless that frost has been polished to make it smooth.

(b) Except for an airplane that has ice protection provisions that meet § 34 of Appendix A, or those for transport category airplane type certificate, no pilot may fly—

(1) Under IFR into known or forecast light or moderate icing conditions; or

(2) Under VFR into known light or moderate icing conditions; unless the aircraft has functioning deicing or anti-icing equipment protecting each rotor blade, propeller, windshield, wing, stabilizing or control surface, and each airspeed, altimeter, rate of climb, or flight attitude instrument system.:

[(c) No pilot may fly a helicopter under IFR into known or forecast icing conditions or under VFR into known icing conditions unless it has been type certificated and appropriately equipped for operations in icing conditions.]

[(d)] Except for an airplane that has ice protection provisions that meet § 34 of Appendix A, or those for transport category airplane type certification, no pilot may fly an aircraft into known or forecast severe icing conditions.

[(e)] If current weather reports and briefing information relied upon by the pilot in command indicate that the forecast icing condition that would otherwise prohibit the flight will not be encountered during the flight because of changed weather conditions since the forecast, [the restrictions in paragraphs (b), (c), and (d)] of this section based on forecast conditions do not apply.

§ 135.229 Airport requirements.

(a) No certificate holder may use any airport unless it is adequate for the proposed operation, considering such items as size, surface, obstructions, and lighting.

(b) No pilot of an aircraft carrying passengers at night may take off from, or land on, an airport unless—

(1) That pilot has determined the wind direction from an illuminated wind direction indicator or local ground communications or, in the case of takeoff, that pilot's personal observations; and

(2) The limits of the area to be used for landing or takeoff are clearly shown—

(i) For airplanes, by boundary or runway marker lights;

(ii) For helicopters, by boundary or runway marker lights or reflective material.

(c) For the purpose of paragraph (b) of this section, if the area to be used for takeoff or landing is marked by flare pots or lanterns, their use must be approved by the Administrator.

Subpart E—Flight Crewmember Requirements

§ 135.241 Applicability.

This subpart prescribes the flight crewmember requirements for operations under this Part.

§ 135.243 Pilot in command qualifications.

(a) No certificate holder may use a person,

nor may any person serve, as pilot in command in passenger-carrying operations of a turbojet airplane, of an airplane having a passenger seating configuration, excluding any pilot seat, of 10 seats or more, or a multiengine airplane being operated by the "Commuter Air Carrier" (as defined in Part 298 of this Title), unless that person holds an airline transport pilot certificate with appropriate category and class ratings and, if required, an appropriate type rating for that airplane.

(b) Except as provided in paragraph (a) of this section, no certificate holder may use a person, nor may any person serve, as pilot in command of an aircraft under VFR unless that person—

(1) Holds at least a commercial pilot certificate with appropriate category and class ratings and, if required, an appropriate type rating for that aircraft; and

(2) Has had at least 500 hours of flight time as a pilot, including at least 100 hours of cross-country flight time, at least 25 hours of which were at night; and

(3) For an airplane, holds an instrument rating or an airline transport pilot certificate with an airplane category rating; or

(4) For helicopter operations conducted VFR over-the-top, holds a helicopter instrument rating, or an airline transport pilot certificate with a category and class rating for that aircraft, not limited to VFR.

(c) Except as provided in paragraph (a) of this section, no certificate holder may use a person, nor may any person serve, as pilot in command of an aircraft under IFR unless that person—

(1) Holds at least a commercial pilot certificate with appropriate category and class ratings and, if required, an appropriate type rating for that aircraft; and

(2) Has had at least 1,200 hours of flight time as a pilot, including 500 hours of cross-country flight time, 100 hours of night flight time, and 75 hours of actual or simulated instrument time at least 50 hours of which were in actual flight; and

(3) For an airplane, holds an instrument rating or an airline transport pilot certificate with an airplane category rating; or

(4) For a helicopter, holds a helicopter instrument rating, or an airline transport pilot certificate with a category and class rating for the aircraft, not limited to VFR.

[(d) Paragraph (b)(3) of this section does not apply when—

[(1) The aircraft used is a single reciprocating-engine-powered airplane;

[(2) The certificate holder does not conduct any operation pursuant to a published flight schedule which specifies five or more round trips a week between two or more points and places between which the round trips are performed, and does not transport mail by air under a contract or contracts with the United States Postal Service having total amount estimated at the beginning of any semiannual reporting period (January 1-June 30; July 1-December 31) to be in excess of $20,000 over the 12 months commencing with the beginning of the reporting period;

[(3) The area, as specified in the certificate holder's operations specifications, is an isolated area, as determined by the Flight Standards district office, if it is shown that—

[(i) The primary means of navigation in the area is by pilotage, since radio navigational aids are largely ineffective; and

[(ii) The primary means of transportation in the area is by air;

[(4) Each flight is conducted under day VFR with a ceiling of not less than 1,000 feet and visibility not less than 3 statute miles;

[(5) Weather reports or forecasts, or any combination of them, indicate that for the period commencing with the planned departure and ending 30 minutes after the planned arrival at the destination the flight may be conducted under VFR with a ceiling of not less than 1,000 feet and visibility of not less than 3 statute miles, except that if weather reports and forecasts are not available, the pilot in command may use that pilot's observations or those of other persons competent to supply weather observations if those observations indicate the flight may be conducted under VFR with the ceiling and visibility required in this paragraph;

[(6) The distance of each flight from the certificate holder's base of operation to destination does not exceed 250 nautical miles for a pilot who holds a commercial pilot certificate with an airplane rating without an instrument rating, provided the pilot's certificate does not contain any limitation to the contrary; and

[(7) The areas to be flown are approved by the certificate-holding FAA Flight Standards district office and are listed in the certificate holder's operations specifications.]

§ 135.244 Operating experience.

(a) No certificate holder may use any person, nor may any person serve, as a pilot in command of an aircraft operated by a Commuter Air Carrier (as defined in § 298.2 of this title) in passenger-carrying operations, unless that person has completed, prior to designation as pilot in command, on that make and basic model aircraft and in that crewmember position, the following operating experience in each make and basic model of aircraft to be flown:

(1) Aircraft, single engine—10 hours.

(2) Aircraft multiengine, reciprocating engine-powered—15 hours.

(3) Aircraft multiengine, turbine engine-powered—20 hours.

(4) Airplane, turbojet-powered—25 hours.

(b) In acquiring the operating experience, each person must comply with the following:

(1) The operating experience must be acquired after satisfactory completion of the appropriate ground and flight training for the aircraft and crewmember position. Approved provisions for the operating experience must be included in the certificate holder's training program.

(2) The experience must be acquired in flight during commuter passenger-carrying operations under this Part. However, in the case of an aircraft not previously used by the

certificate holder in operations under this Part, operating experience acquired in the aircraft during proving flights or ferry flights may be used to meet this requirement.

(3) Each person must acquire the operating experience while performing the duties of a pilot in command under the supervision of a qualified check pilot.

(4) The hours of operating experience may be reduced to not less than 50 percent of the hours required by this section by the substitution of one additional takeoff and landing for each hour of flight.

§ 135.245 Second in command qualifications.

(a) Except as provided in paragraph (b), no certificate holder may use any person, nor may any person serve, as second in command of an aircraft unless that person holds at least a commercial pilot certificate with appropriate category and class ratings and an instrument rating. For flight under IFR, that person must meet the recent instrument experience requirements of Part 61 of this chapter.

(b) A second in command of a helicopter operated under VFR, other than over-the-top, must have at least a commercial pilot certificate with an appropriate aircraft category and class rating.

§ 135.247 Pilot qualifications: recent experience.

(a) No certificate holder may use any person, nor may any person serve, as pilot in command of an aircraft carrying passengers unless, within the preceding 90 days, that person has—

(1) Made three takeoffs and three landings as the sole manipulator of the flight controls in an aircraft of the same category and class and, if a type rating is required, of the same type in which that person is to serve; or

(2) For operation during the period beginning 1 hour after sunset and ending 1 hour before sunrise (as published in the Air Almanac), made three takeoffs and three landings during that period as the sole manipulator of the flight controls in an aircraft of the same category and class and, if a type rating is required, of the same type in which that person is to serve.

A person who complies with paragraph (a) (2) of this paragraph need not comply with paragraph (a) (1) of this paragraph.

(b) For the purpose of paragraph (a) of this section, if the aircraft is a tailwheel airplane, each takeoff must be made in a tailwheel airplane and each landing must be made to a full stop in a tailwheel airplane.

§ 135.249 Use of prohibited drugs.

(a) This section applies to persons who perform a function listed in Appendix I to Part 121 of this chapter for a certificate holder or an operator. For the purpose of this section, a person who performs such a function pursuant to a contract with the certificate holder or the operator is considered to be performing that function for the certificate holder or the operator.

(b) No certificate holder or operator may knowingly use any person to perform, nor may any person perform for a certificate holder or an operator, either directly or by contract, any function listed in Appendix I to Part 121 of this chapter while that person has a prohibited drug, as defined in that appendix, in his or her system.

(c) Except as provided in paragraph (d) of this section, no certificate holder or operator may knowingly use any person to perform, nor may any person perform for a certificate holder or an operator, either directly or by contract, any function listed in Appendix I to Part 121 of this chapter if that person has failed a test or refused to submit to a test required by that appendix given by any certificate holder or any operator.

(d) Paragraph (c) of this section does not apply to a person who has received a recommendation to be hired or to return to duty from a medical review officer in accordance with Appendix I to Part 121 of this chapter or who has received a special issuance medical certificate after evaluation by the Federal Air Surgeon for drug dependency in accordance with Part 67 of this chapter.

§ 135.251 Testing for prohibited drugs.

(a) Each certificate holder or operator shall test each of its employees who performs a function listed in Appendix I to Part 121 of this chapter in accordance with that appendix.

(b) No certificate holder or operator may use any contractor to perform a function listed in Appendix I to Part 121 of this chapter unless that contractor tests each employee performing such a function for the certificate holder or operator in accordance with that appendix.

[Subpart F—Flight Crewmember Flight Time Limitations and Rest Requirements

[§ 135.261 Applicability.

[Sections 135.263 through 135.271 prescribe flight time limitations and rest requirements for operations conducted under this Part as follows:

[(a) Section 135.263 applies to all operations under this subpart.

[(b) Section 135.265 applies to:

[(1) Scheduled passenger-carrying operations except those conducted solely within the state of Alaska. "Scheduled passenger-carrying operations" means passenger-carrying operations that are conducted in accordance with a published schedule which covers at least five round trips per week on at least one route between two or more points, includes dates or times (or both), and is openly advertised or otherwise made readily available to the general public, and

[(2) Any other operation under this Part, if the operator elects to comply with § 135.265 and obtains an appropriate operations specification amendment.

[(c) Sections 135.267 and 135.269 apply to any operation that is not a scheduled passenger-carrying operation and to any operation conducted solely within the State of Alaska, unless the operator elects to comply with § 135.265 as authorized under paragraph (b)(2) of this section.

[(d) Section 135.271 contains special daily flight time limits for operations conducted under the helicopter emergency medial evacuation service (HEMES).

[§ 135.263   Flight time limitations and rest requirements: all certificate holders.

[(a) A certificate holder may assign a flight crewmember and a flight crewmember may accept an assignment for flight time only when the applicable requirements of §§ 135.263 through 135.271 are met.

[(b) No certificate holder may assign any flight crewmember to any duty with the certificate holder during any required rest period.

[(c) Time spent in transportation, not local in character, that a certificate holder requires of a flight crewmember and provides to transport the crewmember to an airport at which he is to serve on a flight as a crewmember, or from an airport at which he was relieved from duty to return to his home station, is not considered part of a rest period.

[(d) A flight crewmember is not considered to be assigned flight time in excess of flight time limitations if the flights to which he is assigned normally terminate within the limitations, but due to circumstances beyond the control of the certificate holder or flight crewmember (such as adverse weather conditions), are not at the time of departure expected to reach their destination within the planned flight time.

[§ 135.265   Flight time limitations and rest requirments: scheduled operations.

[(a) No certificate holder may schedule any flight crewmember, and no flight crewmember may accept an assignment, for flight time in scheduled operations or in other commercial flying if that crewmember's total flight time in all commercial flying will exceed—

[(1) 1,200 hours in any calendar year.

[(2) 120 hours in any calendar month.

[(3) 34 hours in any 7 consecutive days.

[(4) 8 hours during any 24 consecutive hours for a flight crew consisting of one pilot.

[(5) 8 hours between required rest periods for a flight crew consisting of two pilots qualified under this Part for the operation being conducted.

[(b) Except as provided in paragraph (c) of this section, no certificate holder may schedule a flight crewmember, and no flight crewmember may accept an assignment, for flight time during the 24 consecutive hours preceding the scheduled completion of any flight segment without a scheduled rest period during that 24 hours of at least the following:

[(1) 9 consecutive hours of rest for less than 8 hours of scheduled flight time.

[(2) 10 consecutive hours of rest for 8 or more but less than 9 hours of scheduled flight time.

[(3) 11 consecutive hours of rest for 9 or more hours of scheduled flight time.

[(c) A certificate holder may schedule a flight crewmember for less than the rest required in paragraph (b) of this section or may reduce a scheduled rest under the following conditions:

[(1) A rest required under paragraph (b)(1) of this section may be scheduled for or reduced to a minimum of 8 hours if the flight crewmember is given a rest period of at least 10 hours that must begin no later than 24 hours after the commencement of the reduced rest period.

[(2) A rest required under paragraph (b)(2) of this section may be scheduled for or reduced to a minimum of 8 hours if the flight crewmember is given a rest period of at least 11 hours that must begin no later than 24 hours after the commencement of the reduced rest period.

[(3) A rest required under paragraph (b)(3) of this section may be scheduled for or reduced to a minimum of 9 hours if the flight crewmember is given a rest period of at least 12 hours that must begin no later than 24 hours after the commencement of the reduced rest period.

[(d) Each certificate holder shall relieve each flight crewmember engaged in scheduled air transportation from all further duty for at least 24 consecutive hours during any 7 consecutive days.

[§ 135.267   Flight time limitations and rest requirments: unscheduled one- and two-pilot crews.

[(a) No certificate holder may assign any flight crewmember, and no flight crewmember may accept an assignment, for flight time as a member of a one- or two-pilot crew if that crewmember's total flight time in all commercial flying will exceed—

[(1) 500 hours in any calendar quarter.

[(2) 800 hours in any two consecutive calendar quarters.

[(3) 1,400 hours in any calendar year.

[(b) Except as provided in paragraph (c) of this section, during any 24 consecutive hours the total flight time of the assigned flight when added to any other commercial flying by that flight crewmember may not exceed—

[(1) 8 hours for a flight crew consisting of one pilot; or

[(2) 10 hours for a flight crew consisting of two pilots qualified under this Part for the operation being conducted.

[(c) A flight crewmember's flight time may exceed the flight time limits of paragraph (b) of this section if the assigned flight time occurs during a regularly assigned duty period of no more than 14 hours and—

[(1) If this duty period is immediately preceded by and followed by a required rest period of at least 10 consecutive hours of rest;

[(2) If flight time is assigned during this period, that total flight time when added to any other commercial flying by the flight crewmember may not exceed—

[(i) 8 hours for a flight crew consisting of one pilot; or

[(ii) 10 hours for a flight crew consisting of two pilots; and

[(3) If the combined duty and rest periods equal 24 hours.

[(d) Each assignment under paragraph (b) of this section must provide for at least 10 consecutive hours of rest during the 24-hour period that precedes the planned completion time of the assignment.

[(e) When a flight crewmember has exceeded the daily flight time limitations in this section, because of circumstances beyond the control of the certificate holder or flight crewmember (such as adverse weather conditions), that flight crewmember must have a rest period before being assigned or accepting an assignment for flight time of at least—

[(1) 11 consecutive hours of rest if the flight time limitation is exceeded by not more than 30 minutes;

[(2) 12 consecutive hours of rest if the flight time limitation is exceeded by more than 30 minutes, but not more than 60 minutes; and

[(3) 16 consecutive hours of rest if the flight time limitation is exceeded by more than 60 minutes.

[(f) The certificate holder must provide each flight crewmember at least 13 rest periods of at least 24 consecutive hours each in each calendar quarter.

[(g) The Director of Flight Operations may issue operations specifications authorizing a deviation from any specific requirement of this section if he finds that the deviation is justified to allow a certificate holder additional time, but in no case beyond October 1, 1987, to bring its operations into full compliance with the requirements of this section. Each application for a deviation must be submitted to the Director of Flight Operations before October 1, 1986. Each applicant for a deviation may continue to operate under the requirements of Subpart F of this Part as in effect on September 30, 1985, until the Director of Flight Operations has responded to the deviation request.

[§ 135.269   Flight time limitations and rest requirments: unscheduled three- and four-pilot crews.

[(a) No certificate holder may assign any flight crewmember, and no flight crewmember may accept an assignment, for flight time as a member of a three- or four-pilot crew if that crewmember's total flight time in all commercial flying will exceed—

[(1) 500 hours in any calendar quarter.

[(2) 800 hours in any two consecutive calendar quarters.

[(3) 1,400 hours in any calendar year.

[(b) No certificate holder may assign any pilot

to a crew of three or four pilots, unless that assignment provides—

〔(1) At least 10 consecutive hours of rest immediately preceding the assignment;

〔(2) No more than 8 hours of flight deck duty in any 24 consecutive hours;

〔(3) No more than 18 duty hours for a three-pilot crew or 20 duty hours for a four-pilot crew in any 24 consecutive hours;

〔(4) No more than 12 hours aloft for a three-pilot crew or 16 hours aloft for a four-pilot crew during the maximum duty hours specified in paragraph (b)(3) of this section;

〔(5) Adequate sleeping facilities on the aircraft for the relief pilot;

〔(6) Upon completion of the assignment, a rest period of at least 12 hours;

〔(7) For a three-pilot crew, a crew which consists of at least the following:

〔(i) A pilot in command (PIC) who meets the applicable flight crewmember requirements of Subpart E of Part 135;

〔(ii) A PIC who meets the applicable flight crewmember requirements of Subpart E of Part 135, except those prescribed in §§ 135.244 and 135.247; and

〔(iii) A second in command (SIC) who meets the SIC qualifications of § 135.245.

〔(8) For a four-pilot crew, at least three pilots who meet the conditions of paragraph (b)(7) of this section, plus a forth pilot who meets the SIC qualifications of § 135.245.

〔(c) When a flight crewmember has exceeded the daily flight deck duty limitation in this section by more than 60 minutes, because of circumstances beyond the control of the certificate holder or flight crewmember, that flight crewmember must have a rest period before the next duty period of at least 16 consecutive hours.

〔(d) A certificate holder must provide each flight crewmember at least 13 rest periods of at least 24 consecutive hours each in each calendar quarter.

〔**§ 135.271   Helicopter hospital emergency medical evacuation service (HEMES).**

〔(a) No certificat holder may assign any flight crewmember, and no flight crewmember may accept an assignment for flight time if that crewmember's total flight time in all commercial flying will exceed—

〔(1) 500 hours in any calendar quarter.

〔(2) 800 hours in any two consecutive calendar quarters.

〔(3) 1,400 hours in any calendar year.

〔(b) No certificate holder may assign a helicopter flight crewmember, and no flight crewmember may accept an assignment, for hospital emergency medical evacuation service helicopter operations unless that assignment provides for at least 10 consecutive hours of rest immediately preceding reporting to the hospital for availability for flight time.

〔(c) No flight crewmember may accrue more than 8 hours of flight time during any 24-consecutive hour period of a HEMES assignment, unless an emergency medical evacuation operation is prolonged. Each flight crewmember who exceeds the daily 8 hour flight time limitation in this paragraph must be relieved of the HEMES assignment immediately upon the completion of that emergency medical operation and must be given a rest period in compliance with paragraph (h) of this section.

〔(d) Each flight crewmember must receive at least 8 consecutive hours of rest during any 24 consecutive hour period of a HEMES assignment. A flight crewmember must be relieved of the HEMES assignment if he or she has not or cannot receive at least 8 consecutive hours of rest during any 24 consecutive hour period of a HEMES assignment.

〔(e) A HEMES assignment may not exceed 72 consecutive hours at the hospital.

〔(f) An adequate place of rest must be provided at, or in close proximity to, the hospital at which the HEMES assignment is being performed.

〔(g) No certificate holder may assign any other duties to a flight crewmember during a HEMES assignment.

〔(h) Each pilot must be given a rest period upon completion of the HEMES assignment and prior to being assigned any further duty with the certificate holder of—

〔(1) At least 12 consecutive hours for an assignment of less than 48 hours.

〔(2) At least 16 consecutive hours for an assignment of more than 48 hours.

〔(i) The certificate holder must provide each flight crewmember at least 13 rest periods of at least 24 consecutive hours each in each calendar quarter.〕

Subpart G—Crewmember Testing Requirements

§ 135.291   Applicability.

This subpart prescribes the tests and checks required for pilot and flight attendant crewmembers and for the approval of check pilots in operations under this Part.

§ 135.293   Initial and recurrent pilot testing requirements.

(a) No certificate holder may use a pilot, nor may any person serve as a pilot, unless, since the beginning of the 12th calendar month before that service, that pilot has passed a written or oral test, given by the Administrator or an authorized check pilot, on that pilot's knowledge in the following areas—

(1) The appropriate provisions of Parts 61, 91, and 135 of this chapter and the operations specifications and the manual of the certificate holder;

(2) For each type of aircraft to be flown by the pilot, the aircraft powerplant, major components and systems, major appliances, performance and operating limitations, standard and emergency operating procedures, and the contents of the approved Aircraft Flight Manual or equivalent, as applicable;

(3) For each type of aircraft to be flown by the pilot, the method of determining compliance with weight and balance limitations for takeoff, landing and en route operations;

(4) Navigation and use of air navigation aids appropriate to the operation or pilot authorization, including, when applicable, instrument approach facilities and procedures;

(5) Air traffic control procedures, including IFR procedures when applicable;

(6) Meteorology in general, including the principles of frontal systems, icing, fog, thunderstorms, and windshear, and, if appropriate for the operation of the certificate holder, high altitude weather;

(7) Procedures for—

(i) Recognizing and avoiding severe weather situations;

(ii) Escaping from severe weather situations, in case of inadvertent encounters, including low-altitude windshear (except that rotorcraft pilots are not required to be tested on escaping from low-altitude windshear); and

(iii) Operating in or near thunderstorms (including best penetrating altitudes), turbulent air (including clear air turbulence), icing, hail, and other potentially hazardous meteorological conditions; and

(8) New equipment, procedures, or techniques, as appropriate.

(b) No certificate holder may use a pilot, nor may any person serve as a pilot, in any aircraft unless, since the beginning of the 12th calendar month before that service, that pilot has passed a competency check given by the Administrator or an authorized check pilot in that class of aircraft, if single-engine airplane other than turbojet, or that type of aircraft, if helicopter, multiengine airplane, or turbojet airplane, to determine the pilot's competence in practical skills and techniques in that aircraft or class of aircraft. The extent of the competency check shall be determined by the Administrator or authorized check pilot conducting the competency check. The competency check may include any of the maneuvers and procedures currently required for the original issuance of the particular pilot certificate required for the operations authorized and appropriate to the category, class and type of aircraft involved. For the purposes of this paragraph, type, as to an airplane, means any one of a group of airplanes determined by the Administrator to have a similar means of propulsion, the same manufacturer, and no significantly different handling or flight characteristics. For the purposes of this paragraph, type, as to a helicopter, means a basic make and model.

(c) The instrument proficiency check required by § 135.297 may be substituted for the competency check required by this section for the type of aircraft used in the check.

(d) For the purpose of this Part, competent performance of a procedure or maneuver by a person to be used as a pilot requires that the pilot be the obvious master of the aircraft, with the successful outcome of the maneuver never in doubt.

(e) The Administrator or authorized check pilot certifies the competency of each pilot who passes the knowledge or flight check in the certificate holder's pilot records.

(f) Portions of a required competency check may be given in an aircraft simulator for other appropriate training device, if approved by the Administrator.

§ 135.295 Initial and recurrent flight attendant crewmember testing requirements.

No certificate holder may use a flight attendant crewmember, nor many any person serve as a flight attendant crewmember unless, since the beginning of the 12th calendar month before that service, the certificate holder has determined by appropriate initial and recurrent testing that the person is knowledgeable and competent in the following areas as appropriate to assigned duties and responsibilities—

(a) Authority of the pilot in command;

(b) Passenger handling, including procedures to be followed in handling deranged persons or other persons whose conduct might jeopardize safety;

(c) Crewmember assignments, functions, and responsibilities during ditching and evacuation of persons who may need the assistance of another person to move expeditiously to an exit in an emergency;

(d) Briefing of passengers;

(e) Location and operation of portable fire extinguishers and other items of emergency equipment;

(f) Proper use of cabin equipment and controls;

(g) Location and operation of passenger oxygen equipment;

(h) Location and operation of all normal and emergency exits, including evacuation chutes and escape ropes; and

(i) Seating of persons who may need assistance of another person to move rapidly to an exit in an emergency as prescribed by the certificate holder's operations manual.

§ 135.297 Pilot in command: instrument proficiency check requirements.

[(a) No certificate holder may use a pilot, nor may any person serve, as a pilot in command of an aircraft under IFR unless, since the beginning of the 6th calendar month before that service, that pilot has passed an instrument proficiency check under this section administered by the Administrator or an authorized check pilot.

[(b) No pilot may use any type of precision instrument approach procedure under IFR unless, since the beginning of the 6th calendar month before that use, the pilot satisfactorily demonstrated that type of approach procedure. No pilot may use any type of nonprecision approach procedure under IFR unless, since the beginning of the 6th calendar month before that use, the pilot has satisfactorily demonstrated either that type of approach procedure or any other two different types of nonprecision approach procedures. The instrument approach procedure or procedures must include at least one straight-in approach, one circling approach, and one missed approach. Each type of approach procedure demonstrated must be conducted to published minimums for that procedure.]

(c) The instrument proficiency check required by paragraph (a) of this section consists of an oral or written equipment test and a flight check under simulated or actual IFR conditions. The equipment test includes questions on emergency procedures, engine operation, fuel and lubrication systems, power settings, stall speeds, best engine-out speed, propeller and supercharger operations, and hydraulic, mechanical, and electrical systems, as appropriate. The flight check includes navigation by instruments, recovery from simulated emergencies, and standard instrument approaches involving navigational facilities which that pilot is to be authorized to use. Each pilot taking the instrument proficiency check must show that standard of competence required by § 135.293(d).

(1) The instrument proficiency check must—

(i) For a pilot in command of an airplane under § 135.243(a), include the procedures and maneuvers for an airline transport pilot certificate in the particular type of airplane, if appropriate; and

(ii) For a pilot in command of an airplane or helicopter under § 135.243(c), include the procedures and maneuvers for a commercial pilot certificate with an instrument rating and, if required, for the appropriate type rating.

(2) The instrument proficiency check must be given by an authorized check airman or by the Administrator.

(d) If the pilot in command is assigned to pilot only one type of aircraft, that pilot must take the instrument proficiency check required by paragraph (a) of this section in that type of aircraft.

(e) If the pilot in command is assigned to pilot more than one type of aircraft, that pilot must take the instrument proficiency check required by paragraph (a) of this section in each type of aircraft to which that pilot is assigned, in rotation, but not more than one flight check during each period described in paragraph (a) of this section.

(f) If the pilot in command is assigned to pilot both single-engine and multiengine aircraft, that pilot must initially take the instrument proficiency check required by paragraph (a) of this section in a multiengine aircraft, and each succeeding check alternately in single-engine and multiengine aircraft, but not more than one flight check during each period described in paragraph (a) of this section. Portions of a required flight check may be given in an aircraft simulator or other appropriate training device, if approved by the Administrator.

(g) If the pilot in command is authorized to use an autopilot system in place of a second in command, that pilot must show, during the required instrument proficiency check, that the pilot is able (without a second in command) both with and without using the autopilot to—

(1) Conduct instrument operations competently; and

(2) Properly conduct air-ground communications and comply with complex air traffic control instructions.

(3) Each pilot taking the autopilot check must show that, while using the autopilot, the airplane can be operated as proficiently as it would be if a second in command were present to handle air-ground communications and air traffic control instructions. The autopilot check need only be demonstrated once every 12 calendar months during the instrument proficiency check required under paragraph (a) of this section.

[(h) [Deleted.]]

§ 135.299 Pilot in command: line checks: routes and airports.

(a) No certificate holder may use a pilot, nor may any person serve as a pilot in command of a flight unless, since the beginning of the 12th calendar month before that service, that pilot has passed a flight check in one of the types of aircraft which that pilot is to fly. The flight check shall—

(1) Be given by an approved check pilot or by the Administrator;

(2) Consist of at least one flight over one route segment; and

(3) Include takeoffs and landings at one or more representative airports. In addition to the requirements of this paragraph, for a pilot authorized to conduct IFR operations, at least one flight shall be flown over a civil airway, an approved off-airway route, or a portion of either of them.

(b) The pilot who conducts the check shall determine whether the pilot being checked satisfactorily performs the duties and responsibilities of a pilot in command in operations under this Part, and shall so certify in the pilot training record.

(c) Each certificate holder shall establish in the manual required by § 135.21 a procedure which will ensure that each pilot who has not flown over a route and into an airport within the preceding 90 days will, before beginning the flight, become familiar with all available information required for the safe operation of that flight.

§ 135.301 Crewmember: tests and checks, grace provisions, training to accepted standards.

(a) If a crewmember who is required to take a test or a flight check under this Part, completes the test or flight check in the calendar month before or after the calendar month in which it is required, that crewmember is considered to have completed the test or check in the calendar month in which it is required.

(b) If a pilot being checked under this subpart fails any of the required maneuvers, the person giving the check may give additional training to the pilot during the course of the check. In addition to repeating the maneuvers failed, the person giving the check may require the pilot being checked to repeat any other maneuvers that are necessary to determine the pilot's proficiency. If the pilot being checked is unable to demonstrate satisfactory performance to the person conducting the check, the certificate holder may not use the pilot, nor may the pilot serve, as a flight crewmember in operations under this Part until the pilot has satisfactorily completed the check.

§ 135.303 Check pilot authorization: application and issue.

Each certificate holder desiring FAA approval of a check pilot shall submit a request in writing to the FAA Flight Standards District Office charged with the overall inspection of the certificate holder. The Administrator may issue an letter of authority to each check pilot if that pilot passes the appropriate oral and flight test. The letter of authority lists the tests and checks in this Part that the check pilot is qualified to give, and the cagetory, class and type aircraft, where appropriate, for which the check pilot is qualified.

Subpart H—Training

§ 135.321 Applicability and terms used.

(a) This subpart prescribes requirements for establishing and maintaining an approved training program for crewmembers, check airmen and instructors, and other operations personnel, and for the other training devices in the conduct of that program.

(b) For the purposes of this subpart, the following terms and definitions apply:

(1) *Initial training.* The training required for crewmembers who have not qualified and served in the same capacity on an aircraft.

(2) *Transition training.* The training required for crewmembers who have qualified and served in the same capacity on another aircraft.

(3) *Upgrade training.* The training required for crewmembers who have qualified and served as second in command on a particular aircraft type, before they serve as pilot in command on that aircraft.

(4) *Differences training.* The training required for crewmembers who have qualified and served on a particular type aircraft, when the Administrator finds differences training is necessary before a crewmember serves in the same capacity on a particular variation of that aircraft.

(5) *Recurrent training.* The training required for crewmembers to remain adequately trained and currently proficient for each aircraft, crewmember position, and type of operation in which the crewmember serves.

(6) *In flight.* The maneuvers, procedures, or functions that must be conducted in the aircraft.

§ 135.323 Training program: general.

(a) Each certificate holder required to have a training program under § 135.341 shall:

(1) Establish, obtain the appropriate initial and final approval of, and provide a training program that meets this subpart and that ensures that each crewmember, flight instructor, check airman, and each person assigned duties for the carriage and handling of hazardous materials (as defined in 49 CFR 171.8) is adequately trained to perform his assigned duties.

(2) Provide adequate ground and flight training facilities and properly qualified ground instructors for the training required by this subpart.

(3) Provide and keep current for each aircraft type used and, if applicable, the particular variations within the aircraft type, appropriate training material, examinations, forms, instructions, and procedures for use in conducting the training and checks required by this subpart.

(4) Provide enough flight instructors, check airmen, and simulator instructors to conduct required flight training and flight checks, and simulator training courses allowed under this subpart.

(b) Whenever a crewmember who is required to take recurrent training under this subpart completes the training in the calendar month before, or the calendar month after, the month in which that training is required, the crewmember is considered to have completed it in the calendar month in which it was required.

(c) Each instructor, supervisor, or check airman who is responsible for a particular ground training subject, segment of flight training, course of training, flight check, or competence check under this Part shall certify as to the proficiency and knowledge of the crewmember, flight instructor, or check airman concerned upon completion of that training or check. That certification shall be made a part of the crewmember's record. When the certification required by this paragraph is made by an entry in a computerized recordkeeping system, the certifying instructor, supervisor, or check airman, must be identified with that entry. However, the signature of the certifying instructor, supervisor, or check airman, is not required for computerized entries.

(d) Training subjects that apply to more than one aircraft or crewmember position and that have been satisfactorily completed during previous training while employed by the certificate holder for another aircraft or another crewmember position, need not be repeated during subsequent training other than recurrent training.

(e) Aircraft simulators and other training devices may be used in the certificate holder's training program if approved by the Administrator.

§ 135.325 Training program and revision: initial and final approval.

(a) To obtain initial and final approval of a training program, or a revision to an approved training program, each certificate holder must submit to the Administrator—

(1) An outline of the proposed or revised curriculum, that provides enough information for a preliminary evaluation of the proposed training program or revision; and

(2) Additional relevant information that may be requested by the Administrator.

(b) If the proposed training program or revision complies with this subpart, the Administrator grants initial approval in writing after which the certificate holder may conduct the training under that program. The Administrator then evaluates the effectiveness of the training program and advises the certificate holder of deficiencies, if any, that must be corrected.

(c) The Administrator grants final approval of the proposed training program or revision if the certificate holder shows that the training conducted under the initial approval in paragraph (b) of this section ensures that each person who successfully completes the training is adequately trained to perform that person's assigned duties.

(d) Whenever the Administrator finds that revisions are necessary for the continued adequacy of a training program that has been granted final approval, the certificate holder shall, after notification by the Administrator, make any changes in the program that are found necessary by the Administrator. Within

30 days after the certificate holder receives the notice, it may file a petition to reconsider the notice with the Administrator. The filing of a petition to reconsider stays the notice pending a decision by the Administrator. However, if the Administrator finds that there is an emergency that requires immediate action in the interest of safety, the Administrator may, upon a statement of the reasons, require a change effective without stay.

§ 135.327 Training program: curriculum.

(a) Each certificate holder must prepare and keep current a written training program curriculum for each type of aircraft for each crewmember required for that type aircraft. The curriculum must include ground and flight training required by this subpart.

(b) Each training program curriculum must include the following:

(1) A list of principal ground training subjects, including emergency training subjects, that are provided.

(2) A list of all the training devices, mockups, systems trainers, procedures trainers, or other training aids that the certificate holder will use.

(3) Detailed descriptions or pictorial displays of the approved normal, abnormal, and emergency maneuvers, procedures and functions that will be performed during each flight training phase or flight check, indicating those maneuvers, procedures and functions that are to be performed during the inflight portions of flight training and flight checks.

§ 135.329 Crewmember training requirements.

(a) Each certificate holder must include in its training program the following initial and transition ground training as appropriate to the particular assignment of the crewmember:

(1) Basic indoctrination ground training for newly hired crewmembers including instruction in at least the—

(i) Duties and responsibilities of crewmembers as applicable;

(ii) Appropriate provisions of this chapter;

(iii) Contents of the certificate holder's operating certificate and operations specifications (not required for flight attendants); and

(iv) Appropriate portions of the certificate holder's operating manual.

(2) The initial and transition ground training in §§ 135.345 and 135.349, as applicable.

(3) Emergency training in § 135.331.

(b) Each training program must provide the initial and transition flight training in § 135.347, as applicable.

(c) Each training program must provide recurrent ground and flight training in § 135.351.

(d) Upgrade training in §§ 135.345 and 135.347 for a particular type aircraft may be included in the training program for crewmembers who have qualified and served as second in command on that aircraft.

(e) In addition to initial, transition, upgrade and recurrent training, each training program must provide ground and flight training, instruction, and practice necessary to ensure that each crewmember—

(1) Remains adequately trained and currently proficient for each aircraft, crewmember position, and type of operation in which the crewmember serves; and

(2) Qualifies in new equipment, facilities, procedures, and techniques, including modifications to aircraft.

§ 135.331 Crewmember emergency training.

(a) Each training program must provide emergency training under this section for each aircraft type, model, and configuration, each crewmember, and each kind of operation conducted, as appropriate for each crewmember and the certificate holder.

(b) Emergency training must provide the following:

(1) Instruction in emergency assignments and procedures, including coordination among crewmembers.

(2) Individual instruction in the location, function, and operation of emergency equipment including—

(i) Equipment used in ditching and evacuation;

(ii) First aid equipment and its proper use; and

(iii) Portable fire extinguishers, with emphasis on the type of extinguisher to be used on different classes of fires.

(3) Instruction in the handling of emergency situations including—

(i) Rapid decompression;

(ii) Fire in flight or on the surface and smoke control procedures with emphasis on electrical equipment and related circuit breakers found in cabin areas;

(iii) Ditching and evacuation;

(iv) Illness, injury, or other abnormal situations involving passengers or crewmembers; and

(v) Hijacking and other unusual situations.

(4) Review of the certificate holder's previous aircraft accidents and incidents involving actual emergency situations.

(c) Each crewmember must perform at least the following emergency drills, using the proper emergency equipment and procedures, unless the Administrator finds that, for a particular drill, the crewmember can be adequately trained by demonstration:

(1) Ditching, if applicable.

(2) Emergency evacuation.

(3) Fire extinguishing and smoke control.

(4) Operation and use of emergency exits, including deployment and use of evacuation chutes, if applicable.

(5) Use of crew and passenger oxygen.

(6) Removal of life rafts from the aircraft, inflation of the life rafts, use of life lines, and boarding of passengers and crew, if applicable.

(7) Donning and inflation of life vests and the use of other individual flotation devices, if applicable.

(d) Crewmembers who serve in operations above 25,000 feet must receive instruction in the following:

(1) Respiration.

(2) Hypoxia.

(3) Duration of consciousness without supplemental oxygen at altitude.

(4) Gas expansion.

(5) Gas bubble formation.

(6) Physical phenomena and incidents of decompression.

§ 135.333 Training requirements: handling and carriage of hazardous materials.

(a) Except as provided in paragraph (d) of this section, no certificate holder may use any person to perform, and no person may perform, any assigned duties and responsibilities for the handling or carriage of hazardous materials (as defined in 49 CFR 171.8), unless within the preceding 12 calendar months that person has satisfactorily completed initial or recurrent training in an appropriate training program established by the certificate holder, which includes instruction regarding—

(1) The proper shipper certification, packaging, marking, labeling, and documentation for hazardous materials; and

(2) The compatibility, loading, storage, and handling characteristics of hazardous materials.

(b) Each certificate holder shall maintain a record of the satisfactory completion of the initial and recurrent training given to crewmembers an ground personnel who perform assigned duties and responsibilities for the handling and carriage of hazardous materials.

(c) Each certificate holder that elects not to accept hazardous materials shall ensure that each crewmember is adequately trained to recognize those items classified as hazardous materials.

(d) If a certificate holder operates into or out of airports at which trained employees or contract personnel are not available, it may use persons not meeting the requirements of paragraphs (a) and (b) of this section to load, offload, or otherwise handle hazardous materials if these persons are supervised by a crewmember who is qualified under paragraphs (a) and (b) of this section.

§ 135.335 Approval of aircraft simulators and other training devices.

(a) Training courses using aircraft simulators

and other training devices may be included in the certificate holder's training program if approved by the Administrator.

(b) Each aircraft simulator and other training device that is used in a training course or in checks required under this subpart must meet the following requirements:

(1) It must be specifically approved for—

[(i) The certificate holder; and

[(ii) The particular maneuver, procedure, or crewmember function involved.]

(2) It must maintain the performance, functional, and other characteristics that are required for approval.

[(3) Additionally, for aircraft simulators, it must be—

[(i) Approved for the type aircraft and, if applicable, the particular variation within type for which the training or check is being conducted; and

[(ii) Modified to conform with any modification to the aircraft being simulated that changes the performance, functional, or other characteristics required for approval.

[(c) A particular aircraft simulator or other training device may be used by more than one certificate holder.

[(d) In granting initial and final approval of training programs or revisions to them, the Administrator considers the training devices, methods, and procedures listed in the certificate holder's curriculum under § 135.327.]

§ 135.337   Training program: check airmen and instructor qualifications.

(a) No certificate holder may use a person, nor may any person serve, as a flight instructor or check airman in a training program established under this subpart unless, for the particular aircraft type involved, that person—

(1) Holds the airman certificate and ratings that must be held to serve as a pilot in command in operations under this Part;

(2) Has satisfactorily completed the appropriate training phases for the aircraft, including recurrent training, required to serve as a pilot in command in operations under this Part;

(3) Has satisfactorily completed the appropriate proficiency or competency checks required to serve as a pilot in command in operations under this Part;

(4) Has satisfactorily completed the applicable training requirements of § 135.339;

(5) Holds a Class I or Class II medical certificate required to serve as a pilot in command in operations under this Part;

(6) In the case of a check airman, has been approved by the Administrator for the airman duties involved; and

(7) In the case of a check airman used in an aircraft simulator only, holds a Class III medical certificate.

(b) No certificate holder may use a person, nor may any person serve, as a simulator instructor for a course of training given in an aircraft simulator under this subpart unless that person—

(1) Holds at least a commercial pilot certificate; and

(2) Has satisfactorily completed the following as evidenced by the approval of a check airman—

(i) Appropriate initial pilot and flight instructor ground training under this subpart; and

(ii) A simulator flight training course in the type simulator in which that person instructs under this subpart.

§ 135.339   Check airmen and flight instructors: initial and transition training.

(a) The initial and transition ground training for pilot check airmen must include the following:

(1) Pilot check airman duties, functions, and responsibilities.

(2) The applicable provisions of this chapter and certificate holder's policies and procedures.

(3) The appropriate methods, procedures, and techniques for conducting the required checks.

(4) Proper evaluation of pilot performance including the detection of—

(i) Improper and insufficient training; and

(ii) Personal characteristics that could adversely affect safety.

(5) The appropriate corrective action for unsatisfactory checks.

(6) The approved methods, procedures, and limitations for performing the required normal, abnormal, and emergency procedures in the aircraft.

(b) The initial and transition ground training for pilot flight instructors, except for the holder of a valid flight instructor certificate, must include the following:

(1) The fundamental principles of the teaching-learning process.

(2) Teaching methods and procedures.

(3) The instructor-student relationship.

(c) The initial and transition flight training for pilot check airmen and pilot flight instructors must include the following:

(1) Enough inflight training and practice in conducting flight checks from the left and right pilot seats in the required normal, abnormal, and emergency maneuvers to ensure that person's competence to conduct the pilot flight checks and flight training under this subpart.

(2) The appropriate safety measures to be taken from either pilot seat for emergency situations that are likely to develop in training.

(3) The potential results of improper or untimely safety measures during training.

The requirements of paragraphs (2) and

(3) of this paragraph may be accomplished in flight or in an approved simulator.

§ 135.341   Pilot and flight attendant crewmember training programs.

[(a) Each certificate holder, other than one who uses only one pilot in the certificate holder's operations,] shall establish and maintain an approved pilot training program, and each certificate holder who uses a flight attendant crewmember shall establish and maintain an approved flight attendant training program, that is appropriate to the operations to which each pilot and flight attendant is to be assigned, and will ensure that they are adequately trained to meet the applicable knowledge and practical testing requirements of §§ 135.293 through 135.301. However, the Administrator may authorize a deviation from this section if the Administrator finds that, because of the limited size and scope of the operation, safety will allow a deviation from these requirements.

(b) Each certificate holder required to have a training program by paragraph (a) of this section shall include in that program ground and flight training curriculums for—(1) Initial training; (2) Transition training; (3) Upgrade training; (4) Differences training; and (5) Recurrent training.

(c) Each certificate holder required to have a training program by paragraph (a) of this section shall provide current and appropriate study materials for use by each required pilot and flight attendant.

(d) The certificate holder shall furnish copies of the pilot and flight attendant crewmember training program, and all changes and additions, to the assigned representative of the Administrator. If the certificate holder uses training facilities of other persons, a copy of those training programs or appropriate portions used for those facilities shall also be furnished. Curricula that follow FAA published curricula may be cited by reference in the copy of the training program furnished to the representative of the Administrator and need not be furnished with the program.

§ 135.343   Crewmember initial and recurrent training requirements.

No certificate holder may use a person, nor may any person serve, as a crewmember in operations under this Part unless that crewmember has completed the appropriate initial or recurrent training phase of the training program appropriate to the type of operation in which the crewmember is to serve since the beginning of the 12th calendar month before that service. [This section does not apply to a certificate holder that uses only one pilot in the certificate holder's operations.]

§ 135.345   Pilots: initial, transition, and upgrade ground training.

Initial, transition, and upgrade ground training for pilots must include instruction in at least the following, as applicable to their duties:

(a) General subjects—

(1) The certificate holder's flight locating procedures;

(2) Principles and methods for determining weight and balance, and runway limitations for takeoff and landing;

(3) Enough meteorology to ensure a practical knowledge of weather phenomena, including the principles of frontal systems, icing, fog, thunderstorms, windshear and, if appropriate, high altitude weather situations;

(4) Air traffic control systems, procedures, and phraseology;

(5) Navigation and the use of navigational aids, including instrument approach procedures;

(6) Normal and emergency communication procedures;

(7) Visual cues before and during descent below DH or MDA; and

(8) Other instructions necessary to ensure the pilot's competence.

(b) For each aircraft type—

(1) A general description;

(2) Performance characteristics;

(3) Engines and propellers;

(4) Major components;

(5) Major aircraft systems (i.e., flight controls, electrical, and hydraulic), other systems, as appropriate, principles of normal, abnormal, and emergency operations, appropriate procedures and limitations;

(6) Procedures for—
(i) Recognizing and avoiding severe weather situations;
(ii) Escaping from severe weather situations, in case of inadvertent encounters, including low-altitude windshear (except that rotorcraft pilots are not required to be trained in escaping from low-altitude windshear); and
(iii) Operating in or near thunderstorms (including best penetrating altitudes), turbulent air (including clear air turbulence), icing, hail, and other potentially hazardous meterological conditions;

(7) Operating limitations;

(8) Fuel consumption and cruise control;

(9) Flight planning;

(10) Each normal and emergency procedure; and

(11) The approved Aircraft Flight Manual, or equivalent.

§ 135.347 Pilots: initial, transition, upgrade, and differences flight training.

(a) Initial, transition, upgrade, and differences training for pilots must include flight and practice in each of the maneuvers and procedures in the approved training program curriculum.

(b) The maneuvers and procedures required by paragraph (a) of this section must be performed in flight, except to the extent that certain maneuvers and procedures may be performed in an aircraft simulator, or an appropriate training device, as allowed by this subpart.

(c) If the certificate holder's approved training program includes a course of training using an aircraft simulator or other training device, each pilot must successfully complete—

(1) Training and practice in the simulator or training device in at least the maneuvers and procedures in this subpart that are capable of being performed in the aircraft simulator or training device; and

(2) A flight check in the aircraft or a check in the simulator or training device to the level of proficiency of a pilot in command or second in command, as applicable, in at least the maneuvers and procedures that are capable of being performed in an aircraft simulator or training device.

§ 135.349 Flight attendants: initial and transition ground training.

Initial and transition ground training for flight attendants must include instruction in at least the following—

(a) General subjects—

(1) The authority of the pilot in command; and

(2) Passenger handling, including procedures to be followed in handling deranged persons or other persons whose conduct might jeopardize safety.

(b) For each aircraft type—

(1) A general description of the aircraft emphasizing physical characteristics that may have a bearing on ditching, evacuation, and inflight emergency procedures and on other related duties;

(2) The use of both the public address system and the means of communicating with other flight crewmembers, including emergency means in the case of attempted hijacking or other unusual situations; and

(3) Proper use of electrical galley equipment and the controls for cabin heat and ventilation.

§ 135.351 Recurrent training.

(a) Each certificate holder must ensure that each crewmember receives recurrent training and is adequately trained and currently proficient for the type aircraft and crewmember position involved.

(b) Recurrent ground training for crewmembers must include at least the following:

(1) A quiz or other review to determine the crewmember's knowledge of the aircraft and crewmember position involved.

(2) Instruction as necessary in the subjects required for initial ground training by this subpart, as appropriate, including low-altitude windshear training as prescribed in § 135.345 of this part and emergency training.

(c) Recurrent flight training for pilots must include, at least, flight training in the maneuvers or procedures in this subpart, except that satisfactory completion of the check required by § 135.293 within the preceding 12 calendar months may be substituted for recurrent flight training.

§ 135.353 Prohibited drugs.

(a) Each certificate holder or operator shall provide each employee performing a function listed in Appendix I to Part 121 of this chapter and his or her supervisor with the training specified in that appendix.

(b) No certificate holder or operator may use any contractor to perform a function specified in Appendix I to Part 121 of this chapter unless that contractor provides each of its employees performing that function for the certificate holder or the operator and his or her supervisor with the training specified in that appendix.

Subpart I—Airplane Performance Operating Limitations

§ 135.361 Applicability.

(a) This subpart prescribes airplane performance operating limitations applicable to the operation of the categories of airplanes listed in § 135.363 when operated under this Part.

(b) For the purpose of this subpart, "effective length of the runway," for landing means the distance from the point at which the obstruction clearance plane associated with the approach end of the runway intersects the centerline of the runway to the far end of the runway.

(c) For the purpose of this subpart, "obstruction clearance plane" means a plane sloping upward from the runway at a slope of 1:20 to the horizontal, and tangent to or clearing all obstructions within a specified area surrounding the runway as shown in a profile view of that area. In the plan view, the centerline of the specified area coincides with the centerline of the runway, beginning at the point where the obstruction clearance plane intersects the centerline of the runway and proceeding to a point at least 1,500 feet from the beginning point. After that the centerline coincides with the takeoff path over the ground for the runway (in the case of takeoffs) or with the instrument approach counterpart (for landings), or, where the applicable one of these paths has not been established, it proceeds consistent with turns of at least 4,000-foot radius until a point is reached beyond which the obstruction clearance plane clears all obstructions. This area extends laterally 200 feet on each side of the centerline at the point where the obstruction clearance plane intersects the runway and continues at this width to the end of the runway; then it increases uniformly to 500 feet on each side of the centerline at a point 1,500 feet from the intersection of the obstruction clearance plane with the runway; after that it extends laterally 500 feet on each side of the centerline.

§ 135.363 General.

(a) Each certificate holder operating a reciprocating engine powered large transport category airplane shall comply with §§ 135.365 through 135.377.

(b) Each certificate holder operating a turbine engine powered large transport category airplane shall comply with §§ 135.379 through 135.387, except that when it operates a turbopropeller-powered large transport category airplane certificated after August 29, 1959, but previously type certificated with the same number of reciprocating engines, it may comply with §§ 135.365 through 135.377.

(c) Each certificate holder operating a large nontransport category airplane shall comply with §§ 135.389 through 135.395 and any determination of compliance must be based only on approved performance data. For the purpose of this subpart, a large nontransport category airplane is an airplane that was type certificated before July 1, 1942.

(d) Each certificate holder operating a small transport category airplane shall comply with § 135.397.

(e) Each certificate holder operating a small nontransport category airplane shall comply with § 135.399.

(f) The performance data in the Airplane Flight Manual applies in determining compliance with §§ 135.365 through 135.387. Where conditions are different from those on which the performance data is based, compliance is determined by interpolation or by computing the effects of change in the specific variables, if the results of the interpolation or computation are substantially as accurate as the results of direct tests.

(g) No person may take off a reciprocating engine powered large transport category airplane at a weight that is more than the allowable weight for the runway being used (determined under the runway takeoff limitations of the transport category operating rules of this subpart) after taking into account the temperature operating correction factors in § 4a.749a-T or § 4b.117 of the Civil Air Regulations in effect on January 31, 1965, and in the applicable Airplane Flight Manual.

(h) The Administrator may authorize in the operations specifications deviations from this subpart if special circumstances make a literal observance of a requirement unnecessary for safety.

(i) The 10-mile width specified in §§ 135.369 through 135.373 may be reduced to 5 miles, for not more than 20 miles, when operating under VFR or where navigation facilities furnish reliable and accurate identification of high ground and obstructions located outside of 5 miles, but within 10 miles, on each side of the intended track.

[(j) Each certificate holder operating a commuter category airplane shall comply with § 135.398.]

§ 135.365 Large transport category airplanes: reciprocating engine powered: weight limitations.

(a) No person may take off a reciprocating engine powered large transport category airplane from an airport located at an elevation outside of the range for which maximum takeoff weights have been determined for that airplane.

(b) No person may take off a reciprocating engine powered large transport category airplane for an airport of intended destination that is located at an elevation outside of the range for which maximum landing weights have been determined for that airplane.

(c) No person may specify, or have specified, an alternate airport that is located at an elevation outside of the range for which maximum landing weights have been determined for the reciprocating engine powered large transport category airplane concerned.

(d) No person may take off a reciprocating engine powered large transport category airplane at a weight more than the maximum authorized takeoff weight for the elevation of the airport.

(e) No person may take off a reciprocating engine powered large transport category airplane if its weight on arrival at the airport of destination will be more than the maximum authorized landing weight for the elevation of that airport, allowing for normal consumption of fuel and oil en route.

§ 135.367 Large transport category airplanes: reciprocating engine powered: takeoff limitations.

(a) No person operating a reciprocating engine powered large transport category airplane may take off that airplane unless it is possible—

(1) To stop the airplane safely on the runway, as shown by the accelerate-stop distance data, at any time during take off until reaching critical engine failure speed;

(2) If the critical engine fails at any time after the airplane reaches critical-engine failure speed V_1, to continue the takeoff and reach a height of 50 feet, as indicated by the takeoff path data, before passing over the end of the runway; and

(3) To clear all obstacles either by at least 50 feet vertically (as shown by the takeoff path data) or 200 feet horizontally within the airport boundaries and 300 feet horizontally beyond the boundaries, without banking before reaching a height of 50 feet (as shown by the takeoff path data) and after that without banking more than 15 degrees.

(b) In applying this section, corrections must be made for any runway gradient. To allow for wind effect, takeoff data based on still air may be corrected by taking into account not more than 50 percent of any reported headwind component and not less than 150 percent of any reported tailwind component.

§ 135.369 Large transport category airplanes: reciprocating engine powered: en route limitations: all engines operating.

(a) No person operating a reciprocating engine powered large transport category airplane may take off that airplane at a weight, allowing for normal consumption of fuel and oil, that does not allow a rate of climb (in feet per minute), with all engines operating, of at least 6.90 V_{S_0} (that is, the number of feet per minute obtained by multiplying the number of knots by 6.90) at an altitude of at least 1,000 feet above the highest ground or obstruction within ten miles of each side of the intended track.

(b) This section does not apply to large transport category airplanes certificated under Part 4a of the Civil Air Regulations.

§ 135.371 Large transport category airplanes: reciprocating engine powered: en route limitations: one engine inoperative.

(a) Except as provided in paragraph (b) of this section, no person operating a reciprocating engine powered large transport category airplane may take off that airplane at a weight, allowing for normal consumption of fuel and oil, that does not allow a rate of climb (in feet per minute), with one engine inoperative, of at least $(0.079–0.106/N) V_{S_0}^2$ (where N is the number of engines installed and V_{S_0} is expressed in knots) at an altitude of at least 1,000 feet above the highest ground or obstruction within 10 miles of each side of the intended track. However, for the purposes of this paragraph the rate of climb for transport category airplanes certificated under Part 4a of the Civil Air Regulations is 0.026 $V_{S_0}^2$.

(b) In place of the requirements of paragraph (a) of this section, a person may, under an approved procedure, operate a reciprocating engine powered large transport category airplane at an all-engine-operating altitude that allows the airplane to continue, after an engine failure, to an alternate airport where a landing can be made under § 135.377, allowing for normal consumption of fuel and oil. After the assumed failure, the flight path must clear the ground and any obstruction within five miles on each side of the intended track by at least 2,000 feet.

(c) If an approved procedure under paragraph (b) of this section is used, the certificate holder shall comply with the following:

(1) The rate of clim (as prescribed in the Airplane Flight Manual for the appropriate weight and altitude) used in calculating the airplane's flight path shall be diminished by an amount in feet per minute, equal to $(0.079–0.106/N) V_{S_0}^2$ (when N is the

number of engines installed and V_{S_0}, is expressed in knots) for airplanes certificated under Part 25 of this chapter and by 0.026 $V_{S_0}^2$ for airplanes certificated under Part 4a of the Civil Air Regulations.

(2) The all-engines-operating altitude shall be sufficient so that in the event the critical engine becomes inoperative at any point along the route, the flight will be able to proceed to a predetermined alternate airport by use of this procedure. In determining the takeoff weight, the airplane is assumed to pass over the critical obstruction following engine failure at a point no closer to the critical obstruction than the nearest approved radio navigational fix, unless the Administrator approves a procedure established on a different basis upon finding that adequate operational safeguards exist.

(3) The airplane must meet the provisions of paragraph (a) of this section at 1,000 feet above the airport used as an alternate in this procedure.

(4) The procedure must include an approved method of accounting for winds and temperatures that would otherwise adversely affect the flight path.

(5) In complying with this procedure, fuel jettisoning is allowed if the certificate holder shows that it has an adequate training program, that proper instructions are given to the flight crew, and all other precautions are taken to ensure a safe procedure.

(6) The certificate holder and the pilot in command shall jointly elect an alternate airport for which the appropriate weather reports or forecasts, or any combination of them, indicate that weather conditions will be at or above the alternate weather minimum specified in the certificate holder's operations specifications for that airport when the flight arrives.

§ 135.373 Part 25 transport category airplanes with four or more engines: reciprocating engine powered: en route limitations: two engines inoperative.

(a) No person may operate an airplane certificated under Part 25 and having four or more engines unless—

(1) There is no place along the intended track that is more than 90 minutes (with all engines operating at cruising power) from an airport that meets § 135.377; or

(2) It is operated at a weight allowing the airplane, with the two critical engines inoperative, to climb at 0.013 V_S^2 feet per minute (that is, the number of feet per minute obtained by multiplying the number of knots squared by 0.013) at an altitude of 1,000 feet above the highest ground or obstruction within 10 miles on each side of the intended track, or at an altitude of 5,000 feet, whichever is higher.

(b) For the purposes of paragraph (a)(2) of this section, it is assumed that—

(1) The two engines fail at the point that is most critical with respect to the takeoff weight;

(2) Consumption of fuel and oil is normal with all engines operating up to the point where the two engines fail with two engines operating beyond that point;

(3) Where the engines are assumed to fail at an altitude above the prescribed minimum altitude, compliance with the prescribed rate of climb at the prescribed minimum altitude need not be shown during the descent from the cruising altitude to the prescribed minimum altitude, if those requirements can be met once the prescribed minimum altitude is reached, and assuming descent to be along a net flight path and the rate of descent to be 0.013 $V_{S_0}^2$ greater than the rate in the approved performance data; and

(4) If fuel jettisoning is provided, the airplane's weight at the point where the two engines fail is considered to be not less than that which would include enough fuel to proceed to an airport meeting § 135.377 and to arrive at an altitude of at least 1,000 feet directly over that airport.

§ 135.375 Large transport category airplanes: reciprocating engine powered: landing limitations: destination airports.

(a) Except as provided in paragraph (b) of this section, no person operating a reciprocating engine powered large transport category airplane may take off that airplane, unless its weight on arrival, allowing for normal consumption of fuel and oil in flight, would allow a full stop landing at the intended destination within 60 percent of the effective length of each runway described below from a point 50 feet directly above the intersection of the obstruction clearance plane and the runway. For the purposes of determining the allowable landing weight at the destination airport the following is assumed:

(1) The airplane is landed on the most favorable runway and in the most favorable direction in still air.

(2) The airplane is landed on the most suitable runway considering the probable wind velocity and direction (forecast for the expected time of arrival), the ground handling characteristics of the type of airplane, and other conditions such as landing aids and terrain, and allowing for the effect of the landing path and roll of not more than 50 percent of the headwind component or not less than 150 percent of the tailwind component.

(b) An airplane that would be prohibited from being taken off because it could not meet paragraph (a)(2) of this section may be taken off if an alternate airport is selected that meets all of this section except that the airplane can accomplish a full stop landing within 70 percent of the effective length of the runway.

§ 135.377 Large transport category airplanes: reciprocating engine powered: landing limitations: alternate airports.

No person may list an airport as an alternate airport in a flight plan unless the airplane (at the weight anticipated at the time of arrival at the airport), based on the assumptions in § 135.375(a)(1) and (2), can be brought to a full stop landing within 70 percent of the effective length of the runway.

§ 135.379 Large transport category airplanes: turbine engine powered: takeoff limitations.

(a) No person operating a turbine engine powered large transport category airplane may take off that airplane at a weight greater than that listed in the Airplane Flight Manual for the elevation of the airport and for the ambient temperature existing at takeoff.

(b) No person operating a turbine engine powered large transport category airplane certificated after August 26, 1957, but before August 30, 1959 (SR422, 422A), may take off that airplane at a weight greater than that listed in the Airplane Flight Manual for the minimum distance required for takeoff. In the case of an airplane certificated after September 30, 1958 (SR422A, 422B), the takeoff distance may include a clearway distance but the clearway distance included may not be greater than one-half of the takeoff run.

(c) No person operating a turbine engine powered large transport category airplane certificated after August 29, 1959 (SR422B), may take off that airplane at a weight greater than that listed in the Airplane Flight Manual at which compliance with the following may be shown:

(1) The accelerate-stop distance, as defined in § 25.109 of this chapter, must not exceed the length of the runway plus the length of any stopway.

(2) The takeoff distance must not exceed the length of the runway plus the length of any clearway except that the length of any clearway included must not be greater than one-half the length of the runway.

(3) The takeoff run must not be greater than the length of the runway.

(d) No person operating a turbine engine powered large transport category airplane may take off that airplane at a weight greater than that listed in the Airplane Flight Manual—

(1) For an airplane certificated after August 26, 1957, but before October 1, 1958 (SR422), that allows a takeoff path that clears all obstacles either by at least (35 + 0.01 D) feet vertically (D is the distance along the intended flight path from the end of the runway in feet), or by at least 200 feet horizontally within the airport boundaries and by at least 300 feet horizontally after passing the boundaries; or

(2) For an airplane certificated after September 30, 1958 (SR422A, 422B), that allows a net takeoff flight path that clears

all obstacles either by a height of at least 35 feet vertically, or by at least 200 feet horizontally within the airport boundaries and by at least 300 feet horizontally after passing the boundaries.

(e) In determining maximum weights, minimum distances and flight paths under paragraphs (a) through (d) of this section, correction must be made for the runway to be used, the elevation of the airport, the effective runway gradient, and the ambient temperature and wind component at the time of takeoff.

(f) For the purposes of this section, it is assumed that the airplane is not banked before reaching a height of 50 feet, as shown by the takeoff path or net takeoff flight path data (as appropriate) in the Airplane Flight Manual, and after that the maximum bank is not more than 15 degrees.

(g) For the purposes of this section, the terms, "takeoff distance," "takeoff run," "net takeoff flight path," have the same meanings as set forth in the rules under which the airplane was certificated.

§ 135.381 Large transport category airplanes: turbine engine powered: en route limitations: one engine inoperative.

(a) No person operating a turbine engine powered large transport category airplane may take off that airplane at a weight, allowing for normal consumption of fuel and oil, that is greater than that which (under the approved, one engine inoperative, en route net flight path data in the Airplane Flight Manual for that airplane) will allow compliance with subparagraph (1) or (2) of this paragraph, based on the ambient temperatures expected en route.

(1) There is a positive slope at an altitude of at least 1,000 feet above all terrain and obstructions within five statute miles on each side of the intended track, and, in addition, if that airplane was certificated after August 29, 1958 (SR422B), there is a positive slope at 1,500 feet above the airport where the airplane is assumed to land after an engine fails.

(2) The net flight path allows the airplane to continue flight from the cruising altitude to an airport where a landing can be made under § 135.387 clearing all terrain and obstructions within five statute miles of the intended track by at least 2,000 feet vertically and with a positive slope at 1,000 feet above the airport where the airplane lands after an engine fails, or, if that airplane was certificated after September 30, 1958 (SR422A, 422B), with a positive slope at 1,500 feet above the airport where the airplane lands after an engine fails.

(b) For the purpose of paragraph (a)(2) of this section, it is assumed that—

(1) The engine fails at the most critical point en route;

(2) The airplane passes over the critical obstruction, after engine failure at a point that is no closer to the obstruction than the approved radio navigation fix, unless the Administrator authorizes a different procedure based on adequate operational safeguards;

(3) An approved method is used to allow for adverse winds;

(4) Fuel jettisoning will be allowed if the certificate holder shows that the crew is properly instructed, that the training program is adequate, and that all other precautions are taken to ensure a safe procedure;

(5) The alternate airport is selected and meets the prescribed weather minimums; and

(6) The consumption of fuel and oil after engine failure is the same as the consumption that is allowed for in the approved net flight path data in the Airplane Flight Manual.

§ 135.383 Large transport category airplanes: turbine engine powered: en route limitations: two engines inoperative.

(a) Airplanes certificated after August 26, 1957, but before October 1, 1958 (SR422). No person may operate a turbine engine powered large transport category airplane along an intended route unless that person complies with either of the following:

(1) There is no place along the intended track that is more than 90 minutes (with all engines operating at cruising power) from an airport that meets § 135.387.

(2) Its weight, according to the two-engine-inoperative, en route, net flight path data in the Airplane Flight Manual, allows the airplane to fly from the point where the two engines are assumed to fail simultaneously to an airport that meets § 135.387, with a net flight path (considering the ambient temperature anticipated along the track) having a positive slope at an altitude of at least 1,000 feet above all terrain and obstructions within five statute miles on each side of the intended track, or at an altitude of 5,000 feet, whichever is higher.

For the purposes of paragraph (2) of this paragraph, it is assumed that the two engines fail at the most critical point en route, that if fuel jettisoning is provided, the airplane's weight at the point where the engines fail includes enough fuel to continue to the airport and to arrive at an altitude of at least 1,000 feet directly over the airport, and that the fuel and oil consumption after engine failure is the same as the consumption allowed for in the net flight path data in the Airplane Flight Manual.

(b) Airplanes certificated after September 30, 1958, but before August 30, 1959 (SR422A). No person may operate a turbine engine powered large transport category airplane along an intended route unless that person complies with either of the following:

(1) There is no place along the intended track that is more than 90 minutes (with all

engines operating at cruising power) from an airport that meets § 135.387.

(2) Its weight, according to the two-engine-inoperative, en route, net flight path data in the Airplane Flight Manual allows the airplane to fly from the point where the two engines are assumed to fail simultaneously to an airport that meets § 135.387 with a net flight path (considering the ambient temperatures anticipated along the track) having a positive slope at an altitude of at least 1,000 feet above all terrain and obstructions within five statute miles on each side of the intended track, or at an altitude of 2,000 feet, whichever is higher.

For the purpose of paragraph (2) of this paragraph, it is assumed that the two engines fail at the most critical point en route, that the airplane's weight at the point where the engines fail includes enough fuel to continue to the airport, to arrive at an altitude of at least 1,500 feet directly over the airport, and after that to fly for 15 minutes at cruise power or thrust, or both, and that the consumption of fuel and oil after engine failure is the same as the consumption allowed for in the net flight path data in the Airplane Flight Manual.

(c) Aircraft certificated after August 29, 1959 (SR422B). No person may operate a turbine engine powered large transport category airplane along an intended route unless that person complies with either of the following:

(1) There is no place along the intended track that is more than 90 minutes (with all engines operating at cruising power) from an airport that meets § 135.387.

(2) Its weight, according to the two-engine-inoperative, en route, net flight path data in the Airplane Flight Manual, allows the airplane to fly from the point where the two engines are assumed to fail simultaneously to an airport that meets § 135.387, with the net flight path (considering the ambient temperatures anticipated along the track) clearing vertically by at least 2,000 feet all terrain and obstructions within five statute miles on each side of the intended track. For the purposes of this paragraph, it is assumed that—

(i) The two engines fail at the most critical point en route;

(ii) The net flight path has a positive slope at 1,500 feet above the airport where the landing is assumed to be made after the engines fail;

(iii) Fuel jettisoning will be approved if the certificate holder shows that the crew is properly instructed, that the training program is adequate, and that all other precautions are taken to ensure a safe procedure;

(iv) The airplane's weight at the point where the two engines are assumed to fail provides enough fuel to continue to the airport, to arrive at an altitude of at least

1,500 feet directly over the airport, and after that to fly for 15 minutes at cruise power or thrust, or both; and

(v) The consumption of fuel and oil after the engines fail is the same as the consumption that is allowed for in the net flight path data in the Airplane Flight Manual.

§ 135.385 Large transport category airplanes: turbine engine powered: landing limitations: destination airports.

(a) No person operating a turbine engine powered large transport category airplane may take off that airplane at a weight that (allowing for normal consumption of fuel and oil in flight to the destination or alternate airport) the weight of the airplane on arrival would exceed the landing weight in the Airplane Flight Manual for the elevation of the destination or alternate airport and the ambient temperature anticipated at the time of landing.

(b) Except as provided in paragraph (c), (d), or (e) of this section, no person operating a turbine engine powered large transport category airplane may take off that airplane unless its weight on arrival, allowing for normal consumption of fuel and oil in flight (in accordance with the landing distance in the Airplane Flight Manual for the elevation of the destination airport and the wind conditions anticipated there at the time of landing), would allow a full stop landing at the intended destination airport within 60 percent of the effective length of each runway described below from a point 50 feet above the intersection of the obstruction clearance plane and the runway. For the purpose of determining the allowable landing weight at the destination airport the following is assumed:

(1) The airplane is landed on the most favorable runway and in the most favorable direction, in still air.

(2) The airplane is landed on the most suitable runway considering the probable wind velocity and direction and the ground handling characteristics of the airplane, and considering other conditions such as landing aids and terrain.

(c) A turbopropeller powered airplane that would be prohibited from being taken off because it could not meet paragraph (b)(2) of this section, may be taken off if an alternate airport is selected that meets all of this section except that the airplane can accomplish a full stop landing within 70 percent of the effective length of the runway.

(d) Unless, based on a showing of actual operating landing techniques on wet runways, a shorter landing distance (but never less than that required by paragraph (b) of this section) has been approved for a specific type and model airplane and included in the Airplane Flight Manual, no person may take off a turbojet airplane when the appropriate weather reports or forecasts, or any combination of them,

indicate that the runways at the destination airport may be wet or slippery at the estimated time of arrival unless the effective runway length at the destination airport is at least 115 percent of the runway length required under paragraph (b) of this section.

(e) A turbojet airplane that would be prohibited from being taken off because it could not meet paragraph (b)(2) of this section may be taken off if an alternate airport is selected that meets all of paragraph (b) of this section.

§ 135.387 Large transport category airplanes: turbine engine powered: landing limitations: alternate airports.

No person may select an airport as an alternate airport for a turbine engine powered large transport category airplane unless (based on the assumption in § 135.385(b)) that airplane, at the weight anticipated at the time of arrival, can be brought to a full stop landing within 70 percent of the effective length of the runway for turbopropeller-powered airplanes and 60 percent of the effective length of the runway for turbojet airplanes, from a point 50 feet above the intersection of the obstruction clearance plane and the runway.

§ 135.389 Large nontransport category airplanes: takeoff limitations.

(a) No person operating a large nontransport category airplane may take off that airplane at a weight greater than the weight that would allow the airplane to be brought to a safe stop within the effective length of the runway, from any point during the takeoff before reaching 105 percent of minimum control speed (the minimum speed at which an airplane can be safely controlled in flight after an engine becomes inoperative) or 115 percent of the power off stalling speed in the takeoff configuration, whichever is greater.

(b) For the purposes of this section—

(1) It may be assumed that takeoff power is used on all engines during the acceleration;

(2) Not more than 50 percent of the reported headwind component, or not less than 150 percent of the reported tailwind component, may be taken into account;

(3) The average runway gradient (the difference between the elevations of the endpoints of the runway divided by the total length) must be considered if it is more than one-half of one percent;

(4) It is assumed that the airplane is operating in standard atmosphere; and

(5) For takeoff, "effective length of the runway" means the distance from the end of the runway at which the takeoff is started to a point at which the obstruction clearance plane associated with the other end of the runway intersects the runway centerline.

§ 135.391 Large nontransport category airplanes: en route limitations: one engine inoperative.

(a) Except as provided in paragraph (b)

of this section, no person operating a large nontransport category airplane may take off that airplane at a weight that does not allow a rate of climb of at least 50 feet a minute, with the critical engine inoperative, at an altitude of at least 1,000 feet above the highest obstruction within five miles on each side of the intended track, or 5,000 feet, whichever is higher.

(b) Without regard to paragraph (a) of this section, if the Administrator finds that safe operations are not impaired, a person may operate the airplane at an altitude that allows the airplane, in case of engine failure, to clear all obstructions within five miles on each side of the intended track by 1,000 feet. If this procedure is used, the rate of descent for the appropriate weight and altitude is assumed to be 50 feet a minute greater than the rate in the approved performance data. Before approving such a procedure, the Administrator considers the following for the route, route segment, or area concerned:

(1) The reliability of wind and weather forecasting.

(2) The location and kinds of navigation aids.

(3) The prevailing weather conditions, particularly the frequency and amount of turbulence normally encountered.

(4) Terrain features.

(5) Air traffic problems.

(6) Any other operational factors that affect the operations.

(c) For the purposes of this section, it is assumed that—

(1) The critical engine is inoperative;

(2) The propeller of the inoperative engine is in the minimum drag position;

(3) The wing flaps and landing gear are in the most favorable position;

(4) The operating engines are operating at the maximum continuous power available;

(5) The airplane is operating in standard atmosphere; and

(6) The weight of the airplane is progressively reduced by the anticipated consumption of fuel and oil.

§ 135.393 Large nontransport category airplanes: landing limitations: destination airports.

(a) No person operating a large nontransport category airplane may take off that airplane at a weight that—

(1) Allowing for anticipated consumption of fuel and oil, is greater than the weight that would allow a full stop landing within 60 percent of the effective length of the most suitable runway at the destination airport; and

(2) Is greater than the weight allowable if the landing is to be made on the runway—

(i) With the greatest effective length in still air; and

(ii) Required by the probable wind, taking into account not more than 50 percent of the headwind component or not less than 150 percent of the tailwind component.

(b) For the purpose of this section, it is assumed that—

(1) The airplane passes directly over the intersection of the obstruction clearance plane and the runway at a height of 50 feet in a steady gliding approach at a true indicated airspeed of at least 1.3 V_{SO};

(2) The landing does not require exceptional pilot skill; and

(3) The airplane is operating in standard atmosphere.

§ 135.395 Large nontransport category airplanes: landing limitations: alternate airports.

No person may select an airport as an alternate airport for a large nontransport category airplane unless that airplane (at the weight anticipated at the time of arrival), based on the assumptions in § 135.393(b), can be brought to a full stop landing within 70 percent of the effective length of the runway.

§ 135.397 Small transport category airplane performance operating limitations.

(a) No person may operate a reciprocating engine powered small transport category airplane unless that person complies with the weight limitations in § 135.365, the takeoff limitations in § 135.367 (except paragraph (a)(3)), and the landing limitations in §§ 135.375 and 135.377.

(b) No person may operate a turbine engine powered small transport category airplane unless that person complies with the weight limitations in § 135.379 (except paragraphs (d) and (f)) and the landing limitations in §§ 135.385 and 135.387.

[§ 135.398 Commuter Category airplane performance operating limitations.

[(a) No person may operate a commuter category airplane unless that person complies with the takeoff weight limitations in the approved Airplane Flight Manual.

[(b) No person may take off an airplane type certificated in the commuter category at a weight greater than that listed in the Airplane Flight Manual that allows a net takeoff flight path that clears all obstacles either by a height of at least 35 feet vertically, or at least 200 feet horizontally within the airport boundaries and by at least 300 feet horizontally after passing the boundaries.

[(c) No person may operate a commuter category airplane unless that person complies with the landing limitations prescribed in §§ 135.385 and 135.387 of this Part. For purposes of this paragraph, §§ 135.385 and 135.387 are applicable to all commuter category airplanes notwithstanding their stated applicability to turbine-engine-powered large transport category airplanes.

[(d) In determining maximum weights, minimum distances and flight paths under paragraphs (a) through (c) of this section, correction must be made for the runway gradient, and ambient temperature, and wind component at the time of takeoff.

[(e) For the purpose of this section, the assumption is that the airplane is not banked before reaching a height of 50 feet as shown by the new takeoff flight path data in the Airplane Flight Manual and thereafter the maximum bank is not more than 15 degrees.]

§ 135.399 Small nontransport category airplane performance operating limitations.

(a) No person may operate a reciprocating engine or turbopropeller-powered small airplane that is certificated under § 135.169(b)(2), (3), (4), (5), or (6) unless that person complies with the takeoff weight limitations in the approved Airplane Flight Manual or equivalent for operations under this Part, and, if the airplane is certificated under § 135.169(b)(4) or (5) with the landing weight limitations in the Approved Airplane Flight Manual or equivalent for operations under this Part.

(b) No person may operate an airplane that is certificated under § 135.169(b)(6) unless that person complies with the landing limitations prescribed in §§ 135.385 and 135.387 of this Part. For purposes of this paragraph, §§ 135.385 and 135.387 are applicable to reciprocating and turbopropeller-powered small airplanes notwithstanding their stated applicability to turbine engine powered large transport category airplanes.

Subpart J—Maintenance, Preventive Maintenance, and Alterations

§ 135.411 Applicability.

(a) This subpart prescribes rules in addition to those in other parts of this chapter for the maintenance, preventive maintenance, and alterations for each certificate holder as follows:

(1) Aircraft that are type certificated for a passenger seating configuration, excluding any pilot seat, of nine seats or less, shall be maintained under Parts 91 and 43 of this chapter and §§ 135.415, 135.417, and 135.421. An approved aircraft inspection program may be used under § 135.419.

(2) Aircraft that are type certificated for a passenger seating configuration, excluding any pilot seat, of ten seats or more, shall be maintained under a maintenance program in §§ 135.415, 135.417, and 135.423 through 135.443.

(b) A certificate holder who is not otherwise required, may elect to maintain its aircraft under paragraph (a)(2) of this section.

§ 135.413 Responsibility for airworthiness.

(a) Each certificate holder is primarily responsible for the airworthiness of its aircraft, including airframes, aircraft engines, propellers, rotors, appliances, and parts, and shall have its aircraft maintained under this chapter, and shall have defects repaired between

required maintenance under Part 43 of this chapter.

(b) Each certificate holder who maintains its aircraft under § 135.411(a)(2) shall—

(1) Perform the maintenance, preventive maintenance, and alteration of its aircraft, including airframe, aircraft engines, propellers, rotors, appliances, emergency equipment and parts, under its manual and this chapter; or

(2) Make arrangements with another person for the performance of maintenance, preventive maintenance or alteration. However, the certificate holder shall ensure that any maintenance, preventive maintenance, or alteration that is performed by another person is performed under the certificate holder's manual and this chapter.

§ 135.415 Mechanical reliability reports.

(a) Each certificate holder shall report the occurrence or detection of each failure, malfunction, or defect in an aircraft concerning—

(1) Fires during flight and whether the related fire-warning system functioned properly;

(2) Fires during flight not protected by related fire-warning system;

(3) False fire-warning during flight;

(4) An exhaust system that causes damage during flight to the engine, adjacent structure, equipment, or components;

(5) An aircraft component that causes accumulation or circulation of smoke, vapor, or toxic or noxious fumes in the crew compartment or passenger cabin during flight;

(6) Engine shutdown during flight because of flameout;

(7) Engine shutdown during flight when external damage to the engine or aircraft structure occurs;

(8) Engine shutdown during flight due to foreign object ingestion or icing;

(9) Shutdown of more than one engine during flight;

(10) A propeller feathering system or ability of the system to control overspeed during flight;

(11) A fuel or fuel-dumping system that affects fuel flow or causes hazardous leakage during flight;

(12) An unwanted landing gear extension or retraction or opening or closing of landing gear doors during flight;

(13) Brake system components that result in loss of brake actuating force when the aircraft is in motion on the ground;

(14) Aircraft structure that requires major repair;

(15) Cracks, permanent deformation, or corrosion of aircraft structures, if more than the maximum acceptable to the manufacturer or the FAA; and

(16) Aircraft components or systems that result in taking emergency actions during

flight (except action to shut-down an engine).

(b) For the purpose of this section, "during flight" means the period from the moment the aircraft leaves the surface of the earth on takeoff until it touches down on landing.

(c) In addition to the reports required by paragraph (a) of this section, each certificate holder shall report any other failure, malfunction, or defect in an aircraft that occurs or is detected at any time if, in its opinion, the failure, malfunction, or defect has endangered or may endanger the safe operation of the aircraft.

(d) Each certificate holder shall send each report required by this section, in writing, covering each 24-hour period beginning at 0900 hours local time of each day and ending at 0900 hours local time on the next day to the FAA Flight Standards District Office charged with the overall inspection of the certificate holder. Each report of occurrences during a 24-hour period must be mailed or delivered to that office within the next 72 hours. However, a report that is due on Saturday or Sunday may be mailed or delivered on the following Monday and one that is due on a holiday may be mailed or delivered on the next work day. For aircraft operated in areas where mail is not collected, reports may be mailed or delivered within 72 hours after the aircraft returns to a point where the mail is collected.

(e) The certificate holder shall transmit the reports required by this section on a form and in a manner prescribed by the Administrator, and shall include as much of the following as is available:

(1) The type and identification number of the aircraft.

(2) The name of the operator.

(3) The date.

(4) The nature of the failure, malfunction, or defect.

(5) Identification of the part and system involved, including available information pertaining to type designation of the major component and time since last overhaul, if known.

(6) Apparent cause of the failure, malfunction or defect (e.g., wear, crack, design deficiency, or personnel error).

(7) Other pertinent information necessary for more complete identification, determination of seriousness, or corrective action.

(f) A certificate holder that is also the holder of a type certificate (including a supplemental type certificate), a Parts Manufacturer Approval, or a Technical Standard Order Authorization, or that is the licensee of a type certificate need not report a failure, malfunction, or defect under this section if the failure, malfunction, or defect has been reported by it under § 21.3 or § 37.17 of this chapter or under the accident reporting provisions of Part 830 of the regulations of the National Transporta-

tion Safety Board.

(g) No person may withhold a report required by this section even though all information required by this section is not available.

(h) When the certificate holder gets additional information, including information from the manufacturer or other agency, concerning a report required by this section, it shall expeditiously submit it as a supplement to the first report and reference the date and place of submission of the first report.

§ 135.417 Mechanical interruption summary report.

Each certificate holder shall mail or deliver, before the end of the 10th day of the following month, a summary report of the following occurrences in multiengine aircraft for the preceding month to the FAA Flight Standards District Office charged with the overall inspection of the certificate holder:

(a) Each interruption to a flight, unscheduled change of aircraft en route, or unscheduled stop or diversion from a route, caused by known or suspected mechanical difficulties or malfunctions that are not required to be reported under § 135.415.

(b) The number of propeller featherings in flight, listed by type of propeller and engine and aircraft on which it was installed. Propeller featherings for training, demonstration, or flight check purposes need not be reported.

§ 135.419 Approved aircraft inspection program.

(a) Whenever the Administrator finds that the aircraft inspections required or allowed under Part 91 of this chapter are not adequate to meet this Part, or upon application by a certificate holder, the Administrator may amend the certificate holder's operations specifications under § 135.17, to require or allow an approved aircraft inspection program for any make and model aircraft of which the certificate holder has the exclusive use of at least one aircraft (as defined in § 135.25(b)).

(b) A certificate holder who applies for an amendment of its operations specifications to allow an approved aircraft inspection program must submit that program with its application for approval by the Administrator.

(c) Each certificate holder who is required by its operations specifications to have an approved aircraft inspection program shall submit a program for approval by the Administrator within 30 days of the amendment of its operations specifications or within any other period that the Administrator may prescribe in the operations specifications.

(d) The aircraft inspection program submitted for approval by the Administrator must contain the following:

(1) Instructions and procedures for the conduct of aircraft inspections (which must include necessary tests and checks), setting forth in detail the parts and areas of the

airframe, engines, propellers, rotors, and appliances, including emergency equipment, that must be inspected.

(2) A schedule for the performance of the aircraft inspections under paragraph (1) of this paragraph expressed in terms of the time in service, calendar time, number of system operations, or any combination of these.

(3) Instructions and procedures for recording discrepancies found during inspections and correction or deferral of discrepancies including form and disposition of records.

(e) After approval, the certificate holder shall include the approved aircraft inspection program in the manual required by § 135.21.

(f) Whenever the Administrator finds that revisions to an approved aircraft inspection program are necessary for the continued adequacy of the program, the certificate holder shall, after notification by the Administrator, make any changes in the program found by the Administrator to be necessary. The certificate holder may petition the Administrator to reconsider the notice to make any changes in a program. The petition must be filed with the representatives of the Administrator assigned to it within 30 days after the certificate holder receives the notice. Except in the case of an emergency requiring immediate action in the interest of safety, the filing of the petition stays the notice pending a decision by the Administrator.

(g) Each certificate holder who has an approved aircraft inspection program shall have each aircraft that is subject to the program inspected in accordance with the program.

(h) The registration number of each aircraft that is subject to an approved aircraft inspection program must be included in the operations specifications of the certificate holder.

§ 135.421 Additional maintenance requirements.

(a) Each certificate holder who operates an aircraft type certificated for a passenger seating configuration, excluding any pilot seat, of nine seats or less, must comply with the manufacturer's recommended maintenance programs, or a program approved by the Administrator, for each aircraft engine, propeller, rotor, and each item of emergency equipment required by this chapter.

(b) For the purpose of this section, a manufacturer's maintenance program is one which is contained in the maintenance manual or maintenance instructions set forth by the manufacturer as required by this chapter for the aircraft, aircraft engine, propeller, rotor or item of emergency equipment.

§ 135.423 Maintenance, preventive maintenance, and alteration organization.

(a) Each certificate holder that performs any of its maintenance (other than required inspections), preventive maintenance, or alterations, and each person with whom it arranges for the performance of that work, must have an organization adequate to perform the work.

(b) Each certificate holder that performs any inspections required by its manual under § 135.427(b)(2) or (3), (in this subpart referred to as "required inspections"), and each person with whom it arranges for the performance of that work, must have an organization adequate to perform that work.

(c) Each person performing required inspections in addition to other maintenance, preventive maintenance, or alterations, shall organize the performance of those functions so as to separate the required inspection functions from the other maintenance, preventive maintenance, and alteration functions. The separation shall be below the level of administrative control at which overall responsibility for the required inspection functions and other maintenance, preventive maintenance, and alteration functions is exercised.

§ 135.425 Maintenance, preventive maintenance, and alteration programs.

Each certificate holder shall have an inspection program and a program covering other maintenance, preventive maintenance, and alterations, that ensures that—

(a) Maintenance, preventive maintenance, and alterations performed by it, or by other persons, are performed under the certificate holder's manual;

(b) Competent personnel and adequate facilities and equipment are provided for the proper performance of maintenance, preventive maintenance, and alterations; and

(c) Each aircraft released to service is airworthy and has been properly maintained for operation under this Part.

§ 135.427 Manual requirements.

(a) Each certificate holder shall put in its manual the chart or description of the certificate holder's organization required by § 135.423 and a list of persons with whom it has arranged for the performance of any of its required inspections, other maintenance, preventive maintenance, or alterations, including a general description of that work.

(b) Each certificate holder shall put in its manual the programs required by § 135.425 that must be followed in performing maintenance, preventive maintenance, and alterations of that certificate holder's aircraft, including airframes, aircraft engines, propellers, rotors, appliances, emergency equipment, and parts, and must include at least the following:

(1) The method of performing routine and nonroutine maintenance (other than required inspections), preventive maintenance, and alterations.

(2) A designation of the items of maintenance and alteration that must be inspected (required inspections) including at least those that could result in a failure, malfunction, or defect endangering the safe opera-

tion of the aircraft, if not performed properly or if improper parts or materials are used.

(3) The method of performing required inspections and a designation by occupational title of personnel authorized to perform each required inspection.

(4) Procedures for the reinspection of work performed under previous required inspection findings ("buy-back procedures").

(5) Procedures, standards, and limits necessary for required inspections and acceptance or rejection of the items required to be inspected and for periodic inspection and calibration of precision tools, measuring devices, and test equipment.

(6) Procedures to ensure that all required inspections are performed.

(7) Instructions to prevent any person who performs any item of work from performing any required inspection of that work.

(8) Instructions and procedures to prevent any decision of an inspector regarding any required inspection from being countermanded by persons other than supervisory personnel of the inspection unit, or a person at the level of administrative control that has overall responsibility for the management of both the required inspection functions and the other maintenance, preventive maintenance, and alterations functions.

(9) Procedures to ensure that required inspections, other maintenance, preventive maintenance, and alterations that are not completed as a result of work interruptions are properly completed before the aircraft is released to service.

(c) Each certificate holder shall put in its manual a suitable system (which may include a coded system) that provides for the retention of the following information—

(1) A description (or reference to data acceptable to the Administrator) of the work performed;

(2) The name of the person performing the work if the work is performed by a person outside the organization of the certificate holder; and

(3) The name or other positive identification of the individual approving the work.

§ 135.429 Required inspection personnel.

(a) No person may use any person to perform required inspections unless the person performing the inspection is appropriately certificated, properly trained, qualified, and authorized to do so.

(b) No person may allow any person to perform a required inspection unless, at the time, the person performing that inspection is under the supervision and control of an inspection unit.

(c) No person may perform a required inspection if that person performed the item of work to be inspected.

[(d) In the case of rotorcraft that operate in remote areas or sites, the Administrator may

approve procedures for the performance of required inspection items by a pilot when no other qualified person is available, provided—

[(1) The pilot is employed by the certificate holder;

[(2) It can be shown to the satisfaction of the Administrator that each pilot authorized to perform required inspections is properly trained and qualified;

[(3) The required inspection is a result of a mechanical interruption and is not a part of a certificate holder's continuous airworthiness maintenance program;

[(4) Each item is inspected after each flight until the item has been inspected by an appropriately certificated mechanic other than the one who originally performed the item of work; and

[(5) Each item of work that is a required inspection item that is part of the flight control system shall be flight tested and reinspected before the aircraft is approved for return to service.]

[(e) Each certificate holder shall maintain, or shall determine that each person with whom it arranges to perform its required inspections maintains, a current listing of persons who have been trained, qualified, and authorized to conduct required inspections. The persons must be identified by name, occupational title and the inspections that they are authorized to perform. The certificate holder (or person with whom it arranges to perform its required inspections) shall give written information to each person so authorized, describing the extent of that person's responsibilities, authorities, and inspectional limitations. The list shall be made available for inspection by the Administrator upon request.

§ 135.431 Continuing analysis and surveillance.

(a) Each certificate holder shall establish and maintain a system for the continuing analysis and surveillance of the performance and effectiveness of its inspection program and the program covering other maintenance, preventive maintenance, and alterations and for the correction of any deficiency in those programs, regardless of whether those programs are carried out by the certificate holder or by another person.

(b) Whenever the Administrator finds that either or both of the programs described in paragraph (a) of this section does not contain adequate procedures and standards to meet this part, the certificate holder shall, after notification by the Administrator, make changes in those programs requested by the Administrator.

(c) A certificate holder may petition the Administrator to reconsider the notice to make a change in a program. The petition must be filed with the FAA Flight Standards District Office charged with the overall inspection of the certificate holder within 30 days after the certificate holder receives the notice. Except in the case of an emergency requiring immediate action in the interest of safety, the filing of the petition stays the notice pending a decision by the Administrator.

§ 135.433 Maintenance and preventive maintenance training program.

Each certificate holder or a person performing maintenance or preventive maintenance functions for it shall have a training program to ensure that each person (including inspection personnel) who determines the adequacy of work done is fully informed about procedures and techniques and new equipment in use and is competent to perform that person's duties.

§ 135.435 Certificate requirements.

(a) Except for maintenance, preventive maintenance, alterations, and required inspections performed by repair stations certificated under the provisions of Subpart C of Part 145 of this chapter, each person who is directly in charge of maintenance, preventive maintenance, or alterations, and each person performing required inspections must hold an appropriate airman certificate.

(b) For the purpose of this section, a person "directly in charge" is each person assigned to a position in which that person is responsible for the work of a shop or station that performs maintenance, preventive maintenance, alterations, or other functions affecting airworthiness. A person who is "directly in charge" need not physically observe and direct each worker constantly but must be available for consultation and decision on matters requiring instruction or decision from higher authority than that of the person performing the work.

§ 135.437 Authority to perform and approve maintenance, preventive maintenance, and alterations.

(a) A certificate holder may perform, or make arrangements with other persons to perform, maintenance, preventive maintenance, and alterations as provided in its maintenance manual. In addition, a certificate holder may perform these functions for another certificate holder as provided in the maintenance manual of the other certificate holder.

(b) A certificate holder may approve any airframe, aircraft engine, propeller, rotor, or appliance for return to service after maintenance, preventive maintenance, or alterations that are performed under paragraph (a) of this section. However, in the case of a major repair or alteration, the work must have been done in accordance with technical data approved by the Administrator.

§ 135.439 Maintenance recording requirements.

(a) Each certificate holder shall keep (using the system specified in the manual required in § 135.427) the following records for the periods specified in paragraph (b) of this section:

(1) All the records necessary to show that all requirements for the issuance of an airworthiness release under § 135.443 have been met.

(2) Records contain the following information:

(i) The total time in service of the airframe, engine, propeller, and rotor.

(ii) The current status of life-limited parts of each airframe, engine, propeller, rotor, and appliance.

(iii) The time since last overhaul of each item installed on the aircraft which are required to be overhauled on a specified time basis.

(iv) The identification of the current inspection status of the aircraft, including the time since the last inspections required by the inspection program under which the aircraft and its appliances are maintained.

(v) The current status of applicable airworthiness directives, including the date and methods of compliance, and, if the airworthiness directive involves recurring action, the time and date when the next action is required.

(vi) A list of current major alterations and repairs to each airframe, engine, propeller, rotor, and appliance.

(b) Each certificate holder shall retain the records required to be kept by this section for the following periods:

(1) Except for the records of the last complete overhaul of each airframe, engine, propeller, rotor, and appliance the records specified in paragraph (a)(1) of this section shall be retained until the work is repeated or superseded by other work or for one year after the work is performed.

(2) The records of the last complete overhaul of each airframe, engine, propeller, rotor, and appliance shall be retained until the work is superseded by work of equivalent scope and detail.

(3) The records specified in paragraph (a)(2) of this section shall be retained and transferred with the aircraft at the time the aircraft is sold.

(c) The certificate holder shall make all maintenance records required to be kept by this section available for inspection by the Administrator or any representative of the National Transportation Safety Board.

§ 135.441 Transfer of maintenance records.

Each certificate holder who sells a United States registered aircraft shall transfer to the purchaser, at the time of the sale, the following records of that aircraft, in plain language form or in coded form which provides for the preservation and retrieval of information in a manner acceptable to the Administrator.

(a) The records specified in § 135.439(a)(2).

(b) The records specified in § 135.439(a)(1) which are not included in the records covered by paragraph (a) of this section, except that the purchaser may allow the seller to keep physical custody of such records. However, custody of records by the seller does not relieve the purchaser of its responsibility under § 135.439(c) to make the records available for inspection by the Administrator or any representative of the National Transportation Safety Board.

§ 135.443 Airworthiness release or aircraft maintenance log entry.

(a) No certificate holder may operate an aircraft after maintenance, preventive maintenance, or alterations are performed on the aircraft unless the certificate holder prepares, or causes the person with whom the certificate holder arranges for the performance of the maintenance, preventive maintenance, or alterations, to prepare—

(1) An airworthiness release; or

(2) An appropriate entry in the aircraft maintenance log.

(b) The airworthiness release or log entry required by paragraph (a) of this section must—

(1) Be prepared in accordance with the procedure in the certificate holder's manual;

(2) Include a certificate that—

(i) The work was performed in accordance with the requirements of the certificate holder's manual;

(ii) All items required to be inspected were inspected by an authorized person who determined that the work was satisfactorily completed;

(iii) No known condition exists that would make the aircraft unairworthy;

(iv) So far as the work performed is concerned, the aircraft is in condition for safe operation; and

(3) Be signed by an authorized certificated mechanic or repairman, except that a certificated repairman may sign the release or entry only for the work for which that person is employed and for which that person is certificated.

Notwithstanding paragraph (b)(3) of this section, after maintenance, preventive maintenance, or alterations performed by a repair station certificated under the provisions of Subpart C of Part 145, the airworthiness release or log entry required by paragraph (a) of this section may be signed by a person authorized by that repair station.

(c) Instead of restating each of the conditions of the certification required by paragraph (b) of this section, the certificate holder may state in its manual that the signature of an authorized certificated mechanic or repairman constitutes that certification.

Appendix A
Additional Airworthiness Standards
for 10 or More Passenger Airplanes

APPLICABILITY

1. *Applicability*. This Appendix prescribes the additional airworthiness standards required by § 135.169.

2. *References.* Unless otherwise provided, references in this Appendix to specific sections of Part 23 of the Federal Aviation Regulations (FAR Part 23) are to those sections of Part 23 in effect on March 30, 1967.

FLIGHT REQUIREMENTS

3. *General.* Compliance must be shown with the applicable requirements of Subpart B of FAR Part 23, as supplemented or modified in §§ 4 through 10.

PERFORMANCE

4. *General.* (a) Unless itherwise prescribed in this Appendix, compliance with each applicable performance requirement in §§ 4 through 7 must be shown for ambient atmospheric conditions and still air.

(b) The performance must correspond to the propulsive thrust available under the particular ambient atmospheric conditions and the particular flight condition. The available propulsive thrust must correspond to engine power or thrust, not exceeding the approved power or thrust less—

(1) Installation losses; and

(2) The power or equivalent thrust absorbed by the accessories and services appropriate to the particular ambient atmospheric conditions and the particular flight condition.

(c) Unless otherwise prescribed in this Appendix, the applicant must select the takeoff, en route, and landing configurations for the airplane.

(3) The airplane configuration may vary with weight, altitude, and temperature, to the extent they are compatible with the operating procedures required by paragraph (e) of this section.

(e) Unless otherwise prescribed in this Appendix, in determining the critical engine inoperative takeoff performance, the accelerate-stop distance, takeoff distance, changes in the airplane's configuration, speed, power, and thrust must be made under procedures established by the applicant for operation in service.

(f) Procedures for the execution of balked landings must be established by the applicant and included in the Airplane Flight Manual.

(g) The procedures established under paragraphs (e) and (f) of this section must—

(1) Be able to be consistently executed in service by a crew of average skill;

(2) Use methods or devices that are safe and reliable; and

(3) Include allowance for any time delays, in the execution of the procedures, that may reasonably be expected in service.

5. *Takeoff—*(a) *General.* Takeoff speeds, the accelerate-stop distance, the takeoff distance, and the one-engine-inoperative takeoff flight path data (described in paragraphs (b), (c), (d), and (f) of this section), must be determined for—

(1) Each weight, altitude, and ambient temperature within the operational limits selected by the applicant;

(2) The selected configuration for takeoff;

(3) The center of gravity in the most unfavorable position;

(4) The operating engine within approved operating limitations; and

(5) Takeoff data based on smooth, dry, hard-surface runway.

(b) *Takeoff speeds.* (1) The decision speed V_1 is the calibrated airspeed on the ground at which, as a result of engine failure or other reasons, the pilot is assumed to have made a decision to continue or discontinue the takeoff. The speed V_1 must be selected by the applicant but may not be less than—

(i) 1.10 V_{S1};

(ii) 1.10 V_{MC};

(iii) A speed that allows acceleration to V_1 and stop under paragraph (c) of this section; or

(iv) A speed at which the airplane can be rotated for takeoff and shown to be adequate to safely continue the takeoff, using normal piloting skill, when the critical engine is suddenly made inoperative.

(2) The initial climb out speed V_2, in terms of calibrated airspeed, must be selected by the applicant so as to allow the gradient of climb required in § 6(b)(2), but it must not be less than V_1 or less than 1.2 V_{S1}.

(3) Other essential take off speeds necessary for safe operation of the airplane.

(c) *Accelerate-stop distance.* (1) The accelerate-stop distance is the sum of the distances necessary to—

(i) Accelerate the airplane from a standing start to V_1; and

(ii) Come to a full stop from the point at which V_1 is reached assuming that in the case of engine failure, failure of the critical engine is recognized b, the pilot at the speed V_1.

(2) Means other than wheel brakes may be used to determine the accelerate-stop distance if that means is available with the critical engine inoperative and—

(i) Is safe and reliable;

(ii) Is used so that consistent results can be expected under normal operating conditions; and

(iii) Is such that exceptional skill is not required to control the airplane.

(d) *All engines operating takeoff distance.* The all engine operating takeoff distance is the horizontal distance required to takeoff and climb to a height of 50 feet above the takeoff surface under the procedures in FAR 23.51(a).

(e) *One-engine-inoperative takeoff.* Determine the weight for each altitude and temperature within the operational limits established for the airplane, at which the airplane has the capability, after failure of the critical engine at V_1, determined under paragraph (b) of this section, to take off and climb at not less than V_2, to a height 1,000 feet above the takeoff surface and attain the speed and configuration at which compliance is shown with the en route one-engine-inoperative gradient of climb specified in § 6(c).

(f) *One-engine-inoperative takeoff flight path data.* The one-engine-inoperative takeoff flight path data consist of takeoff flight paths extending from a standing start to a point in the takeoff at which the airplane reaches a height 1,000 feet above the takeoff surface under paragraph (e) of this section.

6. *Climb—*(a) *Landing climb: All-engines-operating.* The maximum weight must be determined with the airplane in the landing configuration, for each altitude. and ambient temperature within the operational limits established for the airplane, with the most unfavorable center of gravity, and out-of-ground effect in free air, at which the steady gradient of climb will not be less than 3.3 percent, with:

(1) The engines at the power that is available 8 seconds after initiation of movement of the power or thrust controls from the minimum flight idle to the takeoff position.

(2) A climb speed not greater than the approach speed established under § 7 and not less than the greater of 1.05 V_{MC} or 1.10 V_{S1}.

(b) *Takeoff climb: one-engine-inoperative.* The maximum weight at which the airplane meets the minimum climb performance specified in subparagraphs (1) and (2) of this paragraph must be determined for each altitude and ambient temperature within the operational limits established for the airplane, out of ground effect in free air, with the airplane in the takeoff configuration, with the most unfavorable center of gravity, the critical engine inoperative, the remaining engines at the maximum takeoff power or thrust, and the propeller of the inoperative engine windmilling with the propeller controls in the normal position except that, if an approved automatic feathering system is installed, the propellers may be in the feathered position:

(1) *Takeoff: landing gear extended.* The minimum steady gradient of climb must be measurably positive at the speed V_1.

(2) *Takeoff: landing gear retracted.* The minimum steady gradient of climb may not be less than 2 percent at speed V_2. For airplanes with fixed landing gear this requirement must be met with the landing gear extended.

(c) *En route climb: one-engine-inoperative.* The maximum weight must be determined for each altitude and ambient temperature within the operational limits established for the airplane, at which the steady gradient of climb is not less than 1.2 percent at an altitude 1,000 feet above the takeoff surface, with the airplane in the en route configuration, the critical engine inoperative, the remaining engine at the maximum continuous power or thrust, and the most unfavorable center of gravity.

7. *Landing.* (a) The landing field length described in paragraph (b) of this section must be determined for standard atmosphere at each weight and altitude within the operational limits established by the applicant.

(b) The landing field length is equal to the landing distance determined under FAR 23.75

(a) divided by a factor of 0.6 for the destination airport and 0.7 for the alternate airport. Instead of the gliding approach specified in FAR 23.75(a)(1), the landing may be preceded by a steady approach down to the 50-foot height at a gradient of descent not greater than 5.2 percent (3°) at a calibrated airspeed not less than 1.3 V_{s1}.

TRIM

8. *Trim*—(a) *Lateral and directional trim.* The airplane must maintain lateral and directional trim in level flight at a speed of V_H or V_{MO}/M_{MO}, whichever is lower, with landing gear and wing flaps retracted.

(b) *Longitudinal trim.* The airplane must maintain longitudinal trim during the following conditions, except that it need not maintain trim at a speed greater than V_{MO}/M_{MO}:

(1) In the approach conditions specified in FAR 23.161(c)(3) through (5), except that instead of the speeds specified in those paragraphs, trim must be maintained with a stick force of not more than 10 pounds down to a speed used in showing compliance with § 7 or 1.4 V_{s1} whichever is lower.

(2) In level flight at any speed from V_H or V_{MO}/M_{MO}, whichever is lower, to either V_x or 1.4 V_{s1}, with the landing gear and wing flaps retracted.

STABILITY

9. *Static longitudinal stability.* (a) In showing compliance with FAR 23.175(b) and with paragraph (b) of this section, the airspeed must return to within ±7½ percent of the trim speed.

(b) *Cruise stability.* The stick force curve must have a stable slope for a speed range of ±50 knots from the trim speed except that the speeds need not exceed V_{FC}/M_{FC} or be less than 1.4 V_{s1}. This speed range will be considered to begin at the outer extremes of the friction band and the stick force may not exceed 50 pounds with—

(1) Landing gear retracted;

(2) Wing flaps retracted;

(3) The maximum cruising power as selected by the applicant as an operating limitation for turbine engines or 75 percent of maximum continuous power for reciprocating engines except that the power need not exceed that required at V_{MO}/M_{MO};

(4) Maximum takeoff weight; and

(5) The airplane trimmed for level flight with the power specified in subparagraph (3) of this paragraph.

V_{FC}/M_{FC} may not be less than a speed midway between V_{MO}/M_{MO} and V_{DF}/M_{DF}, except that, for altitudes where Mach number is the limiting factor, M_{FC} need not exceed the Mach number at which effective speed warning occurs.

(c) *Climb stability (turbopropeller powered airplanes only).* In showing compliance with FAR 23.175(a), an applicant must, instead of

the power specified in FAR 23.175(a)(4), use the maximum power or thrust selected by the applicant as an operating limitation for use during climb at the best rate of climb speed, except that the speed need not be less than 1.4 V_{s1}.

STALLS

10. *Stall warning.* If artificial stall warning is required to comply with FAR 23.207, the warning device must give clearly distinguishable indications under expected conditions of flight. The use of a visual warning device that requires the attention of the crew within the cockpit is not acceptable by itself.

CONTROL SYSTEMS

11. *Electric trim tabs.* The airplane must meet FAR 23.677 and in addition it must be shown that the airplane is safely controllable and that a pilot can perform all the maneuvers and operations necessary to effect a safe landing following any probable electric trim tab runaway which might be reasonably expected in service allowing for appropriate time delay after pilot recognition of the runaway. This demonstration must be conducted at the critical airplane weights and center of gravity positions.

INSTRUMENTS: INSTALLATION

12. *Arrangement and visibility.* Each instrument must meet FAR 23.1321 and in addition:

(a) Each flight, navigation, and powerplant instrument for use by any pilot must be plainly visible to the pilot from the pilot's station with the minimum practicable deviation from the pilot's normal position and line of vision when the pilot is looking forward along the flight path.

(b) The flight instruments required by FAR 23.1303 and by the applicable operating rules must be grouped on the instrument panel and centered as nearly as practicable about the vertical plane of each pilot's forward vision. In addition—

(1) The instrument that most effectively indicates the attitude must be in the panel in the top center position;

(2) The instrument that most effectively indicates the airspeed must be on the panel directly to the left of the instrument in the top center position;

(3) The instrument that most effectively indicates altitude must be adjacent to and directly to the right of the instrument in the top center position; and

(4) The instrument that most effectively indicates direction of flight must be adjacent to and directly below the instrument in the top center position.

13. *Airspeed indicating system.* Each airspeed indicating system must meet FAR 23.1323 and in addition:

(a) Airspeed indicating instruments must be of an approved type and must be calibrated

to indicate true airspeed at sea level in the standard atmosphere with a minimum practicable instrument calibration error when the corresponding pitot and static pressures are supplied to the instruments.

(b) The airspeed indicating system must be calibrated to determine the system error, i.e., the relation between IAS and CAS, in flight and during the accelerate-takeoff ground run. The ground run calibration must be obtained between 0.8 of the minimum value of V_1 and 1.2 times the maximum value of V_1, considering the approved ranges of altitude and weight. The ground run calibration is determined assuming an engine failure at the minimum value of V_1.

(c) The airspeed error of the installation excluding the instrument calibration error, must not exceed 3 percent or 5 knots whichever is greater, throughout the speed range from V_{MO} to 1.3 V_{s1} with flaps retracted and from 1.3 V_{SO} to V_{FE} with flaps in the landing position.

(d) Information showing the relationship between IAS and CAS must be shown in the Airplane Flight Manual.

14. *Static air vent system.* The static air vent system must meet FAR 23.1325. The altimeter system calibration must be determined and shown in the Airplane Flight Manual.

OPERATING LIMITATIONS AND INFORMATION

15. *Maximum operating limit speed V_{MO}/M_{MO}.* Instead of establishing operating limitations based on V_{NE}/V_{NO}, the applicant must establish a maximum operating limit speed V_{MO}/M_{MO} as follows:

(a) The maximum operating limit speed must not exceed the design cruising speed V_C and must be sufficiently below V_D/M_D or V_{DF}/M_{DF} to make it highly improbable that the latter speeds will be inadvertently exceeded in flight.

(b) The speed V_{MO} must not exceed 0.8V_D/M_D or 0.8V_{DF}/M_{DF} unless flight demonstrations involving upsets as specified by the Administrator indicates a lower speed margin will not result in speeds exceeding V_D/M_D or V_{DF}. Atmospheric variations, horizontal gusts, system and equipment errors, and airframe production variations are taken into account.

16. *Minimum flight crew.* In addition to meeting FAR 23.1523, the applicant must establish the minimum number and type of qualified flight crew personnel sufficient for safe operation of the airplane considering—

(a) Each kind of operation for which the applicant desires approval;

(b) The workload on each crewmember considering the following:

(1) Flight path control.

(2) Collision avoidance.

(3) Navigation.

(4) Communications.

(5) Operation and monitoring of all essential aircraft systems.

(6) Command decisions; and

(c) The accessibility and ease of operation of necessary controls by the appropriate crewmember during all normal and emergency operations when at the crewmember flight station.

17. *Airspeed indicator.* The airspeed indicator must meet FAR 23.1545 except that, the airspeed notations and markings in terms of V_{NO} and V_{NH} must be replaced by the V_{MO}/M_{MO} notations. The airspeed indicator markings must be easily read and understood by the pilot. A placard adjacent to the airspeed indicator is an acceptable means of showing compliance with FAR 23.1545(c).

AIRPLANE FLIGHT MANUAL

18. *General.* The Airplane Flight Manual must be prepared under FARs 23.1583 and 23.1587, and in addition the operating limitations and performance information in §§ 19 and 20 must be included.

19. *Operating limitations.* The Airplane Flight Manual must include the following limitations—

(a) *Airspeed limitations.* (1) The maximum operating limit speed V_{MO}/M_{MO} and a statement that this speed limit may not be deliberately exceeded in any regime of flight (climb, cruise, or descent) unless a higher speed is authorized for flight test or pilot training;

(2) If an airspeed limitation is based upon compressibility effects, a statement to this effect and information as to any symptoms, the probable behavior of the airplane, and the recommended recovery procedures; and

(3) The airspeed limits, shown in terms of V_{MO}/M_{MO} instead of V_{NO} and V_{NE}.

(b) *Takeoff weight limitations.* The maximum takeoff weight for each airport elevation, ambient temperature, and available takeoff runway length within the range selected by the applicant may not exceed the weight at which—

(1) The all-engine-operating takeoff distance determined under § 5(b) or the accelerate-stop distance determined under § 5(c), whichever is greater, is equal to the available runway length;

(2) The airplane complies with the one-engine-inoperative takeoff requirements specified in § 5(e); and

(3) The airplane complies with the one-engine-inoperative takeoff and en route climb requirements specified in §§ 6(b) and (c).

(c) *Landing weight limitations.* The maximum landing weight for each airport elevation (standard temperature) and available landing runway length, within the range selected by the applicant. This weight may not exceed the weight at which the landing field length determined under § 7(b) is equal to the available runway length. In showing compliance with this operating limitation, it is acceptable to assume that the landing weight at the destination will be equal to the takeoff weight reduced by the normal consumption of fuel and oil en route.

20. *Performance information.* The Airplane Flight Manual must contain the performance information determined under the performance requirements of this Appendix. The information must include the following:

(a) Sufficient information so that the takeoff weight limits specified in § 19(b) can be determined for all temperatures and altitudes within the operation limitations selected by the applicant.

(b) The conditions under which the performance information was obtained, including the airspeed at the 50-foot height used to determine landing distances.

(c) The performance information (determined by extrapolation and computed for the range of weights between the maximum landing and takeoff weights) for—

(1) Climb in the landing configuration; and

(2) Landing distance.

(d) Procedure established under § 4 related to the limitations and information required by this section in the form of guidance material including any relevant limitations or information.

(e) An explanation of significant or unusual flight or ground handling characteristics of the airplane.

(f) Airspeeds, as indicated airspeeds, corresponding to those determined for takeoff under § 5(b).

21. *Maximum operating altitudes.* The maximum operating altitude to which operation is allowed, as limited by flight, structural, powerplant, functional, or equipment characteristics, must be specified in the Airplane Flight Manual.

22. *Stowage provision for airplane flight manual.* Provision must be made for stowing the Airplane Flight Manual in a suitable fixed container which is readily accessible to the pilot.

23. *Operating procedures.* Procedures for restarting turbine engines in flight (including the effects of altitude) must be set forth in the Airplane Flight Manual.

AIRFRAME REQUIREMENTS FLIGHT LOADS

24. *Engine Torque.* (a) Each turbopropeller engine mount and its supporting structure must be designed for the torque effects of:

(1) The conditions in FAR 23.361(a).

(2) The limit engine torque corresponding to takeoff power and propeller speed multiplied by a factor accounting for propeller control system malfunction, including quick feathering action, simultaneously with 1g level flight loads. In the absence of a rational analysis, a factor of 1.6 must be used.

(b) The limit torque is obtained by multiplying the mean torque by a factor of 1.25.

25. *Turbine engine gyroscopic loads.* Each turobpropeller engine mount and its supporting structure must be designed for the gyroscopic loads that result, with the engines at maximum continuous r.p.m. under either—

(a) The conditions in FARs 23.351 and 23.423; or

(b) All possible combinations of the following:

(1) A yaw velocity of 2.5 radians per second.

(2) A pitch velocity of 1.0 radians per second.

(3) A normal load factor of 2.5.

(4) Maximum continuous thrust.

26. *Unsymmetrical loads due to engine failure.* (a) Turbopropeller powered airplanes must be designed for the unsymmetrical loads resulting from the failure of the critical engine including the following conditions in combination with a single malfunction of the propeller drag limiting system, considering the probable pilot corrective action on the flight controls:

(1) At speeds between V_{MO} and V_D, the loads resulting from power failure because of fuel flow interruption are considered to be limit loads.

(2) At speeds between V_{MO} and V_c, the loads resulting from the disconnection of the engine compressor from the turbine or from loss of the turbine blades are considered to be ultimate loads.

(3) The time history of the thrust decay and drag buildup occurring as a result of the prescribed engine failures must be substantiated by test or other data applicable to the particular engine-propeller combination.

(4) The timing and magnitude of the probable pilot corrective action must be conservatively estimated, considering the characteristics of the particular engine-propeller-airplane combination.

(b) Pilot corrective action may be assumed to be initiated at the time maximum yawing velocity is reached, but not earlier than 2 seconds after the engine failure. The magnitude of the corrective action may be based on the control forces in FAR 23.397 except that lower forces may be assumed where it is shown by analysis or test that these forces can control the yaw and roll resulting from the prescribed engine failure conditions.

GROUND LOADS

27. *Dual wheel landing gear units.* Each dual wheel landing wear unit and its supporting structure must be shown to comply with the following:

(a) *Pivoting.* The airplane must be assumed to pivot about one side of the main gear with the brakes on that side locked. The limit vertical load factor must be 1.0 and the coefficient of friction 0.8. This condition need apply only to the main gear and its supporting structure.

(b) *Unequal tire inflation.* A 60–40 percent distribution of the loads established under FAR 23.471 through FAR 23.483 must be applied to the dual wheels.

(c) *Flat tire.* (1) Sixty percent of the loads in FAR 23.471 through FAR 23.483 must be applied to either wheel in a unit.

(2) Sixty percent of the limit drag and side loads and 100 percent of the limit vertical load established under FARs 23.493 and 23.485 must be applied to either wheel in a unit except that the vertical load need not exceed the maximum vertical load in paragraph (c)(1) of this section.

FATIGUE EVALUATION

28. *Fatigue evaluation of wing and associated structure.* Unless it is shown that the structure, operating stress levels, materials and expected use are comparable from a fatigue standpoint to a similar design which has had substantial satisfactory service experience, the strength, detail design, and the fabrication of those parts of the wing, wing carrythrough, and attaching structure whose failure would be catastrophic must be evaluated under either—

(a) A fatigue strength investigation in which the structure is shown by analysis, tests, or both to be able to withstand the repeated loads of variable magnitude expected in service; or

(b) A fail-safe strength investigation in which it is shown by analysis, tests, or both that catastrophic failure of the structure is not probable after fatigue, or obvious partial failure, of a principal structural element, and that the remaining structure is able to withstand a static ultimate load factor of 75 percent of the critical limit load factor at V_c. These loads must be multiplied by a factor of 1.15 unless the dynamic effects of failure under static load are otherwise considered.

DESIGN AND CONSTRUCTION

29. *Flutter.* For multiengine turbopropeller powered airplanes, a dynamic evaluation must be made and must include—

(a) The significant elastic, inertia, and aerodynamic forces associated with the rotations and displacements of the plane of the propeller; and

(b) Engine-propeller-nacelle stiffness and damping variations appropriate to the particular configuration.

LANDING GEAR

30. *Flap operated landing gear warning device.* Airplanes having retractable landing gear and wing flaps must be equipped with a warning device that functions continuously when the wing flaps are extended to a flap position that activates the warning device to give adequate warning before landing, using normal landing procedures, if the landing gear is not fully extended and locked. There may not be a manual shut off for this warning device. The flap position sensing unit may be installed at any suitable location. The system for this device may use any part of the system (including the aural warning device) provided for other landing gear warning devices.

PERSONNEL AND CARGO ACCOMMODATIONS

31. *Cargo and baggage compartments.* Cargo and baggage compartments must be designed to meet FAR 23.787 (a) and (b), and in addition means must be provided to protect passengers from injury by the contents of any cargo or baggage compartment when the ultimate forward inertia force is $9g$.

32. *Doors and exits.* The airplane must meet FAR 23.783 and FAR 23.807 (a)(3), (b), and (c), and in addition:

(a) There must be a means to lock and safeguard each external door and exit against opening in flight either inadvertently by persons, or as a result of mechanical failure. Each external door must be operable from both the inside and the outside.

(b) There must be means for direct visual inspection of the locking mechanism by crewmembers to determine whether external doors and exits, to determine whether external doors ment is outward, are fully locked. In addition, there must be a visual means to signal to crewmembers when normally used external doors are closed and fully locked.

(c) The passenger entrance door must qualify as a floor level emergency exit. Each additional required emergency exit except floor level exits must be located over the wing or must be provided with acceptable means to assist the occupants in descending to the ground. In addition to the passenger entrance door:

(1) For a total seating capacity of 15 or less, an emergency exit as defined in FAR 23.807(b) is required on each side of the cabin.

(2) For a total seating capacity of 16 through 23, three emergency exits as defined in FAR 23.807(b) are required with one on the same side as the door and two on the side opposite the door.

(d) An evacuation demonstration must be conducted utilizing the maximum number of occupants for which certification is desired. It must be conducted under simulated night conditions utilizing only the emergency exits on the most critical side of the aircraft. The participants must be representative of average airline passengers with no previous practice or rehearsal for the demonstration. Evacuation must be completed within 90 seconds.

(e) Each emergency exit must be marked with the word "Exit" by a sign which has white letters 1 inch high on a red background 2 inches high, be self-illuminated or independently internally electrically illuminated, and have a minimum luminescence (brightness) of at least 160 microlamberts. The colors may be reserved if the passenger compartment illumination is essentially the same.

(f) Access to window type emergency exits must not be obstructed by seats or seat backs.

(g) The width of the main passenger aisle at any point between seats must equal or exceed the values in the following table:

Total seating capacity	Minimum main passenger aisle width	
	Less than 25 inches from floor	25 inches and more from floor
10 through 23	9 inches	15 inches.

MISCELLANEOUS

33. *Lightning strike protection.* Parts that are electrically insulated from the basic airframe must be connected to it through lightning arrestors unless a lightning strike on the insulated part—

(a) Is improbable because of shielding by other parts; or

(b) Is not hazardous.

34. *Ice protection.* It certification with ice protection provisions is desired, compliance with the following must be shown:

(a) The recommended procedures for the use of the ice protection equipment must be set forth in the Airplane Flight Manual.

(b) An analysis must be performed to establish, on the basis of the airplane's operational needs, the adequacy of the ice protection system for the various components of the airplane. In addition, tests of the ice protection system must be conducted to demonstrate that the airplane is capable of operating safely in continuous maximum and intermittent maximum icing conditions as described in Appendix C of Part 25 of this chapter.

(c) Compliance with all or portions of this section may be accomplished by reference, where applicable because of similarity of the designs, to analysis and tests performed by the applicant for a type certificated model.

35. *Maintenance information.* The applicant must make available to the owner at the time of delivery of the airplane the information the applicant considers essential for the proper maintenance of the airplane. That information must include the following:

(a) Description of systems, including electrical hydraulic, and fuel controls.

(b) Lubrication instructions setting forth the frequency and the lubricants and fluids which are to be used in the various systems.

(c) Pressures and electrical loads applicable to the various systems.

(d) Tolerances and adjustments necessary for proper functioning.

(e) Methods of leveling, raising, and towing.

(f) Methods of balancing control surfaces.

(g) Identification of primary and secondary structures.

(h) Frequency and extent of inspections necessary to the proper operation of the airplane.

(i) Special repair methods applicable to the airplane.

(j) Special inspection techniques, such as X-ray, ultrasonic, and magnetic particle inspection.

(k) List of special tools.

PROPULSION

GENERAL

36. *Vibration characteristics.* For turbopropeller powered airplanes, the engine installation must not result in vibration characteristics of the engine exceeding those established during the type certification of the engine.

37. *In flight restarting of engine.* If the engine on turbopropeller powered airplanes cannot be restarted at the maximum cruise altitude, a determination must be made of the altitude below which restarts can be consistently accomplished. Restart information must be provided in the Airplane Flight Manual.

38. *Engines.* (a) *For turopropeller powered airplanes.* The engine installation must comply with the following:

(1) *Engine isolation.* The powerplants must be arranged an isolated from each other to allow operation, in at least one configuration, so that the failure or malfunction of any engine, or of any system that can affect the engine, will not—

(i) Prevent the continued safe operation of the remaining engines; or

(ii) Require immediate action by any crewmember for continued safe operation.

(2) *Control of engine rotation.* There must be a means to individually stop and restart the rotation of any engine in flight except that engine rotation need not be stopped if continued rotation could not jeopardize the safety of the airplane. Each component of the stopping and restarting system on the engine side of the firewall, and that might be exposed to fire, must be at least fire resistant. If hydraulic propeller feathering systems are used for this purpose, the feathering lines must be at least fire resistant under the operating conditions that may be expected to exist during feathering.

(3) *Engine speed and gas temperature control devices.* The powerplant systems associated with engine control devices, systems, and instrumentation must provide reasonable assurance that those engine operating limitations that adversely affect turbine rotor structural integrity will not be exceeded in service.

(b) *For reciprocating engine powered airplanes.* To provide engine isolation, the powerplants must be arranged and isolated from each other to allow operation, in at least one configuration, so that the failure or malfunction of any engine, or of any system that can affect that engine, will not—

(1) Prevent the continued safe operation of the remaining engines; or

(2) Require immediate action by any crewmember for continued safe operation.

39. *Turbopropeller reversing systems.* (a) Turbopropeller reversing systems intended for ground operation must be designed so that no single failure or malfunction of the system will result in unwanted reverse thrust under any expected operating condition. Failure of structural elements need not be considered if the probability of this kind of failure is extremely remote.

(b) Turbopropeller reversing systems intended for in flight use must be designed so that no unsafe condition will result during normal operation of the system, or from any failure (or reasonably likely combination of failures) of the reversing system, under any anticipated condition of operation of the airplane. Failure of structural elements need not be considered if the probability of this kind of failure is extremely remote.

(c) Compliance with this section may be shown by failure analysis, testing, or both for propeller systems that allow propeller blades to move from the flight low-pitch position to a position that is substantially less than that at the normal flight low-pitch stop position. The analysis may include or be supported by the analysis made to show compliance with the type certification of the propeller and associated installation components. Credit will be given for pertinent analysis and testing completed by the engine and propeller manufacturers.

40. *Turbopropeller drag-limiting systems.* Turbopropeller drag-limiting systems must be designed so that no single failure or malfunction of any of the systems during normal or emergency operation results in propeller drag in excess of that for which the airplane was designed. Failure of structure elements of the drag-limiting systems need not be considered if the probability of this kind of failure is extremely remote.

41. *Turbine engine powerplant operating characteristics.* For turbopropeller powered airplanes, the turbine engine powerplant operating characteristics must be investigated in flight to determine that no adverse characteristics (such as stall, surge, or flameout) are present to a hazardous degree, during normal and emergency operation within the range of operating limitations of the airplane and of the engine.

42. *Fuel flow.* (a) For turbopropeller powered airplanes—

(1) The fuel system must provide for continuous supply of fuel to the engines for normal operation without interruption due to depletion of fuel in any tank other than the main tank; and

(2) The fuel flow rate for turbopropeller engine fuel pump systems must not be less than 125 percent of the fuel flow required to develop the standard sea level atmospheric conditions takeoff power selected and included as an operating limitation in the Airplane Flight Manual.

(b) For reciprocating engine powered airplanes, it is acceptable for the fuel flow rate for each pump system (main and reserve supply) to be 125 percent of the takeoff fuel consumption of the engine.

FUEL SYSTEM COMPONENTS

43. *Fuel pumps.* For turbopropeller powered airplanes, a reliable and independent power source must be provided for each pump used with turbine engines which do not have provisions for mechanically driving the main pumps. It must be demonstrated that the pump installations provide a reliability and durability equivalent to that in FAR 23.991(a).

44. *Fuel strainer or filter.* For turbopropeller powered airplanes, the following apply:

(a) There must be a fuel strainer or filter between the tank outlet an dthe fuel metering device of the engine. In addition, the fuel strainer or filter must be—

(1) Between the tank outlet and the engine-driven positive displacement pump inlet, if there is an engine-driven positive displacement pump;

(2) Accessible for drainage and cleaning and, for the strainer screen, easily removable; and

(3) Mounted so that its weight is not supported by the connecting lines or by the inlet or outlet connections of the strainer or filter itself.

(b) Unless there are means in the fuel system to prevent the accumulation of ice on the filter, there must be means to automatcially maintain the fuel-flow if ice-clogging of the filter occurs; and

(c) The fuel strainer or filter must be of adequate capacity (for operating limitations established to ensure proper service) and of appropriate mesh to ensure proper engine operation, with the fuel contaminated to a degree (for particle size and density) that can be reasonably expected in service. The degree of fuel filtering may not be less than that established for the engine type certification.

45. *Lightning strike protection.* Protection must be provided against the ignition of flammable vapors in the fuel vent system due to lightning strikes.

COOLING

46. *Cooling test procedures for turbopropeller powered airplanes.* (a) Turbopropeller

powered airplanes must be shown to comply with FAR 23.1041 during takeoff, climb, en route, and landing stages of flight that correspond to the applicable performance requirements. The cooling tests must be conducted with the airplane in the configuration, and operating under the conditions that are critical relative to cooling during each stage of flight. For the cooling tests a temperature is "stabilized" when its rate of change is less than 2° F. per minute.

(b) Temperatures must be stabilized under the conditions from which entry is made into each stage of flights being investigated unless the entry condition is not one during which component and engine fluid temperatures would stabilize, in which case, operation through the full entry condition must be conducted before entry into the stage of flight being investigated to allow temperatures to reach their natural levels at the time of entry. The takeoff cooling test must be preceded by a period during which the powerplant component and engine fluid temperatures are stabilized with the engines at ground idle.

(c) Cooling tests for each stage of flight must be continued until—

(1) The component and engine fluid temperatures stabilize;

(2) The stage of flight is completed; or

(3) An operating limitation is reached.

Induction System

47. *Air induction.* For turbopropeller powered airplanes—

(a) There must be means to prevent hazardous quantities of fuel leakage or overflow from drains, vents, or other components of flammable fluid systems from entering the engine intake systems; and

(b) The air inlet ducts must be located or protected so as to minimize the ingestion of foreign matter during takeoff, landing, and taxiing.

48. *Induction system icing protection.* For turbopropeller powered airplanes, each turbine engine must be able to operate throughout its flight power range without adverse effect on engine operation or serious loss of power or thrust, under the icing conditions specified in Appendix C of Part 25 of this chapter. In addition, there must be means to indicate to appropriate flight crewmembers the functioning of the powerplant ice protection system.

49. *Turbine engine bleed air systems.* Turbine engine bleed air systems of turbopropeller powered airplanes must be investigated to determine—

(a) That no hazard to the airplane will result if a duct rupture occurs. This condition must consider that a failure of the duct can occur anywhere between the engine port and the airplane bleed service; and

(b) That, if the bleed air system is used for direct cabin pressurization, it is not possible

for hazardous contamination of the cabin air system to occur in event of lubrication system failure.

Exhaust System

50. *Exhaust system drains.* Turbopropeller engine exhaust systems having low spots or pockets must incorporate drains at those locations. These drains must discharge clear of the airplane in normal and ground attitudes to prevent the accumulation of fuel after the failure of an attempted engine start.

Powerplant Controls and Accessories

51. *Engine controls.* If throttles or power levers for turbopropeller powered airplanes are such that any position of these controls will reduce the fuel flow to the engine(s) below that necessary for satisfactory and safe idle operation of the engine while the airplane is in flight, a means must be provided to prevent inadvertent movement of the control into this position. The means provided must incorporate a positive lock or stop at this idle position and must require a separate and distinct operation by the crew to displace the control from the normal engine operating range.

52. *Reverse thrust controls.* For turbopropeller powered airplanes, the propeller reverse thrust controls must have a means to prevent their inadvertent operation. The means must have a positive lock or stop at the idle position and must require a separate and distinct operation by the crew to displace the control from the flight regime.

53. *Engine ignition systems.* Each turbopropeller airplane ignition system must be considered an essential electrical load.

54. *Powerplant accessories.* The powerplant accessories must meet FAR 23.1163, and if the continued rotation of any accessory remotely driven by the engine is hazardous when malfunctioning occurs, there must be means to prevent rotation without interfering with the continued operation of the engine.

Powerplant Fire Protection

55. *Fire detector system.* For turbopropeller powered airplanes, the following apply:

(a) There must be a means that ensures prompt detection of fire in the engine compartment. An overtemperature switch in each engine cooling air exit is an acceptable method of meeting this requirement.

(b) Each fire detector must be constructed and installed to withstand the vibration, inertia, and other loads to which it may be subjected in operation.

(c) No fire detector may be affected by any oil, water, other fluids, or fumes that might be present.

(d) There must be means to allow the flight crew to check, in flight, the functioning of each fire detector electric circuit.

(e) Wiring and other components of each fire detector system in a fire zone must be at least fire resistant.

56. *Fire protection, cowling and nacelle skin.* For reciprocating engine powered airplanes, the engine cowling must be designed and constructed so that no fire originating in the engine compartment can enter either through openings or by burn through, any other region where it would create additional hazards.

57. *Flammable fluid fire protection.* If flammable fluids or vapors might be liberated by the leakage of fluid systems in areas other than engine compartments, there must be means to—

(a) Prevent the ignition of those fluids or vapors by any other equipment; or

(b) Control any fire resulting from that ignition.

Equipment

58. *Powerplant instruments.* (a) The following are required for turbopropeller airplanes:

(1) The instruments required by FAR 23.1305(a)(1) through (4), (b)(2) and (4).

(2) A gas temperature indicator for each engine.

(3) Free air temperature indicator.

(4) A fuel flowmeter indicator for each engine.

(5) Oil pressure warning means for each engine.

(6) A torque indicator or adequate means for indicating power output for each engine.

(7) Fire warning indicator for each engine.

(8) A means to indicate when the propeller blade angle is below the low-pitch position corresponding to idle operation in flight.

(9) A means to indicate the functioning of the ice protection system for each engine.

(b) For turbopropeller powered airplanes, the turbopropeller blade position indicator must begin indicating when the blade has moved below the flight low-pitch position.

(c) The following instruments are required for reciprocating engine powered airplanes:

(1) The instruments required by FAR 23.1305.

(2) A cylinder head temperature indicator for each engine.

(3) A manifold pressure indicator for each engine.

Systems and Equipments

General

59. *Function and installation.* The systems and equipment of the airplane must meet FAR 23.1301, and the following:

(a) Each item of additional installed equipment must—

(1) Be of a kind and design appropriate to its intended function;

(2) Be labeled as to its identification, function, or operating limitations, or any applicable

combination of these factors, unless misuse or inadvertent actuation cannot create a hazard;

(3) Be installed according to limitations specified for that equipment; and

(4) Function properly when installed.

(b) Systems and installations must be designed to safeguard against hazards to the aircraft in the event of their malfunction or failure.

(c) Where an installation, the functioning of which is necessary in showing compliance with the applicable requirements, requires a power supply, that installation must be considered an essential load on the power supply, and the power sources and the distribution system must be capable of supplying the following power loads in probable operation combinations and for probable durations:

(1) All essential loads after failure of any prime mover, power converter, or energy storage device.

(2) All essential loads after failure of any one engine on two-engine airplanes.

(3) In determining the probable operating combinations and durations of essential loads for the power failure conditions described in subparagraphs (1) and (2) of this paragraph, it is permissible to assume that the power loads are reduced in accordance with a monitoring procedure which is consistent with safety in the types of operations authorized.

60. *Ventilation.* The ventilation system of the airplane must meet FAR 23.831, and in addition, for pressurized aircraft, the ventilating air in flight crew and passenger compartments must be free of harmful or hazardous concentrations of gases and vapors in normal operation and in the event of reasonably probable failures or malfunctioning of the ventilating, heating, pressurization, or other systems, and equipment. If accumulation of hazardous quantities of smoke in the cockpit area is reasonably probable, smoke evacuation must be readily accomplished.

ELECTRICAL SYSTEMS AND EQUIPMENT

61. *General.* The electrical systems and equipment of the airplane must meet FAR 23.1351, and the following:

(a) *Electrical system capacity.* The required generating capacity, and number and kinds of power sources must—

(1) Be determined by an electrical load analysis; and

(2) Meet FAR 23.1301.

(b) *Generating system.* The generating system includes electrical power sources, main power busses, transmission cables, and associated control, regulation and protective devices. It must be designed so that—

(1) The system voltage and frequency (as applicable) at the terminals of all essential load equipment can be maintained within the limits for which the equipment is designed, during any probable operating conditions;

(2) System transients due to switching, fault clearing, or other causes do not make essential loads inoperative, and do not cause a smoke or fire hazard;

(3) There are means, accessible in flight to appropriate crewmembers, for the individual and collective disconnection of the electrical power sources from the system; and

(4) There are means to indicate to appropriate crewmembers the generating system quantities essential for the safe operation of the system, including the voltage and current supplied by each generator.

62. *Electrical equipment and installation.* Electrical equipment, controls, and wiring must be installed so that operation of any one unit or system of units will not adversely affect the simultaneous operation of any other electrical unit or system essential to the safe operation.

63. *Distribution system.* (a) For the purpose of complying with this section, the distribution system includes the distribution busses, their associated feeders, and each control and protective device.

(b) Each system must be designed so that essential load circuits can be supplied in the event of reasonably probable faults or open circuits, including faults in heavy current carrying cables.

(c) If two independent sources of electrical power for particular equipment or systems are required under this Appendix, their electrical energy supply must be ensured by means such as duplicate electrical equipment, throwover switching, or multichannel or loop circuits separately routed.

64. *Circuit protective devices.* The circuit protective devices for the electrical circuits of the airplane must meet FAR 23.1357, and in addition circuits for loads which are essential to safe operation must have individual and exclusive circuit protection.

APPENDIX B—AIRPLANE FLIGHT RECORDER SPECIFICATIONS

Parameters	Range	Installed system [1] minimum accuracy (to recovered data)	Sampling interval (per second)	Resolution [4] read out
Relative time (from recorded on prior to takeoff).	8 hr minimum..........	±0.125% per hour..........	1..........	1 sec.
Indicated airspeed..........	V$_{so}$ to V$_D$ (KIAS)..........	±5% or ±10 kts., whichever is greater. Resolution 2 kts. below 175 KIAS.	1..........	1% [3]
Altitude..........	− 1,000 ft. to max cert. alt. of A/C..........	±100 to ±700 ft. (see Table 1, TSO C51–a).	1..........	25 to 150
Magnetic heading..........	360°..........	±5°.	1..........	1°
Vertical acceleration..........	−3g to +6g..........	±0.2g in addition to ±0.3g maximum datum.	4 (or 1 per second where peaks, ref. to 1g are recorded).	0.03g
Longitudinal acceleration..........	±1.0g..........	±1.5% max. range excluding datum error of ±5%.	2..........	0.01g
Pitch attitude..........	100% of usable..........	±2°.	1..........	0.8°
Roll attitude..........	±60° or 100% of usable range, whichever is greater.	±2°.	1..........	0.8°
Stabilizer trim position. Or	Full range..........	±3% unless higher uniquely required..........	1..........	1% [3]
Pitch control position.	Full range..........	±3% unless higher uniquely required..........	1..........	1% [3]
Engine Power, Each Engine				
Fan or N$_1$ speed or EPR or cockpit indications used for aircraft certification. Or	Maximum range..........	±5%.	1..........	1% [3]
Prop. speed and torque (sample once/sec as close together as practicable).			1 (prop speed), 1 (torque).	
Altitude rate [2] (need depends on altitude resolution).	±8,000 fpm..........	±10%. Resolution 250 fpm below 12,000 ft. indicated.	1..........	250 fpm. Below 12,000.

APPENDIX B—AIRPLANE FLIGHT RECORDER SPECIFICATIONS—Continued

Parameters	Range	Installed system [1] minimum accuracy (to recovered data)	Sampling interval (per second)	Resolution [4] read out
Angle of attack [2] (need depends on altitude resolution).	−20° to 40° or of usable range	±2°	1	0.8% [3]
Radio transmitter keying (discrete)	On/off		1	
TE flaps (discrete or analog)	Each discrete position (U, D, T/O, AAP)		1	
	Or			
	Analog 0-100% range	±3°	1	1% [3]
LE flaps (discrete or analog)	Each discrete position (U, D, T/O, AAP)		1	
	Or			
	Analog 0-100% range	±3°	1	1% [3]
Thrust reverser, each engine (Discrete)	Stowed or full reverse		1	
Spoiler/speedbrake (discrete)	Stowed or out		1	
Autopilot engaged (discrete)	Engaged or disengaged		1	

[1] When data sources are aircraft instruments (except altimeters) of acceptable quality to fly the aircraft the recording system excluding these sensors (but including all other characteristics of the recording system) shall contribute no more than half of the values in this column.

[2] If data from the altitude encoding altimeter (100 ft. resolution) is used, then either one of these parameters should also be recorded. If however, altitude is recorded at a minimum resolution of 25 feet, then these two parameters can be omitted.

[3] Per cent of full range.

[4] This column applies to aircraft manufacturing after October 11, 1991

APPENDIX C—HELICOPTER FLIGHT RECORDER SPECIFICATIONS

Parameters	Range	Installed system [1] minimum accuracy (to recovered data)	Sampling interval (per second)	Resolution [3] read out
Relative time (from recorded on prior to takeoff).	8 hr minimum	±0.125% per hour	1	1 sec.
Indicated airspeed	V_{min} to V_D (KIAS) (minimum airspeed signal attainable with installed pilot-static system).	±5% or ±10 kts., whichever is greater	1	1 kt.
Altitude	−1,000 ft. to 20,000 ft. pressure altitude	±100 to ±700 ft. (see Table 1, TSO C51-a).	1	25 to 150 ft.
Magnetic heading	360°	±5°	1	1°
Vertical acceleration	−3g to +6g	±0.2g in addition to ±0.3g maximum datum.	4 (or 1 per second where peaks, ref. to 1g are recorded).	0.05g.
Longitudinal acceleration	±1.0g	±1.5% max. range excluding datum error of ±5%.	2	0.03g.
Pitch attitude	100% of usable range	±2°	1	0.8°
Roll attitude	±60° or 100% of usable range, whichever is greater.	±2°	1	0.8°
Altitude rate	±8,000 fpm	±10% Resolution 250 fpm below 12,000 ft. indicated.	1	250 fpm below 12,000.
Engine Power, Each Engine				
Main rotor speed	Maximum range	±5%	1	1% [2]
Free or power turbine	Maximum range	+5%	1	1% [2]
Engine torque	Maximum range	±5%	1	1% [2]
Flight Control—Hydraulic Pressure				
Primary (discrete)	High/low		1	
Secondary—if applicable (discrete)	High/low		1	
Radio transmitter keying (discrete)	On/off		1	
Autopilot engaged (discrete)	Engaged or disengaged		1	
SAS status—engaged (discrete)	Engaged/disengaged		1	
SAS fault status (discrete)	Fault/OK		1	
Flight Controls				
Collective	Full range	±3%	2	1% [2]
Pedal position	Full range	±3%	2	1% [2]
Lat. cyclic	Full range	±3%	2	1% [2]
Long. cyclic	Full range	±3%	2	1% [2]
Controllable stabilator position	Full range	±3%	2	1% [2]

[1] When data sources are aircraft instruments (except altimeters) of acceptable quality to fly the aircraft the recording system excluding these sensors (but including all other characteristics of the recording system) shall contribute no more than half of the values in this column.

[2] Per cent of full range.

[3] This column applies to aircraft manufactured after October 11, 1991

APPENDIX D—AIRPLANE FLIGHT RECORDER SPECIFICATION

Parameters	Range	Accuracy sensor input to DFDR readout	Sampling interval (per second)	resolution [4] read out
Time (GMT or Frame Counter) (range 0 to 4095, sampled 1 per frame).	24 Hrs	±0.125% Per Hour	0.25 (1 per 4 seconds).	1 sec.
Altitude	−1,000 ft to max certificated altitude of aircraft.	±100 to ±700 ft (See Table 1, TSO-C51a).	1	5′ to 35′ [1]
Airspeed	50 KIAS to V_{so}, and V_{so} to 1.2 V_D	±5%, ±3%	1	1kt.
Heading	360°	±2°	1	0.5°
Normal Acceleration (Vertical)	−3g to +6g	±1% of max range excluding datum error of ±5%.	8	0.01g
Pitch Attitude	±75°	±2°	1	0.5°
Roll Attitude	±180°	±2°	1	0.5°.
Radio Transmitter Keying	On-Off (Discrete)		1	
Thrust/Power on Each Engine	Full range forward	±2%	1 (per engine)	0.2% [2]
Trailing Edge Flap or Cockpit Control Selection.	Full range or each discrete position	±3° or as pilot's indicator	0.5	0.5% [2]
Leading Edge Flap on or Cockpit Control Selection.	Full range or each discrete position	±3° or as pilot's indicator	0.5	0.5% [2].
Thrust Reverser Position	Stowed, in transit, and reverse (discrete).		1 (per 4 seconds per engine).	
Ground Spoiler Position/Speed Brake Selection.	Full range or each discrete position	±2% unless higher accuracy uniquely required.	1	0.22 [2]
Marker Beacon Passage	Discrete		1	
Autopilot Engagement	Discrete		1	
Longitudinal Acceleration	±1g	±1.5% max range excluding datum error of ±5%.	4	0.01g.
Pilot Input And/or Surface Position-Primary Controls (Pitch, Roll, Yaw) [3].	Full range	±2° unless higher accuracy required.	1	0.2% [2]
Lateral Acceleration	±1g	±1.5% max range excluding datum error of ±5%.	4	0.01g.
Pitch Trim Position	Full range	±3% unless higher accuracy uniquely required.	1	0.3% [2]
Glideslope Deviation	±400 Microamps	±3%	1	0.3% [2]
Localizer Deviation	±400 Microamps	±3%	1	0.3% [2]
AFCS Mode And Engagement Status	Discrete		1	
Radio Altitude	−20 ft to 2,500 ft	±2 Ft or ±3% whichever is greater below 500 ft and ±5% above 500 ft.	1	1 ft + 5% [2] above 500′
Master Warning	Discrete		1	
Main Gear Squat Switch Status	Discrete		1	
Angle of Attack (if recorded directly)	As installed	As installed	2	0.3% [2]
Outside Air Temperature or Total Air Temperature.	−50°C to +90°c	±2°c.	0.5	0.3°c
Hydraulics, Each System Low Pressure	Discrete		0.5	or 0.5% [2]
Groundspeed	As installed	Most accurate systems installed (IMS equipped aircraft only).	1	0.2% [2]

If additional recording capacity is available, recording of the following parameters is recommended. The parameters are listed in order of significance:

Parameters	Range	Accuracy sensor input to DFDR readout	Sampling interval (per second)	resolution read out
Drift Angle	When available. As installed	As installed	4	
Wind Speed and Direction	When available. As installed	As installed	4	
Latitude and Longitude	When available. As installed	As installed	4	
Brake pressure/Brake pedal position	As installed	As installed	1	
Additional engine parameters:				
EPR	As installed	As installed	1 (per engine)	
N1	As installed	As installed	1 (per engine)	
N2	As installed	As installed	1 (per engine)	
EGT	As installed	As installed	1 (per engine)	
Throttle Lever Position	As installed	As installed	1 (per engine)	
Fuel Flow	As installed	As installed	1 (per engine)	
TCAS:				
TA	As installed	As installed	1	
RA	As installed	As installed	1	
Sensitivity level (as selected by crew).	As installed	As installed	2	
GPWS (ground proximity warning system)	Discrete		1	
Landing gear or gear selector position	Discrete		0.25 (1 per 4 seconds).	
DME 1 and 2 Distance	0–200 NM;	As installed	0.25	1mi.
Nav 1 and 2 Frequency Selection	Full range	As installed	0.25	

[1] When altitude rate is recorded. Altitude rate must have sufficient resolution and sampling to permit the derivation of altitude to 5 feet.
[2] Per cent of full range.
[3] For airplanes that can demonstrate the capability of deriving either the control input on control movement (one from the other) for all modes of operation and flight regimes, the "or" applies. For airplanes with non-mechanical control systems (fly-by-wire) the "and" applies. In airplanes with split surfaces, suitable combination of inputs is acceptable in lieu of recording each surface separately.
[4] This column applies to aircraft manufactured after October 11, 1991

APPENDIX E—HELICOPTER FLIGHT RECORDER SPECIFICATIONS

Parameters	Range	Accuracy sensor input to DFDR readout	Sampling interval (per second)	Resolution [2] read out
Time (GMT)	24 Hrs	±0.125% Per Hour	0.25 (1 per 4 seconds).	1 sec.
Altitude	−1,000 ft to max certificated altitude of aircraft.	±100 to ±700 ft (See Table 1, TSO–C51a).	1	5' to 30'
Airspeed	As the installed measuring system	±3%	1	1 kt.
Heading	360°	±2°	1	0.5°
Normal Acceleration (Vertical)	−3g to +6g	±1% of max range excluding datum error of ±5%.	8	0.01g
Pitch Attitude	±75°	±2°	2	0.5°
Roll Attitude	±180°	±2°	2	0.5°
Radio Transmitter Keying	On-Off (Discrete)		1	0.25 sec.
Power in Each Engine: Free Power Turbine Speed *and* Engine Torque.	0–130% (power Turbine Speed) Full range (Torque).	±2%	1 speed 1 torque (per engine).	0.2% [1] to 0.4% [1]
Main Rotor Speed	0–130%	±2%	2	0.3% [1]
Altitude Rate	±6,000 ft/min	As installed	2	0.2% [1]
Pilot Input—Primary Controls (Collective, Longitudinal Cyclic, Lateral Cyclic, Pedal).	Full range	±3%	2	0.5% [1]
Flight Control Hydraulic Pressure Low	Discrete, each circuit		1	
Flight Control Hydraulic Pressure Selector Switch Position, 1st and 2nd stage.	Discrete		1	
AFCS Mode and Engagement Status	Discrete (5 bits necessary)		1	
Stability Augmentation System Engage	Discrete		0.25	
SAS Fault Status	Discrete		0.25	
Main Gearbox Temperature Low	As installed	As installed	0.25	0.5% [1]
Main Gearbox Temperature High	As installed	As installed	0.5	0.5% [1]
Controllable Stabilator Position	Full Range	±3%	2	0.4% [1]
Longitudinal Acceleration	±1g	±1.5% max range excluding datum error of ±5%.	4	0.01g.
Lateral Acceleration	±1g	±1.5% max range excluding datum of ±5%.	4	0.01g.
Master Warning	Discrete		1	
Nav 1 and 2 Frequency Selection	Full range	As installed	0.25	
Outside Air Temperature	−50°C to +90°C	±2°c.	0.5	0.3°c.

[1] Per cent of full range.
[2] This column applies to aircraft manufactured after October 11, 1991.

[Special Federal Aviation Regulation 38–2

Certification and Operating Requirements]

[Contrary provisions of Parts 121, 125, 127, 129, and 135 of the Federal Aviation Regulations notwithstanding—

1. *Applicability.*

(a) This Special Federal Aviation Regulation applies to persons conducting commercial passenger operations, cargo operations, or both, and prescribes—

(1) The types of operating certificates issued by the Federal Aviation Administration;

(2) The certification requirements an operator must meet in order to obtain and hold operations specifications for each type of operation conducted and each class and size of aircraft operated; and

(3) The operating requirements an operator must meet in conducting each type of operation and in operating each class and size of aircraft authorized in its operations specifications.

A person shall be issued only one certificate and all operations shall be conducted under that certificate, regardless of the type of operation or the class or size of aircraft operated. A person holding an air carrier operating certificate may not conduct any operations under the rules of Part 125.

(b) Persons conducting operations under more than one paragraph of this SFAR shall meet the certification requirements specified in each paragraph and shall conduct operations in compliance with the requirements of the Federal Aviation Regulations specified in each paragraph for the operation conducted under that paragraph.

(c) Except as provided under this SFAR, no person may operate as an air carrier or as a commercial operator without, or in violation of, a certificate and operations specifications issued under this SFAR.

2. *Certificates and foreign air carrier operations specifications.*

(a) Persons authorized to conduct operations as an air carrier will be issued an Air Carrier Operating Certificate.

(b) Persons who are not authorized to conduct air carrier operations, but who are authorized to conduct passenger, cargo, or both, operations as a commercial operator will be issued an Operating Certificate.

(c) FAA certificates are not issued to foreign air carriers. Persons authorized to conduct operations in the United States as a foreign air carrier who hold a permit issued under Section 402 of the Federal Aviation Act of 1958, as amended (49 U.S.C. 1372), or other appropriate economic or exemption authority issued by the appropriate agency of the United States of America will be issued operations specifications in accordance with the requirements of Part 129 and shall conduct their operations within the United States in accordance with those requirements.

3. *Operations specifications.*

The operations specifications associated with a certificate issued under paragraph 2(a) or (b) and the operations specifications issued under paragraph 2(c) of this SFAR will prescribe the authorizations, limitations and certain procedures under which each type of operation shall be conducted and each class and size of aircraft shall be operated.

4. *Air carriers, and those commercial operators engaged in scheduled intrastate common carriage.*

Each person who conducts operations as an air carrier or as a commercial operator engaged in scheduled intrastate common carriage of persons or property for compensation or hire in air commerce with—

(a) Airplanes having a passenger seating configuration of more than 30 seats, excluding any required crewmember seat, or a payload capacity of more than 7,500 pounds, shall comply with the certification requirements in Part 121, and conduct its—

(1) Scheduled operations within the 48 contiguous states of the United States and the District of Columbia, including routes that extend outside the United States that are specifically authorized by the Administrator, with those airplanes in accordance with the[requirements of Part 121 applicable to domestic air carriers, and shall be issued operations specifications for those operations in accordance with those requirements.

(2) Scheduled operations to points outside the 48 contiguous states of the United States and the District of Columbia with those airplanes in accordance with the requirements of Part 121 applicable to flag air carriers, and shall be issued operations specifications for those operations in accordance with those requirements.

(3) All-cargo operations and operations that are not scheduled with those airplanes in accordance with the requirements of Part 121 applicable to supplemental air carriers, and shall be issued operations specifications for those operations in accordance with those requirements; except the Administrator may authorize those operations to be conducted under paragraph (4) (a) (1) or (2) of this paragraph.

(b) Airplanes having a maximum passenger seating configuration of 30 seats or less, excluding any required crewmember seat, and a maximum payload capacity of 7,500 pounds or less, shall comply with the certification requirements in Part 135, and conduct its operations with those airplanes in accordance with the requirements of Part 135, and shall be issued operations specifications for those operations in accordance with those requirements; except that the Administrator may authorize a person conducting operations in transport category airplanes to conduct those operations in accordance with the requirements of paragraph 4(a) of this paragraph.

(c) Rotorcraft having a maximum passenger seating configuration of 30 seats or less and a maximum payload capacity of 7,500 pounds or less shall comply with the certification requirements in Part 135, and conduct its operations with those aircraft in accordance with the requirements of Part 135, and shall be issued operations specifications for those operations in accordance with those requirements.

(d) Rotorcraft having a passenger seating configuration of more than 30 seats or a payload capacity of more than 7,500 pounds shall comply with the certification requirements in Part 135, and conduct its operations with those aircraft in accordance with the requirements of Part 135, and shall be issued special operations specifications for thos.. operations in accordance with those requirements and this SFAR.

5. *Operations conducted by a person who is not engaged in air carrier operations, but is engaged in passenger operations, cargo operations, or both, as a commercial operator.*

Each person, other than a person conducting operations under paragraph 2(c) or 4 of this SFAR, who conducts operations with—

(a) Airplanes having a passenger seating configuration of 20 or more, excluding any required crewmember seat, or a maximum payload capacity of 6,000 pounds or more, shall comply with the certification requirements in Part 125, and conduct its operations with those airplanes in accordance with the requirements of Part 125, and shall be issued operations specifications in accordance with those requirements, or shall comply with an appropriate deviation authority.

(b) Airplanes having a maximum passenger seating configuration of less than 20 seats, excluding any required crewmember seat, and a maximum payload capacity of less than 6,000 pounds shall comply with the certification requirements in Part 135, and conduct its operations in those airplanes in accordance with the requirements of Part 135, and shall be issued operations specifications in accordance with those requirements.

(c) Rotorcraft having a maximum passenger seating configuration of 30 seats or less and a maximum payload capacity of 7,500 pounds or less shall comply with the certification requirements in Part 135, and conduct its operations in those aircraft in accordance with the requirements of Part 135, and shall be issued operations specifications for those operations in accordance with those requirements.

(d) Rotorcraft having a passenger seating configuration of more than 30 seats or a payload capacity of more than 7,500 pounds shall comply with the certification requirements in Part 135, and conduct its operations with those aircraft in accordance with the requirements of Part 135, and shall be issued special operations specifications for those operations in accordance with those requirements and this SFAR.]

6. *Definitions.*

(a) Wherever in the Federal Aviation Regulations the terms—

(1) "Domestic air carrier operating certificate," "flag air carrier operating certificate," "supplemental air carrier operating certificate," or "commuter air carrier" (in the context of Air Carrier Operating Certificate) appears, it shall be deemed to mean an "Air Carrier Operating Certificate" issued and maintained under this SFAR.

(2) "ATCO operating certificate," appears, it shall be deemed to mean either an "Air Carrier Operating Certificate" or "Operating Certificate," as is appropriate to the context of the regulation. All other references to an operating certificate shall be deemed to mean an "Operating Certificate" issued under this SFAR unless the context indicates the reference is to an Air Carrier Operating Certificate.

(b) Wherever in the Federal Aviation Regulations a regulation applies to—

(1) "Domestic air carriers," it will be deemed to mean a regulation that applies to scheduled operations solely within the 48 contiguous states of the United States and the District of Columbia conducted by persons described in paragraph 4(a)(1) of this SFAR.

(2) "Flag air carriers," it will be deemed to mean a regulation that applies to scheduled operations to any point outside the 48 contiguous states of the United States and the District of Columbia conducted by persons described in paragraph 4(a)(2) of this SFAR.

(3) "Supplemental air carriers," it will be deemed to mean a regulation that applies to charter and all-cargo operations conducted by persons described in paragraph 4(a)(3) of this SFAR.

(4) "Commuter air carriers," it will be deemed to mean a regulation that applies to scheduled passenger carrying operations, with a frequency of operations of at least five round trips per week on at least one route between two or more points according to the published flight schedules, conducted by persons described in paragraph 4(b) or (c) of this SFAR. This definition does not apply to Part 93 of this chapter.

(c) For the purpose of this SFAR, the term—

(1) "Air carrier" means a person who meets the definition of an air carrier as defined in the Federal Aviation Act of 1958, as amended.

(2) "Commercial operator" means a person, other than an air carrier, who conducts operations in air commerce carrying persons or property for compensation or hire.

(3) "Foreign air carrier" means any person other than a citizen of the United States, who undertakes, whether directly or indirectly or by lease or any other arrangement, to engage in foreign air transportation.

(4) "Scheduled operations" means operations that are conducted in accordance with a published schedule for passenger operations which includes dates or times (or both) that is openly advertised or otherwise made readily available to the general public.

(5) "Size of aircraft" means an aircraft's size as determined by its seating configuration or payload capacity, or both.

(6) "Maximum payload capacity" means:

(i) For an aircraft for which a maximum zero fuel weight is prescribed in FAA technical specifications, the maximum zero fuel weight, less empty weight, less all justifiable aircraft equipment, and less the operating load (consisting of minimum flight crew, foods and beverages, and supplies and equipment related to foods and beverages, but not including disposable fuel or oil).

(ii) For all other aircraft, the maximum certificated takeoff weight of an aircraft, less the empty weight, less all justifiable aircraft equipment, and less the operating load (consisting of minimum fuel load, oil, and flightcrew). The allowance for the weight of the crew, oil, and fuel is as follows:

(A) Crew—200 pounds for each crewmember required by the Federal Aviation Regulations.]

(B) Oil—350 pounds.

(C) Fuel—the minimum weight of fuel required by the applicable Federal Aviation Regulations for a flight between domestic points 174 nautical miles apart under VFR weather conditions that does not involve extended overwater operations.

(7) "Empty weight" means the weight of the airframe, engines, propellers, rotors, and fixed equipment. Empty weight excludes the weight of the crew and payload, but includes the weight of all fixed ballast, unusable fuel supply, undrainable oil, total quantity of engine coolant, and total quantity of hydraulic fluid.

(8) "Maximum zero fuel weight" means the maximum permissible weight of an aircraft with no disposable fuel or oil. The zero fuel weight figure may be found in either the aircraft type certificate data sheet, or the approved Aircraft Flight Manual, or both.

(9) "Justifiable aircraft equipment" means any equipment necessary for the operation of the aircraft. It does not include equipment or ballast specifically installed, permanently or otherwise, for the purpose of altering the empty weight of an aircraft to meet the maximum payload capacity.

This Special Federal Aviation Regulation No. 38-2 terminates [June 1, 1989], or the effective date of the codification of SFAR 38-2 into the Federal Aviation Regulations, whichever occurs first.

FEDERAL AVIATION REGULATIONS

Part 141—Pilot Schools

Contents

Subpart A—General

§ 141.1 Applicability.

This Part prescribes the requirements for issuing pilot school certificates, provisional pilot school certificates, and associated ratings and the general operating rules for the holders of those certificates and ratings.

§ 141.3 Certificate required.

No person may operate as a certificated pilot school without, or in violation of, a pilot school certificate or provisional pilot school certificate issued under this Part.

§ 141.5 Pilot school certificate.

An applicant is issued a pilot school certificate with associated ratings for that certificate if—

(a) It meets the pertinent requirements of Subparts A through C of this Part; and

(b) Within the 24 months before the date of application, it has trained and recommended for pilot certification and rating tests, at least 10 applicants for pilot certificates and ratings and at least 8 of the 10 most recent graduates tested by an FAA inspector or designated pilot examiner, passed that test the first time.

§ 141.7 Provisional pilot school certificate.

An applicant is issued a provisional pilot school certificate with associated ratings if it meets the pertinent requirements of Subparts A through C of this Part, but does not meet the recent training activity requirement specified in § 141.5(b).

§ 141.9 Examining authority.

An applicant is issued an examining authority for its pilot school certificate if it meets the requirements of Subpart D of this Part.

§ 141.11 Pilot school ratings.

Associated ratings are issued with a pilot school certificate or a provisional pilot school certificate, specifying each of the following courses that the school is authorized to conduct:

(a) *Certification courses.*

(1) Private pilot.

(2) Private test course.

(3) Instrument rating.

(4) Commercial pilot.

(5) Commercial test course.

(6) Additional aircraft rating.

(b) *Pilot ground school course.*

(1) Pilot ground school.

(c) *Test preparation courses.*

(1) Flight instructor certification.

(2) Additional flight instructor rating.

(3) Additional instrument rating.

(4) Airline transport pilot certification.

(5) Pilot refresher course.

(6) Agricultural aircraft operations course.

(7) Rotorcraft external load operations course.

§ 141.13 Application for issuance, amendment, or renewal.

(a) Application for an original certificate and rating, for an additional rating, or for the renewal of a certificate under this Part is made on a form and in a manner prescribed by the Administrator.

(b) An application for the issuance or amendment of a certificate or rating must be accompanied by three copies of the proposed training course outline for each course for which approval is sought.

§ 141.15 Location of facilities.

Neither a pilot school certificate nor a provisional pilot school certificate is issued for a school having a base or other facilities located outside the United States unless the Administrator finds that the location of the base or facilities at that place is needed for the training of students who are citizens of the United States.

§ 141.17 Duration of certificates.

(a) Unless sooner surrendered, suspended, or revoked, a pilot school certificate or a provisional pilot school certificate expires—

(1) At the end of the twenty-fourth month after the month in which it was issued or renewed; or

(2) Except as provided in paragraph (b) of this section, on the date that any change in ownership of the school or the facilities upon which its certification is based occurs; or

(3) Upon notice by the Administrator that the school has failed for more than 60 days to maintain the facilities, aircraft, and personnel required for at least one of its approved courses.

(b) A change in the ownership of a certificated pilot school or provisional pilot school does not terminate that certificate if within 30 days after the date that any change in ownership of the school occurs, application is made for an appropriate amendment to the certificate and no change in the facilities, instructor, personnel or training course is involved.

(c) An examining authority issued to the holder of a pilot school certificate expires on the date that the pilot school certificate expires, or is surrendered, suspended, or revoked.

§ 141.18 Carriage of narcotic drugs, marahuana, and depressant or stimulant drugs or substances.

If the holder of a certificate issued under this Part permits any aircraft owner or leased by that holder to be engaged in any operation that the certificate holder knows to be in violation of § 91.12(a) of this chapter, that operation is a basis for suspending or revoking the certificate.

§ 141.19 Display of certificate.

(a) Each holder of a pilot school certificate or a provisional pilot school certificate shall display that certificate at a place in the school that is normally accessible to the public and is not obscured.

(b) A certificate shall be made available for inspection upon request by the Administrator, or an authorized representative of the National Transportation Safety Board, or of any Federal, State, or local law enforcement officer.

§ 141.21 Inspections.

Each holder of a certificate issued under this Part shall allow the Administrator to inspect its personnel, facilities, equipment, and records to determine its compliance with the Federal Aviation Act of 1958, and the Federal Aviation Regulations, and its eligibility to hold its certificate.

§ 141.23 Advertising limitations.

(a) The holder of a pilot school certificate or a provisional pilot school certificate may not make any statement relating to its certification and ratings which is false or designed to mislead any person contemplating enrollment in that school.

(b) The holder of a pilot school certificate or a provisional pilot school certificate may not advertise that the school is certificated unless it clearly differentiates between courses that have been approved and those that have not.

(c) The holder of a pilot school certificate or a provisional pilot school certificate—

(1) That has relocated its school shall promptly remove from the premises it has vacated all signs indicating that the school was certificated by the Administrator; or

(2) Whose certificate has expired, or has been surrendered, suspended, or revoked shall promptly remove all indications (including signs), wherever located, that the school is certificated by the Administrator.

§ 141.25 Business office and operations base.

(a) Each holder of a pilot school or a provisional pilot school certificate shall maintain a principal business office with a mailing address in the name shown on its certificate. The business office shall have facilities and equipment that are adequate to maintain the required school files and records and to operate the business of the school. The office may not be shared with, or used by, another pilot school.

(b) Each certificate holder shall, before changing the location of its business office or base of operations, notify the FAA District Office having jurisdiction over the area of the new location. The notice shall be submitted in writing at least 30 days before the change. For a change in the holder's base of operations, the notice shall be accompanied by any amendments needed for the holder's approved training course outline.

(c) No certificate holder may conduct training at an operations base other than the one specified in its certificate, until—

(1) The base has been inspected and approved by the FAA District Office having jurisdiction over the school for use by the certificate holder; and

(2) The course of training and any needed amendments thereto have been approved for training at that base.

§ 141.27 Renewal of certificates and ratings.

(a) *Pilot school certificates.* The holder of a pilot school certificate may apply for a renewal of the certificate not less than 30 days before the certificate expires. If the school meets the requirements of this Part for the issuance of the certificate, its certificate is renewed for 24 months.

(b) *Pilot school ratings.* Each pilot school rating on a pilot school certificate may be renewed with that certificate for another 24 months if the Administrator finds that the school meets the requirements prescribed in this Part for the issuance of the rating.

(c) *Provisional pilot school certificates.*

(1) A provisional pilot school certificate and any ratings on that certificate may not be renewed. However, the holder of that certificate may apply for a pilot school certificate with appropriate ratings not less than 30 days before the provisional certificate expires. The school is issued a pilot school certificate with appropriate ratings, if it meets the appropriate requirements of this Part.

(2) The holder of a provisional pilot school certificate may not reapply for a provisional pilot school certificate for at least 180 days after the date of its expiration.

§ 141.29 Existing pilot school certificates; validity.

(a) A pilot school certificate issued before November 1, 1974, remains in effect until the expiration date on that certificate, unless it is sooner surrendered, suspended, or revoked.

(b) A pilot school certificate issued before November 1, 1974, may be renewed under the requirements of Part 141 in effect before November 1, 1974, to allow those students enrolled in a school's approved course of training prior to November 1, 1974, to complete that training. The renewal is issued for a period long enough for the students to complete their training in the enrolled course, but not for more than 24 months after November 1, 1974.

Subpart B—Personnel, Aircraft, and Facilities Requirements

§ 141.31 Applicability.

This subpart prescribes the personnel and aircraft requirements for a pilot school or a provisional pilot school certificate. It also prescribes the facilities an applicant must have available to him on a continuous use basis to hold a pilot school or provisional pilot school certificate. As used in this subpart, a person has the continuous use of a facility, including an airport, if it has the use of the facility when needed as the owner, or under a written agreement giving it that use for at least 6 calendar months from the date of the application for the initial certificate or a renewal of that certificate.

§ 141.33 Personnel.

(a) An applicant for a pilot school or provisional pilot school certificate must show that—

(1) It has adequate personnel and authorized instructors, including a chief instructor for each course of training, who are qualified and competent to perform the duties to which they are assigned;

(2) Each dispatcher, aircraft handler, line crewman, and serviceman to be used

has been instructed in the procedures and responsibilities of his employment. (Qualified operations personnel may serve in more than one capacity with a pilot school or provisional pilot school); and

(3) Each instructor to be used for ground or flight instruction holds a flight or ground instructor certificate, as appropriate, with ratings for the course of instruction and any aircraft used in that course.

(b) An applicant for a pilot school or a provisional pilot school certificate shall designate a chief instructor for each course of training who meets the requirements of a § 141.35 of this Part. Where necessary, the applicant shall also designate at least one instructor to assist the chief instructor and serve for the chief instructor in his absence. A chief instructor or his assistant may be designated to serve in that capacity for more than one approved course but not for more than one school.

§ 141.35 Chief instructor qualifications.

(a) To be eligible for a designation as a chief flight instructor or an assistant chief flight instructor for a course of training, a person must meet the following requirements:

(1) He must pass—

(i) An oral test on this Part and on Parts 61 and 91 of this chapter and on the training standards and objectives of the course for which he is designated; and

(ii) A flight test on the flight procedures and maneuvers appropriate to that course.

(2) He must meet the applicable requirements of paragraphs (b), (c), and (d) of this section. However, a chief flight instructor or an assistant chief flight instructor for a course of training for gliders, free balloons or airships is only required to have 40 percent of the hours required in paragraphs (b) and (c) of this section.

(b) For a course of training leading to the issuance of a private pilot certificate or rating, a chief flight instructor or an assistant chief flight instructor must have—

(1) At least a commercial pilot certificate and a flight instructor certificate, each with a rating for the category and class of aircraft used in the course;

(2) At least 1,000 hours as pilot in command;

(3) Primary flight instruction experience, acquired as either a certificated flight instructor or an instructor in a military pilot primary flight training program, or a combination thereof, consisting of at least—

(i) Two years and a total of 500 flight hours; or

(ii) 1,000 flight hours; and

(4) Within the year preceding designation, at least 100 hours of flight instruction as a certificated flight instructor in the category of aircraft used in the course.

(c) For a course of training leading to the issuance of an instrument rating or a rating with instrument privileges, a chief flight instructor or an assistant chief flight instructor must have—

(1) At least a commercial pilot certificate and a flight instructor certificate, each with an appropriate instrument rating;

(2) At least 100 hours of flight time under actual or simulated instrument conditions;

(3) At least 1,000 hours as pilot in command;

(4) Instrument flight instructor experience, acquired as either a certificated instrument flight instructor or an instructor in a military pilot basic or instrument flight training program, or a combination thereof, consisting of at least—

(i) Two years and a total of 250 flight hours; or

(ii) 400 flight hours; and

(5) Within the year preceding designation, at least—

(i) 100 hours of instrument flight instruction as a certificated instrument flight instructor; or

(ii) One year of active service as an FAA designated instrument rating examiner.

(d) For a course of training other than those that lead to the issuance of a private pilot certificate or rating, or an instrument rating or a rating with instrument privileges, a chief flight instructor or an assistant chief flight instructor must have—

(1) At least a commercial pilot certificate and a flight instructor certificate, each with a rating for the category and class of aircraft used in the course of training and, for a course of training using airplanes or airships, an instrument rating on his commercial pilot certificate;

(2) At least 2,000 hours as pilot in command;

(3) Flight instruction experience, acquired as either a certificated flight instructor or an instructor in a military pilot primary or basic flight training program or a combination thereof, consisting of at least—

(i) Three years and a total of 1,000 flight hours; or

(ii) 1,500 flight hours; and

(4) Within the year preceding designation, at least—

(i) 100 hours of pilot instruction as a certificated flight instructor in the category of aircraft used in the course;

(ii) One year of active service as chief flight instructor of an approved course of training; or

(iii) One year of active service as an FAA designated pilot examiner.

(e) To be eligible for designation as a chief instructor or an assistant chief instructor for a ground school course, a person must have one year of experience as a ground school instructor in a certificated pilot school.

§ 141.37 Airports.

(a) An applicant for a pilot school certificate or a provisional pilot school certificate must show that it has continuous use of each airport at which training flights originate.

(b) Each airport used for airplanes and gliders must have at least one runway or take-off area that allows training aircraft to make a normal takeoff or landing at full gross weight—

(1) Under calm wind (not more than five miles per hour) conditions and temperatures equal to the mean high temperature for the hottest month of the year in the operating area;

(2) Clearing all obstacles in the takeoff flight path by at least 50 feet;

(3) With the powerplant operation and landing gear and flap operation, if applicable, recommended by the manufacturer; and

(4) With smooth transition from liftoff to the best rate of climb speed without exceptional piloting skills or techniques.

(c) Each airport must have a wind direction indicator that is visible from the ends of each runway at ground level.

(d) Each airport must have a traffic direction indicator when the airport has no operating control tower and UNICOM advisories are not available.

(e) Each airport used for night training flights must have permanent runway lights.

§ 141.39 Aircraft. ☞ F11

An applicant for a pilot school or provisional pilot school certificate must show that each aircraft used by that school for flight instruction and solo flights meets the following requirements:

(a) It must be registered as a civil aircraft of the United States.

(b) Except for aircraft used for flight instruction and solo flights in a course of training for agricultural aircraft operations, external load operations and similar aerial work operations, it must be certificated in the standard airworthiness category.

(c) It must be maintained and inspected in accordance with the requirements of Part 91 of this chapter that apply to aircraft used to give flight instruction for hire.

(d) For use in flight instruction, it must be at least a two place aircraft having engine power controls and flight controls that are easily reached and that operate in a normal manner from both pilot stations.

(e) For use in IFR en route operations and instrument approaches, it must be equipped and maintained for IFR operations. However, for instruction in the control and precision maneuvering of an aircraft by reference to instruments, the aircraft may be equipped as provided in the approved course of training.

§ 141.41 Ground trainers and training aids.

An applicant for a pilot school or a provisional pilot school certificate must show that its ground trainers, and training aids and equipment meet the following requirements:

(a) *Pilot ground trainers.*

(1) Each pilot ground trainer used to obtain the maximum flight training credit allowed for ground trainers in an approved pilot training course curriculum must have—

(i) An enclosed pilot's station or cockpit which accommodates one or more flight crewmembers;

(ii) Controls to simulate the rotation of the trainer about three axes;

(iii) The minimum instrumentation and equipment required for powered aircraft in § 91.33 of this chapter, for the type of flight operations simulated;

(iv) For VFR instruction, a means for simulating visual flight conditions, including motion of the trainer, or projections, or models operated by the flight controls; and

(v) For IFR instruction, a means for recording the flight path simulated by the trainer.

(2) Pilot ground trainers other than those covered under subparagraph (1) of this paragraph must have—

(i) An enclosed pilot's station or cockpit, which accommodates one or more flight crewmembers;

(ii) Controls to simulate the rotation of the trainer about three axes; and

(iii) The minimum instrumentation and equipment required for powered aircraft in § 91.33 of this chapter, for the type of flight operations simulated.

(b) *Training aids and equipment.*

Each training aid, including any audio-visuals, mockup, chart, or aircraft component listed in the approved training course outline must be accurate and appropriate to the course for which it is used.

§ 141.43 Pilot briefing areas.

(a) An applicant for a pilot school or provisional pilot school certificate must show that it has the continuous use of a briefing area located at each airport at which training flights originate, that is—

(1) Adequate to shelter students waiting to engage in their training flights;

(2) Arranged and equipped for the conduct of pilot briefings; and

(3) For a school with an instrument or commercial pilot course rating, equipped with private landline or telephone communication to the nearest FAA Flight Service Station, except that this communication equipment is not required if the briefing area and the flight service station are located on the same airport and are readily accessible to each other.

(b) A briefing area required by paragraph (a) of this section may not be used by the applicant if it is available for use by any other

pilot school during the period it is required for use by the applicant.

§ 141.45 Ground training facilities.

An applicant for a pilot school or provisional pilot school certificate must show that each room, training booth, or other space used for instructional purposes is heated, lighted, and ventilated to conform to local building, sanitation, and health codes. In addition, the training facility must be so located that the students in that facility are not distracted by the instruction conducted in other rooms, or by flight and maintenance operations on the airport.

Subpart C—Training Course Outline and Curriculum

§ 141.51 Applicability.

This subpart prescribes the curriculum and course outline requirements for the issuance of a pilot school or provisional pilot school certificate and ratings.

§ 141.53 Training course outline: General.

(a) *General.* An applicant for a pilot school or provisional pilot school certificate must obtain the Administrator's approval of the outline of each training course for which certification and rating is sought.

(b) *Application.* An application for the approval of an initial or amended training course outline is made in triplicate to the FAA District Office having jurisdiction over the area in which the operations base of the applicant is located. It must be made at least 30 days before any training under that course, or any amendment thereto, is scheduled to begin. An application for an amendment to an approved training course must be accompanied by three copies of the pages in the course outline for which an amendment is requested.

§ 141.55 Training course outline: Contents.

(a) *General.* The outline for each course of training for which approval is requested must meet the minimum curriculum for that course prescribed in the appropriate appendix of this Part, and contain the following information:

(1) A description of each room for ground training, including its size and the maximum number of students that may be instructed in the room at one time.

(2) A description of each type of audio-visual aid, projector, tape recorder, mockup, aircraft component and other special training aid used for ground training.

(3) A description of each pilot ground trainer used for instruction.

(4) A listing of the airports at which training flights originate and a description of the facilities, including pilot briefing areas that are available for use by the students and operating personnel at each of those airports.

(5) A description of the type of aircraft including any special equipment, used for each phase of instruction.

(6) The minimum qualifications and ratings for each instructor used for ground or flight training.

(b) *Training syllabus.* In addition to the items specified in paragraph (a) of this section, the course outline must include a training syllabus for each course of training that includes at least the following information:

(1) The pilot certificate and ratings, if any; the medical certificate, if necessary; and the training, pilot experience and knowledge, required for enrollment in the course.

(2) A description of each lesson, including its objectives and standards and the measurable unit of student accomplishment or learning to be derived from the lesson or course.

(3) The stage of training (including the standards therefor) normally accomplished within each training period of not more than 90 days.

(4) A description of the tests and checks used to measure a student's accomplishment for each stage of training.

§ 141.57 Special curricula.

An applicant for a pilot school or provisional pilot school certificate may apply for approval to conduct a special course of pilot training for which a curriculum is not prescribed in the appendixes to this Part, if it shows that the special course of pilot training contains features which can be expected to achieve a level of pilot competency equivalent to that achieved by the curriculum prescribed in the appendixes to this Part or the requirements of Part 61 of this chapter.

Subpart D—Examining Authority

§ 141.61 Applicability.

This subpart prescribes the requirements for the issuance of an examining authority to the holder of a pilot school certificate and the privileges and limitations of that authority.

§ 141.63 Application and qualification.

(a) Application for an examining authority is made on a form and in a manner prescribed by the Administrator.

(b) To be eligible for an examining authority an applicant must hold a pilot school certificate. In addition, the applicant must show that—

(1) It has actively conducted a certificated pilot school for at least 24 months before the date of application; and

(2) Within the 24 months before the date of application for the examining authority, at least 10 students were graduated from the course for which the authority is requested, and at least 9 of the most recent 10 graduates

of that course, who were given an interim or final test by an FAA inspector or a designated pilot examiner, passed that test the first time.

§ 141.65 Privileges.

The holder of an examining authority may recommend graduates of the school's approved certification courses for pilot certificates and ratings except flight instructor certificates, airline transport pilot certificates and ratings, and turbojet type ratings, without taking the FAA flight or written test, or both, in accordance with the provisions of this subpart.

§ 141.67 Limitations and reports.

(a) The holder of an examining authority may not recommend any person for the issuance of a pilot certificate or rating without taking the FAA written or flight test unless that person has—

(1) Been enrolled by the holder of the examining authority in its approved course of training for the particular pilot certificate or rating recommended; and

(2) Satisfactorily completed all of that course of training at its school.

(b) Each final written or flight test given by the holder of an examining authority to a person who has completed the approved course of training must be at least equal in scope, depth, and difficulty to the comparable written or flight test prescribed by the Administrator under Part 61 of this chapter.

(c) A final ground school written test may not be given by the holder of an examining authority to a student enrolled in its approved course of training unless the test has been approved by the FAA District Office having jurisdiction over the area in which the holder of the examining authority is located. In addition, an approved test may not be given by the holder of an examining authority when—

(1) It knows or has reason to believe that the test has been compromised; or

(2) It has been notified that the District Office knows or has reason to believe that the test has been compromised.

(d) The holder of an examining authority shall submit to the FAA District Office a copy of the appropriate training record for each person recommended by it for a pilot certificate or rating.

Subpart E—Operating Rules

§ 141.71 Applicability.

This subpart prescribes the operating rules that are applicable to a pilot school or provisional pilot school certificated under the provisions of this Part.

§ 141.73 Privileges.

(a) The holder of a pilot school or a provisional pilot school certificate may advertise and conduct approved pilot training courses in accordance with the certificate and ratings that it holds.

(b) A certificated pilot school holding an examining authority for a certification course may recommend each graduate of that course for the issuance of a pilot certificate and rating appropriate to that course without the necessity of taking an FAA written or flight test from an FAA inspector or designated pilot examiner.

§ 141.75 Aircraft requirements.

(a) A pretakeoff and prelanding checklist, and the operator's handbook for the aircraft (if one is furnished by the manufacturer) or copies of the handbook if furnished to each student using the aircraft, must be carried on each aircraft used for flight instruction and solo flights.

(b) Each aircraft used for flight instruction and solo flight must have a standard airworthiness certificate, except that an aircraft certificated in the restricted category may be used for flight training and solo flights conducted under special courses for agricultural aircraft operation, external load operations, and similar aerial work operations if its use for training is not prohibited by the operating limitations for the aircraft.

§ 141.77 Limitations.

(a) The holder of a pilot school or a provisional pilot school certificate may not issue a graduation certificate to a student, nor may a certificated pilot school recommend a student for a pilot certificate or rating, unless the student has completed the training therefor specified in the school's course of training and passed the required final tests.

(b) The holder of a pilot school or a provisional pilot school certificate may not graduate a student from a course of training unless he has completed all of the curriculum requirements of that course. A student may be credited, but not for more than one-half of the curriculum requirements, with previous pilot experience and knowledge, based upon an appropriate flight check or test by the school. Course credits may be transferred from one certificated school to another. The receiving school shall determine the amount to be transferred, based on a flight check or written test, or both, of the student. Credit for training and instruction received in another school may not be given unless—

(1) The other school holds a certificate issued under this Part and certifies to the kind and amount of training and to the result of each stage and final test given to that student;

(2) The training and instruction was conducted by the other school in accordance with that school's approved training course; and

(3) The student was enrolled in the other school's approved training course before he received the instruction and training.

§ 141.79 Flight instruction.

(a) No person other than a flight instructor who has the ratings and the minimum qualifications specified in the approved training course outline may give a student flight instruction under an approved course of training.

(b) No student pilot may be authorized to start a solo practice flight from an airport until the flight has been approved by an authorized flight instructor who is present at that airport.

(c) Each chief flight instructor must complete at least once each 12 months, a flight instructor refresher course consisting of not less than 24 hours of ground or flight instruction, or both.

(d) Each flight instructor for an approved course of training must satisfactorily accomplish a flight check given to him by the designated chief flight instructor for the school by whom he is employed. He must also satisfactorily accomplish this flight check each 12 months from the month in which the initial check is given. In addition, he must satisfactorily accomplish a flight check in each type of aircraft in which he gives instruction.

(e) An instructor may not be used in an approved course of training until he has been briefed in regard to the objectives and standards of the course by the designated chief instructor or his assistant.

§ 141.81 Ground training.

(a) Except as provided in paragraph (b) of this section, each instructor used for ground training in an approved course of training must hold a flight or ground instructor certificate with an appropriate rating for the course of training.

(b) A person who does not meet the requirements of paragraph (a) of this section may be used for ground training in an approved course of training if—

(1) The chief instructor for that course of training finds him qualified to give that instruction; and

(2) The instruction is given under the direct supervision of the chief instructor or the assistant chief instructor who is present at the base when the instruction is given.

(c) An instructor may not be used in an approved course of training until he has been briefed in regard to the objectives and standards of that course by the designated chief instructor or his assistant.

§ 141.83 Quality of instruction.

(a) Each holder of a pilot school or provisional pilot school certificate must comply with the approved course of training and must provide training and instruction of such quality that at least 8 out of the 10 students or graduates of that school most recently tested by an FAA inspector or designated pilot examiner, passed on their first attempt either of the following tests:

(1) A test for a pilot certificate or rating, or for an operating privilege appropriate to the course from which the student graduated; or

(2) A test given to a student to determine his competence and knowledge of a completed stage of the training course in which he is enrolled.

(b) The failure of a certificated pilot school or provisional pilot school to maintain the quality of instruction specified in paragraph (a) of this section is considered to be the basis for the suspension or revocation of the certificate held by that school.

(c) The holder of a pilot school or provisional pilot school certificate shall allow the Administrator to make any test, flight check, or examination of its students to determine compliance with its approved course of training and the quality of its instruction and training. A flight check conducted under the provisions of this paragraph is based upon the standards prescribed in the school's approved course of training. However, if the student has completed a course of training for a pilot certificate or rating, the flight test is based upon the standards prescribed in Part 61 of this chapter.

§ 141.85 Chief instructor responsibilities.

(a) Each person designated as a chief instructor for a certificated pilot school or provisional pilot school shall be responsible for—

(1) Certifying training records, graduation certificates, stage and final test reports, and student recommendations;

(2) Conducting an initial proficiency check of each instructor before he is used in an approved course of instruction and, thereafter, at least once each 12 months from the month in which the initial check was conducted;

(3) Conducting each stage or final test given to a student enrolled in an approved course of instruction; and

(4) Maintaining training techniques, procedures, and standards for the school that are acceptable to the Administrator.

(b) The chief instructor or his designated assistant shall be available at the school's base of operations during the time that instruction is given for an approved course of training.

§ 141.87 Change of chief instructor.

(a) The holder of a pilot school or provisional pilot school certificate shall immediately notify in writing the FAA District Office having jurisdiction over the area in which the school is located, of any change in its designation of a chief instructor of an approved training course.

(b) The holder of a pilot school or provisional pilot school certificate may, after providing the notification required in paragraph (a) and pending the designation and approval

of another chief instructor, conduct training or instruction without a chief instructor for that course of training for a period of not more than 60 days. However, during that time each stage or final test of a student enrolled in that approved course of training must be given by an FAA inspector, or a designated pilot examiner.

§ 141.89 Maintenance of personnel, facilities, and equipment.

The holder of a pilot school or provisional pilot school certificate may not give instruction or training to a student who is enrolled in an approved course of training unless—

(a) Each airport, aircraft, and facility necessary for that instruction or training meets the standards specified in the holder's approved training course outline and the appropriate requirements of this Part; and

(b) Except as provided in § 141.87, each instructor or chief instructor meets the qualifications specified in the holder's approved course of training and the appropriate requirements of this Part.

§ 141.91 Satellite bases.

The holder of a pilot school or provisional pilot school certificate may conduct ground or flight training and instruction in an approved course of training at a base other than its main operations base if—

(a) The satellite base is located not more than 25 nautical miles from its main operations base;

(b) The airport, facilities, and personnel used at the satellite base meets the appropriate requirements of Subpart B of this Part and its approved training course outline;

(c) The instructors are under the direct supervision of the chief instructor for the appropriate training course, who is readily available for consultation; and

(d) The FAA District Office having jurisdiction over the area in which the school is located is notified in writing if training or instruction is conducted there for more than seven consecutive days.

§ 141.93 Enrollment.

(a) The holder of a pilot school or a provisional pilot school certificate shall furnish each student, at the time he is enrolled in each approved training course, with the following:

(1) A certificate of enrollment containing—

(i) The name of the course in which he is enrolled; and

(ii) The date of that enrollment.

(2) A copy of the training syllabus required under § 141.55(b).

(3) A copy of the safety procedures and practices developed by the school covering the use of its facilities and the operation of its aircraft, including instructions on the following:

(i) The weather minimums required by the school for dual and solo flights.

(ii) The procedures for starting and taxiing aircraft on the ramp.

(iii) Fire precautions and procedures.

(iv) Redispatch procedures after unprogrammed landings, on and off airports.

(v) Aircraft discrepancies and write offs.

(vi) Securing of aircraft when not in use.

(vii) Fuel reserves necessary for local and cross-country flights.

(viii) Avoidance of other aircraft in flight and on the ground.

(ix) Minimum altitude limitations and simulated emergency landing instructions.

(x) Description and use of assigned practice areas.

(b) The holder of a pilot school or provisional pilot school certificate shall, within 5 days after the date of enrollment, forward a copy of each certificate of enrollment required by paragraph (a)(1) of this section to the FAA District Office having jurisdiction over the area in which the school is located.

§ 141.95 Graduation certificate. ☞ F9

(a) The holder of a pilot school or provisional pilot school certificate shall issue a graduation certificate to each student who completes its approved course of training.

(b) The certificate shall be issued to the student upon his completion of the course of training and contain at least the following information:

(1) The name of the school and the number of the school certificate.

(2) The name of the graduate to whom it was issued.

(3) The course of training for which it was issued.

(4) The date of graduation.

(5) A statement that the student has satisfactorily completed each required stage of the approved course of training including the tests for those stages.

(6) A certification of the information contained in the certificate by the chief instructor for that course of training.

(7) A statement showing the cross-country training the student received in the course of training.

Subpart F—Records

§ 141.101 Training records.

(a) Each holder of a pilot school or provisional pilot school certificate shall establish and maintain a current and accurate record of the participation and accomplishment of each student enrolled in an approved course of training conducted by the school (the student's logbook is not acceptable for this record). The record shall include—

(1) The date the student was enrolled;

(2) A chronological log of the student's attendance, subjects, and flight operations covered in his training and instruction, and the names and grades of any tests taken by the student; and

(3) The date the student graduated, terminated his training, or transferred to another school.

(b) Whenever a student graduates, terminates his training, or transfers to another school, his record shall be certified to that effect by the chief instructor.

(c) The holder of a certificate for a pilot school or a provisional pilot school shall retain each student record required by this section for at least 1 year from the date that the student graduates from the course to which the record pertains, terminates his enrollment in that course, or transfers to another school.

(d) The holder of a certificate for a pilot school or a provisional pilot school shall, upon request of a student, make a copy of his record available to him.

Appendix A

Private Pilot
Certification Course (Airplanes)

1. *Applicability.* This Appendix prescribes the minimum curriculum for a private pilot certification course (airplanes) required by § 141.55.

2. *Ground training.* The course must consist of at least 35 hours of ground training in the following subjects:

(a) The Federal Aviation Regulations applicable to private pilot privileges, limitations, and flight operations; the rules of the National Transportation Safety Board pertaining to accident reporting; the use of the Airman's Information Manual; and the FAA Advisory Circular System.

(b) VFR navigation using pilotage, dead reckoning, and radio aids.

(c) The recognition of critical weather situations from the ground and in flight and the procurement and use of aeronautical weather reports and forecasts.

(d) The safe and efficient operation of airplanes, including high density airport operations, collision avoidance precautions, and radio communication procedures.

3. *Flight training.*

(a) The course must consist of at least 35 hours of the flight training listed in this section and section 4 of this Appendix. Instruction in a pilot ground trainer that meets the requirements of § 141.41(a)(1) may be credited for not more than 5 of the required 35 hours of flight time. Instruction in a pilot ground trainer that meets the requirement of § 141.41(a)(2) may be credited for not more than 2.5 of the required 35

hours of flight time.

(b) Each training flight must include a preflight briefing and a postflight critique of the student by the instructor assigned to that flight.

(c) Flight training must consist of at least 20 hours of instruction in the following subjects:

(1) Preflight operations, including weight and balance determination, line inspection, starting and runups, and airplane servicing.

(2) Airport and traffic pattern operations, including operations at controlled airports, radio communications, and collision avoidance precautions.

(3) Flight maneuvering by reference to ground objects.

(4) Flight at critically slow airspeeds, recognition of imminent stalls, and recovery from imminent and full stalls.

(5) Normal and crosswind takeoffs and landings.

(6) Control and maneuvering an airplane solely by reference to instruments, including emergency descents and climbs using radio aids or radar directives.

(7) Cross-country flying using pilotage, dead reckoning, and radio aids, including a two-hour dual flight at least part of which must be on Federal airways.

(8) Maximum performance takeoffs and landings.

(9) Night flying, including 5 takeoffs and landings as sole manipulator of the controls, and VFR navigation.

(10) Emergency operations, including simulated aircraft and equipment malfunctions, lost procedures, and emergency go-arounds.

[4. *Solo flights.* The course must provide at least 15 hours of solo flights, including:]

(a) *Solo practice.*

[Directed solo practice on all VFR flight operations for which flight instruction is required (except simulated emergencies) to develop proficiency, resourcefulness, and self-reliance.]

(b) *Cross-country flights.*

[(1) Ten hours of cross-country flights, each flight with a landing at a point more than 50 nautical miles from the original departure point. One flight must be of at least 300 nautical miles with landings at a minimum of three points, one of which is at least 100 nautical miles from the original departure point.]

(2) If a pilot school or a provisional pilot school shows that it is located on an island from which cross-country flights cannot be accomplished without flying over water more than 10 nautical miles from the nearest shoreline, it need not include cross-country flights under subparagraph (1) of this paragraph. However, if other airports that permit civil operations are available to

which a flight may be made without flying over water more than 10 nautical miles from the nearest shoreline, the school must include in its course, two round trip solo flights between those airports that are farthest apart, including a landing at each airport on both flights.

5. *Stage and final tests.*

(a) Each student enrolled in a private pilot certification course must satisfactorily accomplish the stage and final test prescribed in this section. The written tests may not be credited for more than 3 hours of the 35 hours of required ground training, and the flight tests may not be credited for more than 4 hours of the 35 hours of required flight training.

(b) Each student must satisfactorily accomplish a written examination at the completion of each stage of training specified in the approved training syllabus for the private certification course and a final test at the conclusion of that course.

(c) Each student must satisfactorily accomplish a flight test at the completion of the first solo flight and at the completion of the first solo cross-country flight and at the conclusion of that course.

Appendix B

Private Test Course (Airplanes)

1. *Applicability.* This Appendix prescribes the minimum curriculum for a private test course (airplanes) required by § 141.55.

2. *Experience.* For enrollment as a student in a private test course (airplanes) an applicant must—

(a) Have logged at least 30 hours of flight time as a pilot; and

(b) Have such experience and flight training that upon completion of his approved private test course (airplanes) he will meet the aeronautical experience requirements prescribed in Part 61 of this chapter for a private pilot certificate.

3. *Ground training.* The course must consist of at least 35 hours of ground training in the subjects listed in § 2 of Appendix A of this Part.

4. *Flight training.*

(a) The course must consist of a total of at least 10 hours of flight instruction in the subjects listed in § 3(c) of Appendix A of this Part.

(b) Each training flight must include a preflight briefing and a postflight critique of the student by the instructor assigned to that flight.

5. *Stage and final tests.* Each student enrolled in the course must satisfactorily accomplish the final tests prescribed in § 5 of Appendix A of this Part. Written tests may not be credited for more than 3 hours of the required 35 hours of ground training, and the

flight tests may not be credited for more than 2 hours of the required 10 hours of flight training.

Appendix C

Instrument Rating Course (Airplanes)

1. *Applicability.* This Appendix prescribes the minimum curriculum for a training course for an Instrument Rating Course (airplanes) required by § 141.55.

2. *Ground training.* The course must consist of at least 30 hours of ground training instruction in the following subjects:

(a) The Federal Aviation Regulations that apply to flight under IFR conditions, the IFR air traffic system and procedures, and the provisions of the Airman's Information Manual pertinent to IFR flights.

(b) Dead reckoning appropriate to IFR navigation, IFR navigation by radio aids using the VOR, ADF, and ILS systems, and the use of IFR charts and instrument approach procedure charts.

(c) The procurement and use of aviation weather reports and forecasts, and the elements of forecasting weather trends on the basis of that information and personal observation of weather conditions.

(d) The function, use, and limitations of flight instruments required for IFR flight, including transponders, radar and radio aids to navigation.

3. *Flight training.* The course must consist of at least 35 hours of instrument flight instruction given by an appropriately rated flight instructor, covering the operations listed in paragraphs (a) through (d) of this section. Instruction given by an authorized instructor in a pilot ground trainer which meets the requirements of § 141.41(a)(1) may be credited for not more than 15 hours of the required flight instruction. Instruction in a pilot ground trainer that meets the requirements of § 141.41(a)(2) may be credited for not more than 7.5 of the required 35 hours of flight time.

(a) Control and accurate maneuvering of an airplane solely by reference to flight instruments.

(b) IFR navigation by the use of VOR and ADF systems, including time, speed and distance computations and compliance with air traffic control instructions and procedures.

(c) Instrument approaches to published minimums using the VOR, ADF, and ILS systems (instruction in the use of the ILS glide slope may be given in an instrument ground trainer or with an airborne ILS simulator).

(d) Cross-country flying in simulated or actual IFR conditions, on Federal airways or as routed by ATC, including one such trip of at least 250 nautical miles including

VOR, ADF, and ILS approaches at different airports.

(e) Emergency procedures appropriate to the maneuvering of an airplane solely by reference to flight instruments.

4. *Stage and final tests.*

(a) Each student must satisfactorily accomplish a written test at the completion of each stage of training specified in the approved training syllabus for the instrument rating course. In addition, he must satisfactorily accomplish a final written test at the conclusion of that course. The written tests may not be credited for more than 5 hours of the 30 hours of required ground training.

(b) Each student must satisfactorily accomplish a flight stage test at the completion of each operation listed in paragraphs (a), (b), and (c) of § 3 of this Appendix. In addition, he must satisfactorily accomplish a final flight test at the completion of the course. The stage and final tests may not be credited for more than 5 hours of the required 35 hours of flight training.

Appendix D

Commercial Pilot Certification Course (Airplanes)

1. *Applicability.* This Appendix prescribes the minimum curriculum for a commercial pilot certification course (airplanes) required by § 141.55.

2. *Ground training.* The course must consist of at least 100 hours of ground training instruction in the following subjects:

(a) The ground training subjects prescribed in § 2 of Appendix A of this Part for a private pilot certification course, except the private pilot privileges and limitations of paragraph (a) of that section.

(b) The ground training subjects prescribed in § 2 of Appendix C of this Part for an Instrument Rating Course.

(c) The Federal Aviation Regulations covering the privileges, limitations, and operations of a commercial pilot, and the operations for which an air taxi/commercial operator, agricultural aircraft operator, and external load operator certificate, waiver, or exemption is equired.

(d) Basic aerodynamics, and the principles of flight which apply to airplanes.

(e) The safe and efficient operation of airplanes, including inspection and certification requirements, operating limitations, high altitude operations and physiological considerations, loading computations, the significance of the use of airplane performance speeds, the computations involved in runway and obstacle clearance and cross-wind component considerations, and cruise control.

3. *Flight training.*

(a) *General.* The course must consist of

at least 190 hours of the flight training and instruction prescribed in this section. Instruction in a pilot ground trainer that meets the requirements of § 141.41(a)(1) may be credited for not more than 40 hours of the required 190 hours of flight time. Instruction in a pilot ground trainer that meets the requirements of § 141.41(a)(2) may be credited for not more than 20 hours of the required 190 hours of flight time.

(b) *Flight instruction.* The course must consist of at least 75 hours of instruction in the operations listed in subparagraphs (1) through (6) of this paragraph. Instruction in a pilot ground trainer that meets the requirements of § 141.41(a)(1) may be credited for not more than 20 hours of the required 75 hours. Instruction in a pilot ground trainer that meets the requirements of § 141.41(a)(2) may be credited for not more than 10 hours of the required 75 hours.

(1) The pilot operations for the Private Pilot Course prescribed in § 3 of Appendix A of this Part.

(2) The IFR operations for the Instrument Rating Course prescribed in § 3 of Appendix C of this Part.

(3) Ten hours of flight instruction in an airplane with retractable gear, flaps, a controllable propeller, and powered by at least 180 hp. engine.

(4) Night flying, including a cross-country night flight with a landing at a point more than 100 miles from the point of departure.

(5) Normal and maximum performance takeoffs and landings using precision approaches and prescribed airplane performance speeds, including operation at maximum authorized takeoff weight.

(6) Emergency procedures appropriate to VFR and IFR flight and to the operation of complex airplane systems.

(c) *Solo practice.* The course must consist of at least 100 hours of the flights listed in subparagraphs (1) through (4) of this paragraph. Flight time as pilot in command of an airplane carrying only those persons who are pilots assigned by the school to specific flight crew duties on the flight may be credited for not more than 50 hours of that requirement.

(1) Directed solo practice on each VFR operation for which flight instruction is required (except simulated emergencies).

[(2) At least 40 hours of solo cross-country flights, each flight with a landing at a point more than 50 nautical miles from the original departure point. One flight must have landings at a minimum of three points, one of which is at least 150 nautical miles from the original departure point if the flight is conducted in Hawaii, or at least 250 nautical miles from the original departure point if it is conducted elsewhere.]

(3) At least 5 hours of pilot in command

time in an airplane described in paragraph (b) (3) of this section, including not less than 10 takeoffs and 10 landings to a full stop.

(4) At least 5 hours of night flight, including at least 10 takeoffs and 10 landings to a full stop.

4. *Stage and final tests.*

(a) *Written examinations.* Each student enrolled in the course must satisfactorily accomplish a written test upon the completion of each stage of training specified in the approved training syllabus for the commercial pilot certification course. In addition, he must satisfactorily accomplish a final stage test at the completion of all of that course. The stage and final tests may be credited for not more than 6 hours of the required 100 hours of ground training.

(b) *Flight tests.* Each student enrolled in a commercial pilot certification course (airplanes) must satisfactorily accomplish a stage flight test at the completion of each of the stages listed in subparagraphs (1), (2), (3), (4), and (5), of this paragraph. In addition, he must satisfactorily accomplish a final test at the completion of all of those stages. The stage and final tests may not be credited for more than 10 hours of the required 190 hours of flight training.

(1) Solo.

(2) Cross-country.

(3) High performance airplane operations.

(4) IFR operations.

(5) Commercial Pilot Course test, VFR and IFR.

Appendix E

Commercial Test Course (Airplanes)

1. *Applicability.* This Appendix prescribes the minimum curriculum for a commercial test course (airplanes) required by § 141.55.

2. *Experience.* For enrollment as a student in a commerical test course (airplanes) an applicant must—

(a) Hold a valid private pilot certificate;

(b) Hold a valid instrument rating, or be enrolled in an approved instrument rating course; and

(c) Have such experience and flight training that upon completion of his approved commercial test course he will meet the aeronautical experience requirements prescribed in Part 61 of this chapter for a commercial pilot certificate.

3. *Ground training.* The course must consist of at least 50 hours of ground training instruction in the following subjects:

(a) A review of the ground training sub-

jects prescribed in § 2 of Appendix A of this Part for a private pilot certification.

(b) A review of the ground training subjects prescribed in § 2 of Appendix C of this Part for an instrument rating course.

(c) The Federal Aviation Regulations covering the privileges, limitations, and operations of a commercial pilot, and the operations for which an air taxi/commercial operator, agricultural aircraft operator, and external load operator certificate, waiver or exemption is required.

(d) Basic aerodynamics, and the principles of flight that apply to airplanes.

(e) The safe and efficient operation of airplanes, including inspection and certification requirements, operating limitations, high altitude operations and physiological considerations, loading computations, the significance and use of airplane performance speeds, and computations involved in runway and obstacle clearance and crosswind component considerations.

4. *Flight training.*

(a) *General.* The course must consist of at least 25 hours of flight training prescribed in this section. Instruction in a pilot ground trainer that meets the requirements of § 141.41 (a) (1) may be credited for not more than 20 percent of the total number of hours of flight time. Instruction in a pilot ground trainer that meets the requirements of § 141.41 (a) (2) may be credited for not more than 10 percent of the total number of hours of flight time.

(b) *Flight instruction.* The course must consist of at least 20 hours of flight instruction in the subjects listed in subparagraphs (1) through (3) of this paragraph. Instruction in a ground trainer that meets the requirements of § 141.41 (a) (1) may be credited for not more than 4 hours of the required 20 hours. Instruction in a ground trainer that meets the requirements of § 141.41 (a) (2) may not be credited for more than 2 hours of the required 20 hours.

(1) A review of the VFR operations prescribed in § 3 of Appendix A of this Part for a private course.

(2) A review of the IFR operations prescribed in § 3 of Appendix C of this Part for an instrument rating course.

(3) A review of the VFR operations prescribed in § 3 (b) (3) through (6) of Appendix D of this Part for a commercial pilot certification course.

(c) *Directed solo practice.* If the course includes directed solo practice necessary to develop the flight proficiency of each student, the practice may not exceed a ratio of 3 hours of directed solo practice for each hour of the flight instruction required by the school's approved course outline.

5. *Stage and final tests.*

(a) *Written tests.* Each student enrolled in the course must satisfactorily accomplish

a stage test upon the completion of each stage of training specified in the approved training syllabus for the commercial test course. In addition, he must satisfactorily accomplish a final test at the conclusion of that course. The stage and final tests may not be credited for more than 4 hours of the required 50 hours of ground training.

(b) *Flight tests.* Each student enrolled in the course must satisfactorily accomplish a final test at the completion of the course. However, if the approved course of training exceeds 35 hours he must be given a test at an appropriate stage prior to completion of 35 hours of flight training. The flight tests may not be credited for more than 3 hours of the required hours of flight training.

(c) *Total flight experience.* The approved training course outline must specify the minimum number of hours of flight instruction and directed solo practice (if any) that is provided for each student under the requirements of paragraphs (b) and (c) of § 4 of this Appendix. The total number of hours of all flight training given to a student under this section and the minimum experience required for enrollment under § 2 of this Appendix must meet the minimum aeronautical experience requirements of § 61.129 of this chapter for the issuance of a commercial pilot certificate.

Appendix F

Rotorcraft, Gliders, Lighter-Than-Air Aircraft and Aircraft Rating Courses

A. *Applicability.* This Appendix prescribes the minimum curriculum for a pilot certification course for a rotorcraft, glider, lighter-than-air aircraft, or aircraft rating, required by § 141.55.

B. *General Requirements.* The course must be comparable in scope, depth, and detail with the curriculum prescribed in Appendixes A through D of this Part for a pilot certification course (airplanes) with the same rating. Each course must provide ground and flight training covering the aeronautical knowledge and skill items required by Part 61 of this chapter for the certificate or rating concerned. In addition, each course must meet the appropriate requirements of this Appendix.

C. *Rotorcraft.*

I. *Kinds of rotorcraft pilot certification courses.* An approved rotorcraft pilot certification course includes—

(a) A helicopter or gyroplane course—private pilots;

(b) A helicopter or gyroplane course—commercial pilots; and

(c) An instrument rating—helicopter.

II. *Helicopter or gyroplane course: private pilots.*

(a) A private pilot certification course

for helicopters or gyroplanes must consist of at least the following:

(1) Ground training—35 hours.

(2) Flight training—35 hours, including the following:

(i) Flight instruction—20 hours.

(ii) Solo practice—10 hours, including a flight with landings at three points, each of which is more than 25 nautical miles from the other two points.

(b) Stage and final tests may be credited for not more than 3 hours of the 35 hours of ground training, and for not more than 4 hours of the 35 hours of flight training required by paragraphs (a)(1) and (a)(2) of this section.

III. *Helicopter or gyroplane course—commercial pilots.*

(a) A commercial pilot certification course of training for helicopters or gyroplanes must consist of at least the following:

(1) Ground training—65 hours.

(2) Flight training—150 hours of flight training at least 50 hours of which must be in helicopters or gyroplanes. The flight training must include the following:

(i) Flight instruction—50 hours.

(ii) Directed solo—100 hours (including a cross-country flight with landings at three points, each of which is more than 50 nautical miles from the other two points).

(b) Stage and final tests may be credited for not more than 5 hours of the required 65 hours of ground training, and for not more than 7 hours of the required 150 hours of flight training prescribed in paragraphs (a)(1) and (a)(2) of this section.

IV. *Instrument rating—helicopter course.*

(a) An instrument rating — helicopter course of training must consist of at least the following:

(1) Ground training—35 hours.

(2) Instrument flight training — 35 hours. Instrument instruction in a pilot ground trainer that meets the requirements of § 141.41(a)(1) may be credited for not more than 10 hours of the required 35 hours of flight training. Instruction in a ground trainer that meets the requirements of § 141.41(a)(2) may be credited for not more than 5 hours of the required 35 hours. The instrument flight instruction must include a 100 mile simulated or actual IFR cross-country flight, and 25 hours of flight instruction.

(3) Stage and final tests may be credited for not more than 5 hours of the 35 hours of required ground training, and not more than 5 hours of the 35 hours of instrument training.

D. *Gliders.*

I. *Kinds of glider pilot certification courses.*

An approved glider certification course includes—

(a) A glider course—private pilots; and

(b) A glider course—commercial pilots.

II. *Glider course: private pilot.*

A private pilot certification course for gliders must consist of at least the following:

(a) Ground training—15 hours.

(b) Flight training—8 hours (including 35 flights if ground tows are used, or 20 flights if aero tows are used). The flight training must include the following:

(1) Flight instruction—2 hours (including 20 flights if ground tows are used or 15 flights if aero tows are used).

(2) Directed solo—5 hours (including at least 15 flights if ground tows are used or 5 flights if aero tows are used).

(c) Stage and flight tests may be credited for not more than one hour of the 15 hours of ground training, and for not more than one-half hour of the 2 hours of flight instruction required by paragraphs (a)(1) and (a)(2) of this section.

III. *Glider course: commercial pilot.*

(a) An approved commercial pilot certification course for gliders must consist of at least the following:

(1) Ground training—25 hours.

(2) Flight training—20 hours of flight time in gliders (consisting of at least 50 flights), including the following:

(i) Flight instruction—8 hours.

(ii) Directed solo—10 hours.

(b) Stage and final tests may be credited for not more than 2 hours of the 25 hours of ground training, and for not more than 2 hours of the 20 hours of flight training required by paragraphs (a)(1) and (a)(2) of this section.

E. *Lighter-than-air aircraft.*

I. *Kinds of lighter-than-air pilot certification courses.* An approved lighter-than-air pilot certification course includes—

(a) An airship course—private pilot;

(b) A free balloon course—private pilot;

(c) An airship course—commercial pilot; and

(d) A free balloon course—commercial pilot.

II. *Airship—private pilot.*

(a) A private pilot certification course for an airship must consist of at least the following:

(1) Ground training—35 hours.

(2) Flight training—50 hours (45 hours must be in airships), including the following:

(i) Flight instruction—20 hours in airships.

(ii) Directed solo, or performing the functions of a pilot in command of an airship for which more than one pilot is required—10 hours.

(b) Stage and final tests may be credited for not more than 5 hours of the 35 hours of ground training, and not more than 5 hours of the 50 hours of flight training required by paragraphs (a)(1) and (a)(2) of this section.

III. *Free balloon course; private pilot.*

(a) A private pilot course for a free balloon must consist of at least the following:

(1) *Ground training*—10 hours.

(2) Flight training—6 free flights, including—

(i) Two flights of one hour duration each if a gas balloon is used, or of 30 minutes duration if a hot air balloon is used;

(ii) At least one solo flight; and

(iii) One ascent under control to 5,000 feet above the point of takeoff if a gas balloon is used, or 3,000 feet above the point of takeoff if a hot air balloon is used.

(b) The written and stage checks may be credited for not more than one hour of the ground training, and not more than one of the 6 flights required by paragraph (a)(1) and (a)(2) of this section.

IV. *Airship course—commercial pilot.*

(a) A commercial pilot course for an airship must consist of at least the following:

(1) Ground training—100 hours.

(2) Flight training—190 hours in airships as follows:

(i) Flight instruction—80 hours, including 30 hours instrument time.

(ii) 100 hours of solo time, or flight time performing the functions or a pilot in command in an airship that requires more than one pilot, including 10 hours of cross-country flying and 10 hours of night flying.

(b) Stage and final tests may be credited for not more than 6 hours of the 100 hours of ground training, and not more than 10 hours of the 190 hours of flight training required by paragraphs (a)(1) and (a)(2), respectively, of this section.

V. *Free balloon course; commercial pilot.*

(a) A commercial pilot certification course for free balloons must consist of the following:

(1) Ground training—20 hours.

(2) Flight training—8 free flights, including—

(i) 2 flights of more than 2 hours duration if a gas balloon is used, or 2 flights of more than 1 hour duration if a hot air balloon is used;

(ii) 1 ascent under control to more than 10,000 feet above the takeoff point if a gas balloon is used, or to more than 5,000 feet above the takeoff point if a hot air balloon is used; and

(iii) 2 solo flights.

(b) Stage and final tests may be credited

for not more than 2 hours of the 20 hours of ground training, and not more than one of the flights required by paragraphs (a)(1) and (a)(2), respectively, of this section.

F. *Aircraft rating course.*

I. *Kinds of aircraft rating courses.* An approved aircraft rating course includes—

(a) An aircraft category rating;

(b) An aircraft class rating; and

(c) An aircraft type rating.

II. *Aircraft category rating.* An aircraft rating course must include at least the ground training and flight instruction required by Part 61 of this chapter for the issuance of a pilot certificate with a category rating appropriate to the course. However, the Administrator may approve a lesser number of hours of ground training, or flight instruction, or both, if the course provides for the use of special training aids, such as ground procedures, trainers, systems mockups, and audiovisual training materials, or requires appropriate aeronautical experience of the students as a prerequisite for enrollment in the course.

III. *Aircraft class rating.* An aircraft class rating course must include at least the flight instruction required by Part 61 of this chapter for the issuance of a pilot certificate with a class rating appropriate to the course.

IV. *Aircraft type rating.* An aircraft type rating course must include at least 10 hours of ground training on the aircraft systems, performance, operation, and loading. In addition, it must include at least 10 hours of flight instruction. Instruction in a pilot ground trainer that meets the requirements of § 141.41(a)(1) may be credited for not more than 5 of the 10 hours of required flight instruction. Instruction in a pilot ground trainer that meets the requirements of § 141.41 (a)(2) may be credited for not more than 2.5 of the 10 hours of required flight instruction.

Appendix G
Pilot Ground School Course

1. *Applicability.* This Appendix prescribes the minimum curriculum for a pilot ground school course required by § 141.55.

2. *General requirements.* An approved course of training for a pilot ground school course must contain the instruction necessary to provide each student with adequate knowledge of those subjects needed to safely exercise the privileges of the pilot certificate sought.

3. *Ground training instruction.* A pilot ground school course must include at least the subjects and the number of hours of ground training specified in the ground training section of the curriculum prescribed in the Appendixes to this Part for the certification or test preparation course to which the ground school course is directed.

4. *Stage and final tests.* Each student must pass a written test at the completion of each stage of training specified in the approved training syllabus for each ground training course in which he is enrolled. In addition, he must pass a final written test at the completion of the course. The stage and final tests may be credited towards the total ground training time required for each certification and test preparation course as provided in the curriculum prescribed in the Appendixes to this Part for that course.

Appendix H
Test Preparation Courses

1. *Applicability.* This Appendix prescribes the minimum curriculum required under § 141.55 for this Part for each test preparation course listed in § 141.11.

2. *General requirements.*

(a) A test preparation course is eligible for approval if the Administrator determines that it is adequate for a student enrolled in that course, upon graduation, to safely exercise the privileges of the certificate, rating, or authority for which the course is conducted.

(b) Each course for a test preparation must be equivalent in scope, depth, and detail with the curriculum for the corresponding test course prescribed in Appendixes A, B, C, and D of this Part. However, the number of hours of ground training and flight training included in the course must meet the curriculum prescribed in this Appendix. (The minimums prescribed in this Appendix for each test preparation course are based upon the amount of training that is required for students who meet the total flight experience requirements prescribed in Part 61 of this chapter at the time of enrollment.)

(c) Minimum experience, knowledge, or skill, requirements necessary as a prerequisite for enrollment are prescribed in the appropriate test preparation courses contained in this Appendix.

3. *Flight instructor certification course.*

(a) An approved course of training for a flight instructor certification course must contain at least the following:

(1) *Ground training*—40 hours.

(2) *Instructor training*—25 hours, including—

(i) 10 hours of flight instruction in the analysis and performance of flight training maneuvers;

(ii) 5 hours of practice ground instruction; and

(iii) 10 hours of practice flight instruction (with the instructor in the aircraft).

(b) *Credit for previous training of ex-*

perience: A student may be credited with the following training and experience acquired before his enrollment in the course.

(1) Satisfactory completion of two years of study on the principles of education in a college or university may be credited for 20 hours of the required 40 hours of ground training prescribed in paragraph (a)(1) of this section.

(2) One year of experience as a full-time instructor in an institution of secondary or advanced education may be credited for 5 hours of the required practice ground instruction prescribed in paragraph (a)(2) of this section.

(c) *Prerequisite for enrollment.* To be eligible for enrollment each student must hold—

(1) A commercial pilot certificate;

(2) A rating for the aircraft used in the course; and

(3) An instrument rating for enrollment in an airplane instructor rating course.

4. *Additional flight instructor rating courses.*

(a) An approved course of training for an additional flight instructor rating course must consist of at least the following:

(1) *Ground training*—20 hours.

(2) *Instructor training (with an instructor in the aircraft).* 20 hours, including—

(i) 10 hours of analysis of flight training maneuvers, or, in the case of a glider instructor rating course, 10 flights in a glider; and

(ii) 10 hours of practice flight instruction, or, in the case of glider instructor rating course, 10 flights in a glider.

5. *Additional instrument rating course (airplane or helicopter).*

(a) An approved training course for an additional instrument rating course must include at least the following:

(1) *Ground training*—15 hours.

(2) *Flight instruction*—15 hours.

(b) *Prerequisites for enrollment.* To be eligible for enrollment each student must hold a valid pilot certificate with an instrument rating, and an aircraft rating for the aircraft used in the course.

6. *Airline transport pilot test course.*

(a) An approved training course for an airline transport pilot test course must include at least the following:

(1) *Ground training*—40 hours.

(2) *Flight instruction*—25 hours, including at least 15 hours of instrument flight instruction.

(b) *Prerequisites for enrollment.* To be eligible for enrollment each student must—

(1) Hold a commercial pilot certificate

with an instrument rating and a rating for the aircraft used in the course; and

(2) Meet the experience requirements of Part 61 of this chapter for the issuance of an airline transport pilot certificate.

7. *Pilot certificate, aircraft or instrument rating refresher course.*

(a) An approved refresher training course for a pilot certificate, aircraft rating, or an instrument rating must contain at least the following:

(1) *Ground training*—4 hours.

(2) *Flight instruction*—6 hours, which may include not more than 2 hours of directed solo or pilot in command practice.

(b) *Prerequisites for enrollment.* To be eligible for enrollment each student must hold a valid pilot certificate with ratings appropriate to the refresher course.

8. *Agricultural aircraft operations course.*

(a) An approved training course for pilots of agricultural aircraft must include at least the following:

(1) *Ground training*—25 hours, including at least 15 hours on the handling of agricultural and industrial chemicals.

(2) *Flight instruction*—15 hours, which may include not more than 5 hours of directed solo practice.

(b) *Prerequisite for enrollment.* To be eligible for enrollment each student must hold a valid commercial pilot certificate with a rating for the aircraft used in the course.

9. *Rotorcraft external-load operations course.*

(a) An approved training course for pilots of a rotorcraft with an external-load must contain at least the following:

(1) *Ground training*—10 hours.

(2) *Flight instruction*—15 hours.

(b) *Prerequisite for enrollment.* To be eligible for enrollment each student must hold a valid commercial pilot certificate with a rating for the rotorcraft used in the course.

FEDERAL AVIATION REGULATIONS

Part 143—Ground Instructors

Contents

---

§ 143.1 Applicability.

This Part prescribes the requirements for issuing ground instructor certificates and associated ratings and the general operating rules for the holders of those certificates and ratings.

§ 143.3 Application and issue.

(a) An application for a certificate and rating, or for an additional rating, under this Part, is made on a form and in a manner prescribed by the Administrator. However, a person whose ground instructor certificate has been revoked may not apply for a new certificate for a period of one year after the effective date of the revocation unless the order of revocation provides otherwise.

(b) An applicant who meets the requirements of this Part is entitled to an appropriate certificate with ratings naming the ground school subjects that he is authorized to teach.

(c) Unless authorized by the Administrator, a person whose ground instructor certificate is suspended may not apply for any rating to be added to that certificate during the period of suspension.

(d) Unless the order of revocation provides otherwise, a person whose ground instructor certificate is revoked may not apply for any ground instructor certificate for one year after the date of revocation.

§ 143.5 Temporary certificate.

A certificate or rating effective for a period of not more than 90 days may be issued to a qualified applicant, pending the issue of the certificate or rating for which he applied.

§ 143.7 Duration of certificate.

(a) A certificate or rating issued under this Part is effective until it is surrendered, suspended, or revoked.

(b) The holder of any certificate issued under this Part that is suspended or revoked shall, upon the Administrator's request, return it to the Administrator.

§ 143.8 Change of name; replacement of lost or destroyed certificate.

(a) An application for a change of name on a certificate issued under this Part must be accompanied by the applicant's current certificate and the marriage license, court order, or other document verifying the change. The documents are returned to the applicant after inspection.

(b) An application for a replacement of a lost or destroyed certificate is made by letter to the Department of Transportation, Federal Aviation Administration, Airman Certification Branch, P.O. Box 25082, Oklahoma City, Okla. 73125. The letter must—

(1) Contain the name in which the certificate was issued, the permanent mailing address (including zip code), social security number (if any), and date and place of birth of the certificate holder, and any available information regarding the grade, number, and date of issue of the certificate, and the ratings on it; and

(2) Be accompanied by a check or money order for $2.00, payable to the Federal Aviation Administration.

(c) A person whose certificate issued under this Part has been lost may obtain a telegram from the FAA confirming that it was issued. The telegram may be carried as a certificate for a period not to exceed 60 days pending his receiving a duplicate certificate under paragraph (b) of this section, unless he has been notified that the certificate has been suspended or revoked. The request for such a telegram may be made by prepaid telegram, stating the date upon which a duplicate certificate was requested, or including the request for a duplicate and a money order for the necessary amount. The request for a telegraphic certificate should be sent to the office prescribed in **paragraph (b) of this section.**

§ 143.9 Eligibility requirements: general.

To be eligible for a certificate under this Part, a person must be at least 18 years of age, be of good moral character, and comply with § 143.11.

§ 143.11 Knowledge requirements.

Each applicant for a ground instructor certificate must show his practical and theoretical knowledge of the subject for which he seeks a rating by passing a written test on that subject.

§ 143.15 Tests: general procedures.

(a) Tests prescribed by or under this Part are given at times and places, and by persons, designated by the Administrator.

(b) The minimum passing grade for each test is 70 percent.

§ 143.17 Retesting after failure.

An applicant for a ground instructor rating who fails a test under this Part may apply for retesting—

(a) After 30 days after the date he failed that test; or

(b) Upon presenting a statement from a certificated ground instructor, rated for the subject of the test failed, certifying that he has given the applicant at least five hours additional instruction in that subject and now considers that he can pass the test.

§ 143.18 Written tests: cheating or other unauthorized conduct.

(a) Except as authorized by the Administrator, no person may—

(1) Copy, or intentionally remove, a written test under this Part;

(2) Give to another, or receive from another, any part or copy of that test;

(3) Give help on that test to, or receive help on that test from, any person during the period that test is being given;

(4) Take any part of that test in behalf of another person;

(5) Use any material or aid during the period that test is being given; or

(6) Intentionally cause, assist, or participate in any act prohibited by this paragraph.

(b) No person who commits an act prohibited by paragraph (a) of this section is eligible for any airman or ground instructor certificate or rating under this chapter for a period of one year after the date of that act. In addition, the commission of that act is a basis for suspending or revoking any airman or ground instructor certificate or rating held by that person.

§ 143.19 Recent experience.

The holder of a ground instructor certificate may not perform the duties of a ground instructor unless, within the 12 months before he intends to perform them—

(a) He has served for at least three months as a ground instructor; or

(b) The Administrator has determined that he meets the standards prescribed in this Part for the certificate and rating.

§ 143.20 Applications, certificates, logbooks, reports, and records: falsification, reproduction, or alteration.

(a) No person may make or cause to be made—

(1) Any fraudulent or intentionally false statement on any application for a certificate or rating under this Part;

(2) Any fraudulent or intentionally false entry in any logbook, record, or report that is required to be kept, made, or used, to show compliance with any requirement for any certificate or rating under this Part;

(3) Any reproduction, for fraudulent purpose, of any certificate or rating under this Part; or

(4) Any alteration of any certificate or rating under this Part.

(b) The commission by any person of an act prohibited under paragraph (a) of this section is a basis for suspending or revoking any airman or ground instructor certificate or rating held by that person.

§ 143.21 Display of certificate.

Each person who holds a ground instructor certificate shall keep it readily available to him while instructing and shall present it for inspection upon the request of the Administrator or an authorized representative of the National Transportation Safety Board, or of any Federal, State, or local law enforcement officer.

§ 143.23 Change of address.

Within 30 days after any change in his permanent mailing address, the holder of a ground instructor certificate shall notify the Department of Transportation, Federal Aviation Administration, Airman Certification Branch, P.O. Box 25082, Oklahoma City, Okla. 73125 in writing, of his new address.

NATIONAL TRANSPORTATION SAFETY BOARD

PART 830—NOTIFICATION AND REPORTING OF AIRCRAFT ACCIDENTS OR INCIDENTS AND OVERDUE AIRCRAFT, AND PRESERVATION OF AIRCRAFT WRECKAGE, MAIL, CARGO, AND RECORDS

Subpart A—General

Sec.
830.1   Applicability.
830.2   Definitions.

Subpart B—Initial Notification of Aircraft Accidents, Incidents, and Overdue Aircraft

830.5   Immediate notification.
830.6   Information to be given in notification.

Subpart C—Preservation of Aircraft Wreckage, Mail, Cargo, and Records

830.10   Preservation of aircraft wreckage, mail, cargo, and records.

Subpart D—Reporting of Aircraft Accidents, Incidents, and Overdue Aircraft

830.15   Reports and statement to be filed.

Subpart E—Reporting of Public Aircraft Accidents and Incidents

830.20   Reports to be filed.

Subpart A—General

§ 830.1   Applicability.

This part contains rules pertaining to:

(a) Notification and reporting aircraft accidents and incidents and certain other occurrences in the operation of aircraft when they involve civil aircraft of the United States wherever they occur, or foreign civil aircraft when such events occur in the United States, its territories or possessions.

(b) Reporting aircraft accidents and listed incidents in the operation of aircraft when they involve certain public aircraft.

(c) Preservation of aircraft wreckage, mail, cargo, and records involving all civil aircraft in the United States, its territories or possessions.

§ 830.2   Definitions.

As used in this part the following words or phrases are defined as follows:

"Aircraft accident" means an occurrence associated with the operation of an aircraft which takes place between the time any person boards the aircraft with the intention of flight and all such persons have disembarked, and in which any person suffers death or serious injury, or in which the aircraft receives substantial damage.

"Civil aircraft" means any aircraft other than a public aircraft.

"Fatal injury" means any injury which results in death within 30 days of the accident.

"Incident" means an occurrence other than an accident, associated with the operation of an aircraft, which affects or could affect the safety of operations.

"Operator" means any person who causes or authorizes the operation of an aircraft, such as the owner, lessee, or bailee of an aircraft.

"Public aircraft" means an aircraft used exclusively in the service of any government or of any political subdivision thereof, including the government of any State, Territory, or possession of the United States, or the District of Columbia, but not including any government-owned aircraft engaged in carrying persons or property for commercial purposes. For purposes of this section "used exclusively in the service of" means, for other than the Federal Government, an aircraft which is owned and operated by a governmental entity for other than commercial purposes or which is exclusively leased by such governmental entity for not less than 90 continuous days.

"Serious injury" means any injury which (1) requires hospitalization for more than 48 hours, commencing within 7 days from the date of the injury was received; (2) results in a fracture of any bone (except simple fractures of fingers, toes, or nose); (3) causes severe hemorrhages, nerve, muscle, or tendon damage; (4) involves any internal organ; or (5) involves second or third-degree burns, or any burns affecting more than 5 percent of the body surface.

"Substantial damage" means damage or failure which adversely affects the structural strength, performance, or flight characteristics of the aircraft, and which would normally require major repair or replacement of the affected component. Engine failure or damage limited to an engine if only one engine fails or is damaged, bent fairings or cowling, dented skin, small punctured holes in the skin or fabric, ground damage to rotor or propeller blades and damage to landing gear, wheels, tires, flaps, engine accessories, brakes, or wingtips are not considered "substantial damage" for the purpose of this part.

Subpart B—Initial Notification of Aircraft Accidents, Incidents, and Overdue Aircraft

§ 830.5   Immediate notification.

The operator of an aircraft shall immediately, and by the most expeditious means available, notify the nearest National Transportation Safety Board (Board), field office[1] when:

(a) An aircraft accident or any of the following listed incidents occur:

(1) Flight control system malfunction or failure;

(2) Inability of any required flight crewmember to perform normal flight duties as a result of injury or illness;

(3) Failure of structural components of a turbine engine excluding compressor and turbine blades and vanes;

(4) In-flight fire; or

(5) Aircraft collide in flight.

(6) Damage to property, other than the aircraft, estimated to exceed $25,000 for repair (including materials and labor) or fair market value in the event of total loss, whichever is less.

(7) For large multiengine aircraft (more than 12,500 pounds maximum certificated takeoff weight):

(i) In-flight failure of electrical systems which requires the sustained use of an emergency bus powered by a back-up source such as a battery, auxiliary power unit, or air-driven generator to retain flight control or essential instruments;

(ii) In-flight failure of hydraulic systems that results in sustained reliance on the sole remaining hydraulic or mechanical system for movement of flight control surfaces;

(iii) Sustained loss of the power or thrust produced by two or more engines; and

(iv) An evacuation of an aircraft in which an emergency egress is utilized.

(b) An aircraft is overdue and is believed to have been involved in an accident.

§ 830.6   Information to be given in notification.

The notification required in § 830.5 shall contain the following information, if available:

(a) Type, nationality, and registration marks of the aircraft;

(b) Name of owner, and operator of the aircraft;

(c) Name of the pilot-in-command;

---

[1] The National Transportation Safety Board field offices are listed under U.S. Government in the telephone directories in the following cities: Anchorage, Alaska; Atlanta, Ga.; Chicago, Ill.; Denver Colo.; Fort Worth, Tex.; Kansas City, Mo.; Los Angeles, Calif.; Miami, Fla.; New York, N.Y.; Seattle, Wash.

(d) Date and time of the accident;

(e) Last point of departure and point of intended landing of the aircraft;

(f) Position of the aircraft with reference to some easily defined geographical point;

(g) Number of persons aboard, number killed and number seriously injured;

(h) Nature of the accident, the weather and the extent of damage to the aircraft, so far as is known; and

(i) A description of any explosives, radioactive materials, or other dangerous articles carried.

Subpart C—Preservation of Aircraft Wreckage, Mail, Cargo, and Record

§ 830.10 Preservation of aircraft wreckage, mail, cargo, and records.

(a) The operator of an aircraft involved in an accident or incident for which notification must be given is responsible for preserving to the extent possible any aircraft wreckage, cargo, and mail aboard the aircraft, and all records, including all recording mediums of flight, maintenance, and voice recorders, pertaining to the operation and maintenance of the aircraft and to the airmen until the Board takes custody thereof or a release is granted pursuant to § 831.10(b).

(b) Prior to the time the Board or its authorized representative takes custody of aircraft wreckage, mail, or cargo, such wreckage, mail, or cargo may not be disturbed or moved except to the extent necessary:

(1) To remove persons injured or trapped;

(2) To protect the wreckage from further damage; or

(3) To protect the public from injury.

(c) Where it is necessary to move aircraft wreckage, mail or cargo, sketches, descriptive notes, and photographs shall be made, if possible, of the original position and condition of the wreckage and any significant impact marks.

(d) The operator of an aircraft involved in an accident or incident shall retain all records, reports, internal documents, and memoranda dealing with the accident or incident, until authorized by the Board to the contrary.

Subpart D—Reporting of Aircraft Accidents, Incidents, and Overdue Aircraft

§ 830.15 Reports and statements to be filed.

(a) *Reports.* The operator of an aircraft shall file a report on Board Form 6120.1 (OMB No. 3147-005) or Board Form 7120.2 (OMB No. 3147-0001)[2] within 10 days after an accident, or after 7 days if an overdue aircraft is still missing. A report on an incident for which notification is required by § 830.5(a) shall be filed only as requested by an authorized representative of the Board.

(b) *Crewmember statement.* Each crewmember, if physically able at the time the report is submitted, shall attach a statement setting forth the facts, conditions and circumstances relating to the accident or incident as they appear to him. If the crewmember is incapacitated, he shall submit the statement as soon as he is physically able.

(c) *Where to file the reports.* The operator of an aircraft shall file any report with the field office of the Board nearest the accident or incident.

Subpart E—Reporting of Public Aircraft Accidents and Incidents

§ 830.20 Reports to be filed.

The operator of a public aircraft other than an aircraft of the Armed Forces or Intelligence Agencies shall file a report on NTSB Form 6120.1 (OMB No. 3147-001)[3] within 10 days after an accident or incident listed in § 830.5(a). The operator shall file the report with the field office of the Board nearest the accident or incident.[4]

[2]Forms are obtainable from the Board field offices (see footnote 1), the National Transportation Safety Board, Washington, DC 20594, and the Federal Aviation Administration, Flight Standards District Office.

[3]To obtain this form see footnote 2.

[4]The location of the Board's field offices are set forth in footnote 1.

Proposed Part 91 Revision

Finding your way through the regulatory maze

The midsection of the Federal Aviation Regulations known as Part 91 is almost as familiar to and frequently consulted by serious pilots as a family Bible. Material covering flight operations was codified in a very logical and systematic form from the pre-existing Civil Aviation Regulations in 1963. The part was assigned 311 rule slots (91.1, 91.2, etc., on up to 91.311) with alternate numbers left blank for later insertions. Like material was grouped into five subparts, covering (A) general information, (B) flight rules, (C) maintenance, (D) large and turbine-powered airplanes, and (E) operating noise limits. After less than 25 years of existence in its present form it is about to be revised: why?

Because many pilots have complained recently that looking up the appropriate rule can be excessively time-consuming, as some rules with related information are no longer adjacently located; some are extremely complicated, and some require the services of an aviation lawyer to interpret.

To some extent this is true, perhaps inevitably so. With the very rapid expansion of aviation in the 60's and 70's, rulemaking quickly filled in vacant number slots. Soon it became necessary to locate new rules at a considerable distance from related material—sometimes even in a different subpart. Consequently, it can be necessary to inspect almost the entire Part 91 in order to be sure of identifying all of the regulatory matter concerning one specific subject. In some cases, alternately, what would have been a new rule normally was tacked on to an existing rule to which it was more or less related. The result of this procedure was often confusing as to the intent of the rule, as well as to its meaning. Rules were "stretched" to a maximum, divided into paragraphs, sections, subsections, etc., in order to avoid locating new rules remotely.

Solving this problem of overcrowding is not an easy matter, given the obligatory standard form of rule codification in the Federal Government. Once enacted, rules cannot be readily abolished or altered numerically because many rules refer to other rules by number. To realign and simplify Part 91 it has been necessary to carry out a complete overhaul, which has taken place over the past two years, after numerous consultations with the general flying public and the aviation industry. The proposed form of this revision was published in the *Federal Register* dated March 20, 1985.

The proposed revised format is very similar to the original, except that there are now 10 subparts instead of five, and 134 rules contained within 899 number slots. All of the even numbers are initially left blank, and there are also generous blocks of unused (reserved) rule numbers for rule development until well into the next century. This should provide ample room for rule development until well into the next century.

The proposed reorganization is a boon to general aviation airmen, since the new subparts A through E would pertain mostly to VFR pilots flying small reciprocating engine airplanes. Material found in subparts F through J would apply primarily to operators of larger aircraft. If enacted, the new Part 91 would outline like this:

- Subpart A—General §91.1–§91.25
- Subpart B—Flight Rules §91.101–§91.193
- Subpart C—Equipment, Instrument, and Certificate Requirements §91.201–§91.219
- Subpart D—Special Flight Operations §91.301–§91.323
- Subpart E—Maintenance, Preventive Maintenance, and Alterations §91.401–§91.421
- Subpart F—Recoverable Expenses for Certain Operators §91.501
- Subpart G—Additional Equipment and Operating Requirements for Large and Transport Category Aircraft §91.601–§91.645
- Subpart H—Foreign Aircraft Operations and Operations of U.S. Registered Civil Aircraft to Areas Outside of the U.S. §91.701–§91.715
- Subpart I—Operating Noise Limits §91.801–§91.821
- Subpart J—Waivers §91.901–§91.905

Current Appendices A, B, and C would remain the same. Note that each subpart is set out in hundreds; i.e., all flight rules would be found in the 91.200's, all maintenance rules in the 91.400's and so on. The final significant difference in the proposal is the use of "he or she" and "him or her" when referring to airmen.

In addition to the realignment, FAA took the opportunity to propose a few changes in some of the rules. Most of these alterations would fall under the category of minor editorial revisions, designed to clarify a term or promote consistency. For example, proposed §91.107(b) (current §91.14—Use of safety belts) inserts the word "approved" before any mention of "seat or berth" to make it unmistakably clear that unapproved seats or berths—such as suitcases or cargo boxes—may not be used by occupants.

One change for the sake of consistency concerns the use of "statute miles" in sections of Part 91 that refer to instrument flight: these references have been changed to "nautical miles" for commonality with instrument charts.

There are also a few significant proposed changes that bear particular notice because of the benefits offered airmen and operators. Proposed §91.117, Aircraft Speed (current §91.70) would increase the maximum speed allowed in an airport traffic area for a reciprocating engine aircraft from 156 knots to 200 knots. This would allow easier mixing of multiengine reciprocating aircraft with the increasing numbers of small, turbine-power business aircraft.

Current §91.97, Positive Control Areas and Route Segments, requires that operators who desire for whatever reasons to deviate from the equipment or other requirements of this section submit such a request in writing at least four days in advance of the operation. Recognizing that current telecommunications and air traffic control equipment are more flexible, the proposed §91.35 would require a 48-hour oral notification.

Proposed §91.409 (replacing current §91.169, Inspections) would allow operators of turbine-powered rotorcraft to use alternate inspection programs—other than the annual or the 100-hour—provided the program is FAA-approved.

A final proposed change of note is that instead of separate references to General Aviation District Offices, Air Carriers District Offices, and Flight Standards District Offices, all FAA field offices would be called (in the rule) Flight Standards district offices. Each of the three offices can perform all of their separate functions, so the FAA felt a little streamlining of language would aid in the simplification of Part 91.

Following is a table, which, when used with a copy of the current Part 91 and the proposed Part 91, will help you understand the proposed reorganization as it now stands. Some changes are likely to be made before final enactment of the rule, as a result of comments, but the table can be updated by hand for subsequent guidance.

[Editor's Note: At press time, this revision was awaiting final adoption by the FAA.]

A COMPARISON OF CURRENT PART 91 AND THE PROPOSED CHANGES

MATERIAL IN OLD SECTION . . .	MOVES TO NEW SECTION . . .	OLD SECTION NUMBER NOW COVERS . . .	MATERIAL IN OLD SECTION . . .	MOVES TO NEW SECTION . . .	OLD SECTION NUMBER NOW COVERS . . .
SUBPART A (General)		**SUBPART A** (General)	§91.9 Careless or reckless operation	§91.13	§91.9 Civil aircraft operating limitations and marking requirements
§91.1 Applicability	§91.1 and §91.703	§91.1 Applicability	§91.10 Careless or reckless operation other than for the purpose of air navigation	§91.13	Not in use
§91.2 Certificate of authorization for Category II operations	§91.193	Not in use	§91.11 Liquor and drugs	§91.17	§91.11 Prohibition against interference with crewmembers
§91.3 Responsibility & authority of the pilot-in-command	Unchanged	Unchanged	§91.12 Carriage of narcotic drugs, marihuana, and depressant or stimulant drugs or substances	§91.19	Not in use
§91.4 Pilot-in-command of aircraft requiring more than one pilot	§91.5	Not in use			
§91.5 Preflight Action	§91.103	§91.5 Pilot-in-command of aircraft requiring more than one pilot	§91.13 Dropping objects	§91.15	§91.13 Careless or reckless operation
§91.6 Category II and III operations: General operating rules	§91.189	Not in use	§91.14 Use of safety belts	§91.107	Not in use
§91.7 Flight crewmembers at stations	§91.105	§91.7 Civil aircraft airworthiness	§91.15 Parachutes and parachuting	§91.307	§91.15 Dropping objects
§91.8 Prohibition against interference with crewmembers	§91.11	Not in use	§91.17 Towing: Gliders	§91.309	§91.17 Liquor and drugs
			§91.18 Towing: Other . . .	§91.311	Not in use

MATERIAL IN OLD SECTION . . .	MOVES TO NEW SECTION . . .	OLD SECTION NUMBER NOW COVERS . . .	MATERIAL IN OLD SECTION . . .	MOVES TO NEW SECTION . . .	OLD SECTION NUMBER NOW COVERS . . .
§91.19 Portable electronic devices	§91.21	§91.19 Carriage of narcotic drugs, marihuana, depressant or stimulant drugs or substances	§91.47 Emergency exits for airplanes carrying passengers for hire	§91.607	Reserved
§91.20 Operations within the North Atlantic Minimum Navigation Performance Specifications Airspace	§91.705	Not in use	§91.49 Aural speed warning device	§91.603	
			§91.50 reserved	Deleted	
§91.21 Flight Instruction: Simulated instrument flight and certain flight tests	§91.109	§91.21 Portable electronic devices	§91.51 Altitude alerting system or device; turbojet powered civil airplanes	§91.219	
§91.22 Fuel requirements for flight under VFR	§91.151	Not in use	§91.52 Emergency locator transmitters	§91.207	
§91.23 Fuel requirements for flight in IFR conditions	§91.167	§91.23 Truth in leasing clause requirement in leases and conditional sales contracts	§91.53 Reserved	Deleted	
			§91.54 Truth in leasing clause requirement in leases and conditional sales contracts	§91.23	
§91.24 ATC transponder and altitude reporting equipment use	§91.215	Not in use	§91.55 Civil aircraft sonic boom	§91.817	
§91.25 VOR equipment check for IFR operations	§91.171	§91.25 Aviation Safety Reporting Program: Prohibition against use of reports for enforcement purposes	§91.56 Agricultural and firefighting airplanes; noise operating limitations	§91.815	
			§91.57 Aviation Safety Reporting Program: Prohibition against using reports for enforcement purposes	§91.25	
§91.27 Civil aircraft: Certifications required	§91.203	§91.27–91.99 Reserved			
§91.28 Special flight authorizations for foreign civil aircraft	§91.715		§91.58 Material for compartment interiors	§91.613	
§91.29 Civil aircraft airworthiness	§91.7		§91.59 Carriage of candidates in Federal elections	§91.321	
§91.30 Inoperable instruments and equipment for multiengine aircraft	§91.213		**SUBPART B** (Flight Rules)		
§91.31 Civil aircraft operating limitations and marking requirements	§91.9		§91.61 Applicability	§91.101	
			§91.63 Waivers	§91.903	
§91.32 Supplemental Oxygen	§91.211		§91.65 Operating near other aircraft	§91.111	
§91.33 Powered civil aircraft with standard category U.S. airworthiness certificates: Instrument and equipment requirements	§91.205		§91.67 Right of way rules; except water operations	§91.113	
			§91.69 Right of way rules; water operations	§91.115	
§91.34 Category II manual	§91.191		§91.70 Aircraft speed	§91.117	
§91.35 Flight recorders and cockpit voice recorders	§91.609		§91.71 Acrobatic flight	§91.303	
§91.36 Data correspondence between automatically reported pressure altitude data and the pilot's altitude reference	§91.217		§91.73 Aircraft lights	§91.209	
			§91.75 Compliance with ATC clearances and instructions	§91.123	
			§91.77 ATC light signals	§91.125	
			§91.79 Minimum safe altitudes	§91.119	
			§91.81 Altimeter settings	§91.121	
§91.37 Transport category civil airplane weight limitations	§91.605		§91.83 Flight plan; information required	§91.153 and 91.169	
§91.38 Increased maximum certificated weights for certain airplanes operating in Alaska	§91.323		§91.84 Flights between Mexico or Canada and the U.S.	§91.707	
			§91.85 Operating on or in the vicinity of an airport	§91.127	
§91.39 Restricted category civil aircraft; operating limitations	§91.313		§91.87 Operation at airports with an operating control tower	§91.129	
§91.40 Limited category civil aircraft; operating limitations	§91.315		§91.89 Operation at airports without control towers	§91.127	
§91.41 Provisionally certificated civil aircraft; operating limitations	§91.317		§91.90 Terminal Control areas	§91.131	
§91.42 Aircraft having experimental certificates; operating limitations	§91.319		§91.91 Temporary flight restrictions	§91.137	
			§91.93 Flight test areas	§91.305	
§91.43 Special rules for foreign civil aircraft	§91.711		§91.95 Restricted and prohibited areas	§91.133	
§91.45 Authorization for ferry flights with one engine inoperative	§91.611		§91.97 Positive control areas and route segments	§91.135	

MATERIAL IN OLD SECTION...	MOVES TO NEW SECTION...	OLD SECTION NUMBER NOW COVERS...
§91.100 Emegency air traffic rules	§91.139	▼
§91.101 Operations to Cuba	§91.709	**SUBPART B** (Flight Rules)
		§91.101 Applicability
§91.102 Flight limitation in the proximity of space flight recovery operations	§91.143	Not in use
§91.103 Operation of civil aircraft of Cuban registry	§91.713	§91.103 Preflight action
§91.104 Flight restrictions in the proximity of the Presidential and other parties	§91.141	Not in use
Visual Flight Rules		
§91.105 Basic VFR weather minimums	§91.155	§91.105 Flight crewmembers at stations
§91.107 Special VFR weather minimums	§91.157	§91.107 Use of safety belts
§91.109 VFR cruising altitude or flight level	§91.159	§91.109 Flight instruction; simulated instrument flight and certain flight tests
§91.111 none		§91.111 Operating near other aircraft
§91.113 none		§91.113 Right of way rules; except water operations
Instrument Flight Rules		
§91.115 ATC clearance and flight plan required	§91.173	§91.115 Right of way rules; water operations
§91.116 Takeoff and landing under IFR	§91.175	Not in use
§91.117 Reserved		§91.117 Aircraft speed
§91.119 Minimum altitudes for IFR operations	§91.177	§91.119 Minimum safe altitudes; general
§91.121 IFR cruising altitude or flight level	§91.179	§91.121 Altimeter settings
§91.123 Course to be flown	§91.181	§91.123 Compliance with ATC clearance and instructions
§91.125 IFR radio communications	§91.183	§91.125 ATC light signals
§91.127 IFR operations; two way radio communications failure	§91.185	§91.127 Operating on or in the vicinity of an airport; general rules
§91.129 Operation under IFR in controlled airspace; malfunction reports	§91.187	§91.129 Operations at airports with operating control towers
§91.131–91.159		§91.131 Terminal Control Areas
		§91.133 Restricted and prohibited areas
		§91.135 Positive control areas and route segments
		§91.137 Temporary flight restrictions
		§91.139 Emergency air traffic rules
		§91.141 Flight restrictions in the proximity of the Presidential and other parties
		§91.143 Flight limitations in the proximity of space flight recovery operations
		§91.145–91.149 Reserved
		Visual Flight Rules
		§91.151 Fuel requirements for flight under VFR

NONEXISTENT

MATERIAL IN OLD SECTION...	MOVES TO NEW SECTION...	OLD SECTION NUMBER NOW COVERS...
		§91.153 VFR flight plan; information required
		§91.155 Basic VFR weather minimums
		§91.157 Special VFR weather minimums
		§91.159 VFR cruising altitude or flight level
SUBPART C (Maintenance, Preventive Maintenance, and Alterations)		
§91.161 Applicability	§91.401	§91.161 Reserved
§91.163 General	§91.403	§91.163 Reserved
§91.165 Maintenance required	§91.405	§91.165 Reserved
§91.167 Operation after maintenance, preventive maintenance, or alterations	§91.407	**Instrument Flight Rules**
		§91.167 Fuel requirements for flight in IFR conditions
§91.169 Inspections	§91.409	§91.169 IFR flight plans; information required
§91.170 Changes to aircraft inspection programs	§91.415	Not in use
§91.171 Altimeter system and altitude reporting equipment tests and inspection	§91.411	§91.171 VOR equipment check for IFR operations
§91.172 ATC transponder tests and inspections	§91.413	Not in use
§91.173 Maintenance records	§91.417	§91.173 ATC clearance and flight plan required
§91.174 Transfer of maintenance records	§91.419	Not in use
§91.175 Rebuilt engine maintenance records	§91.421	§91.175 Takeoff and landing under IFR
§91.177–91.179		§91.177 Minimum altitudes for IFR operations
Nonexistent		§91.179 IFR cruising altitude or flight level
SUBPART D (Large and Turbine-powered Multiengine Airplanes)		
§91.181 Applicability	§91.501	§91.181 Course to be flown
§91.183 Flying equipment and operating information	§91.615	§91.183 IFR radio communications
§91.185 Familiarity with operating limitations and emergency equipment	§91.617	§91.185 IFR operations; two way radio communications failure
§91.187 Equipment requirements; over the top or night VFR operations	§91.619	§91.187 Operation under IFR in controlled airspace; malfunction reports
§91.189 Survival equipment for overwater operations	§91.621	§91.189 Category II and III operations; general operating rules
§91.191 Radio equipment for overwater operations	§91.623	§91.191 Category II manual
§91.193 Emergency equipment	§91.625	§91.193 Certificate of authorization for certain Category II operations
§91.195 Flight altitude rules	§91.627	§91.195–91.199 Reserved
§91.197 Smoking and safety belt signs	§91.629	
§91.199 Passenger briefing	§91.631	
§91.200 Shoulder harness	§91.633	
§91.201 Carry-on baggage	§91.635	**SUBPART C** (Equipment, Instrument, and Certificate Requirements)
		§91.201 Applicability

MATERIAL IN OLD SECTION ...	MOVES TO NEW SECTION ...	OLD SECTION NUMBER NOW COVERS ...
§91.203 Carriage of cargo	§91.637	§91.203 Civil aircraft; certifications required
§91.205 Transport category airplane weight limitations	Deleted	§91.205 Powered civil aircraft with standard category U.S. airworthiness certificates; instrument and equipment requirements
§91.207 None		§91.207 Emergency locator transmitters
§91.209 Operating in icing conditions	§91.639	§91.209 Aircraft lights
§91.211 Flight engineer requirements	§91.641	§91.211 Supplemental oxygen
§91.213 Second-in-command requirements	§91.643	§91.213 Inoperable instruments and equipment for multiengine aircraft
§91.215 Flight attendant requirements	§91.645	§91.215 ATC transponder and altitude reporting equipment use
§91.217–91.299 None		§91.217 Data correspondence between automatically reported pressure altitude data and the pilot's altitude reference
Nonexistent		§91.219 Altitude alerting system or device; turbo jet-powered civil airplanes
		§91.221–91.299 Reserved

MATERIAL IN OLD SECTION ...	MOVES TO NEW SECTION ...	OLD SECTION NUMBER NOW COVERS ...
SUBPART E (Operating Noise Limits)		SUBPART D (Special Flight Operations)
§91.301 Applicability; relation to Part 36	§91.801	§91.301 Applicability
§91.302 Part 125 operators; designation of applicable regulations	§91.803	Not in use
§91.303 Final compliance; subsonic airplanes	§91.805	§91.303 Aerobatic flight
§91.305 Phased compliance under Parts 121 and 135; subsonic airplanes	§91.807	§91.305 Flight test areas
§91.306 Replacement airplanes	§91.809	Not in use
§91.307 Service to small communities exemption; two-engine, subsonic airplanes	§91.811	§91.307 Parachutes and parachuting
§91.308 Compliance plans and status; U.S. operators of subsonic airplanes	§91.813	Not in use
§91.309 Civil supersonic airplanes that do not comply with Part 36	§91.819	§91.309 Towing: Gliders
§91.311 Civil supersonic airplanes: Noise limits	§91.821	§91.311 Towing: Other than §91.309
§91.311 **ENDS CURRENT FAR PART 91**		

SECTION NUMBERS NOT FOUND IN EXISTING PART 91

NEW SECTION NUMBERS...

§91.313 Restricted category civil aircraft; operating limitations

§91.315 Limited category civil aircraft; operating limitations

§91.317 Provisionally certified civil aircraft operating limitations

§91.319 Aircraft having experimental certificates

§91.321 Carriage of candidates in Federal elections

§91.323 Increased maximum certificated weights for certain airplanes operated in Alaska

§91.325–91.399 Reserved

SUBPART E—Maintenance, Preventive Maintenance, and Alterations
§91.401 Applicabiity

§91.403 General

§91.405 Maintenance required

§91.407 Operatons after maintenance

§91.409 Inspections

§91.411 Altimeter system and altitude reporting equipment tests and inspections

§91.413 ATC transponder tests and inspections

§91.415 Changes to aircraft inspection programs

§91.417 Maintenance records

§91.419 Transfer of maintenance records

§91.421 Rebuilt engine maintenance records

§91.423–91.499 Reserved

SUBPART F—Recoverable expenses for certain operations
§91.501 Applicability

§91.503–91.599 Reserved

SUBPART G—Additional Equipment and Operating Requirements for Large and Transport Category Aircraft
§91.601 Applicability

§91.603 Aural speed warning device

§91.605 Transport category civil airplane weight limitations

§91.607 Emergency exits for airplanes carrying passengers for hire

§91.609 Flight recorders and cockpit voice recorders

§91.611 Authorization for ferry flight with one engine inoperative

§91.613 Materials for compartment interiors

§91.615 Flying equipment and operating information

§91.617 Familiarity with operating limitations and emergency equipment

§91.619 Equipment requirements; over the top or night VFR operations

§91.621 Survival equipment for overwater operations

§91.623 Radio equipment for overwater operations

§91.625 Emergency equipment

§91.627 Flight altitude rules

§91.629 Smoking and safety belts

§91.631 Passenger briefing

§91.633 Shoulder harness

§91.635 Carry-on baggage

§91.637 Carriage of cargo

§91.639 Operating in icing conditions

§91.641 Flight engineer requirements

§91.643 Second-in-command requirements

§91.645 Flight attendant requirements

§91.647–91.699 Reserved

SUBPART H—Foreign Aircraft Operations and Operations of U.S. Registered Civil Aircraft to Areas Outside of the United States
§91.701 Applicability

§91.703 Operations of civil aircraft of U.S. registry outside of the U.S.

§91.705 Operations within North Atlantic MNPS airspace

§91.707 Flights between Mexico or Canada and the U.S.

§91.709 Operations to Cuba

§91.711 Special rules for foreign civil aircraft

§91.713 Operation of civil aircraft of Cuban registry

§91.715 Special flight authorization for foreign civil aircraft

§91.717–91.799 Reserved

SUBPART I—Operating Noise Limits
§91.801 Applicability: Relation to Part 36

§91.803 Part 125 operators; designation of applicable regulations

§91.805 Final compliance; subsonic airplanes

§91.807 Phased compliance under Parts 121, 125, and 135; subsonic airplanes

§91.809 Replacement airplanes

§91.811 Service to small communities; two-engine subsonic airplanes

§91.813 Compliance plans and status; U.S. operations of subsonic airplanes

§91.815 Agricultural and firefighting airplanes

§91.817 Civil aircraft sonic boom

§91.819 Civil supersonic airplanes that do not comply with Part 36

§91.821 Civil supersonic airplanes; noise limits

§91.823–91.899 Reserved

SUBPART J—Waivers
§91.901 Applicability

§91.903 Policy and procedures

§91.905 List of rules subject to waiver

§91.907–91.999 Reserved

END OF PROPOSED PART 91 ∎

FAR Violation Sanctions*

Extracted from *FAA Compliance and Enforcement Program Handbook* (4/4/88)

Individuals and General Aviation - Owners, Pilots, Repair Stations, Maintenance Personnel.

A. *Owners and Operators Other Than Required Crewmembers.*

1. Failure to comply with airworthiness directives—*Moderate to maximum civil penalty.*
2. Failure to perform or improper performance of maintenance, including required maintenance—*Moderate to maximum civil penalty.*
3. Failure to make proper entries in aircraft logs—*Minimum to moderate civil penalty.*
4. Operation of aircraft beyond annual, 100-hour, or progressive inspection—*Minimum to moderate civil penalty.*
5. Operation of unairworthy aircraft—*Moderate to maximum civil penalty.*
6. Falsification of any record—*Revocation.*

B. *Repair Stations*

1. Failure to provide adequately for proper servicing, maintenance repairs, and inspection—*Moderate to maximum civil penalty.*
2. Failure to provide adequate personnel who can perform, supervise, and inspect work for which the station is rated—*Maximum civil penalty to 7 day suspension and thereafter until adequate personnel are provided.*
3. Failure to have enough qualified personnel to keep up with the volume of work—*Maximum civil penalty to 7 day suspension and thereafter until certificate holder has enough qualified personnel.*
4. Failure to maintain records of supervisory and inspection personnel—*Moderate to maximum civil penalty.*
5. Failure to maintain performance records and reports—*Moderate to maximum civil penalty.*
6. Failure to insure correct calibration of all inspection and test equipment is accomplished at prescribed intervals—*Minimum to maximum civil penalty.*
7. Failure to set forth adequate description of work performed—*Minimum to moderate civil penalty.*
8. Failure of mechanic to make log entries, records, or reports—*Moderate to maximum civil penalty.*
9. Failure to sign or complete maintenance release—*Minimum to moderate civil penalty.*
10. Inspection of work performed and approval for return to service by other than a qualified inspector—*Maximum civil penalty to 30 day suspension.*
11. Failure to have an adequate inspection system that produces satisfactory quality control—*Moderate civil penalty to 30 day suspension and thereafter until an adequate inspection system is attained.*

12. Maintaining or altering an article for which it is rated, without using required technical data, equipment, or facilities—*Maximum civil penalty to 30 day suspension.*
13. Failure to perform or properly perform maintenance, repairs, alterations, and required inspections—*Moderate civil penalty to 30 day suspension.*
14. Maintaining or altering an airframe, powerplant, propeller, instrument, radio, or accessory for which it is not rated—*Maximum civil penalty to revocation.*
15. Failure to report defects or unairworthy conditions to FAA in a timely manner—*Moderate to maximum civil penalty.*
16. Failure to satisfy housing and facility requirements—*Moderate civil penalty to suspension until housing and facility requirements are satisfied.*
17. Change of location, housing, or facilities without advance written approval—*Moderate civil penalty to suspension until approval is given.*
18. Operating as a certificated repair station without a repair station certificate—*Maximum civil penalty.*
19. Failure to permit FAA to inspect—*Maximum civil penalty to suspension until FAA is permitted to inspect.*

C. *General Aviation Maintenance Personnel.*

1. Failure to revise aircraft data after major repairs or alterations—*30 to 60 day suspension.*
2. Failure to perform or improper performance of maintenance—*30 to 120 day suspension.*
3. Failure of mechanic to properly accomplish inspection—*30 to 60 day suspension.*
4. Failure of mechanic to record inspection—*Minimum civil penalty to 30 day suspension.*
5. Failure of IA holder to properly accomplish inspection—*60 to 180 day suspension of IA.*
6. Failure of IA holder to record inspection—*Moderate civil penalty to 30 day suspension of IA.*
7. Maintenance performed by person without a certificate—*Moderate to maximum civil penalty.*
8. Maintenance performed by person who exceeded certificate limitations—*15 to 60 day suspension.*
9. Improper approval for return to service—*Moderate civil penalty to 60 day suspension.*
10. Failure to make maintenance record entries—*Moderate civil penalty to 60 day suspension.*
11. Failure to set forth adequate description of work performed—*Minimum civil penalty to 30 day suspension.*

12. Falsification of maintenance records—*Revocation.*

D. *Student Operations.*

1. Carrying passengers—*Revocation.*
2. Solo flight without required endorsement—*45 to 90 day suspension.*
3. Operation on international flight—*60 to 90 day suspension.*
4. Use of aircraft in business—*90 to 120 day suspension.*
5. Operation for compensation or hire—*Revocation.*

E. *Flight Instructors.*

1. False endorsement of student pilot certificate—*Revocation.*
2. Exceeding flight time limitations—*30 to 90 day suspension.*
3. Instruction in aircraft for which he/she is not rated—*30 to 90 day suspension.*

F. *Operational Violations.*

1. Operation without valid airworthiness or registration certificate—*30 to 90 day suspension.*
2. Failure to close flight plan or file arrival notice—*Administrative action to minimum civil penalty.*
3. Operation without valid pilot certificate (no certificate)—*Maximum civil penalty.*
4. Operation while pilot certificate is suspended—*Emergency revocation.*
5. Operation without pilot or medical certificate in personal possession—*Administrative action to 15 day suspension.*
6. Operation without valid medical certificate—*30 to 180 day suspension.*
7. Operation for compensation or hire without commercial pilot certificate—*180 day suspension to revocation.*
8. Operation without type or class rating—*60 to 120 day suspension.*

---

per violation. These are FAA's internal guidelines only. Sanctions may be imposed outside of the normal range listed, depending on circumstances, including degree of hazard; whether violation was inadvertent or deliberate; nature of activity (private, public, commercial); and alleged violator's attitude, level of experience, ability to absorb the sanction, and demonstrated lack of qualifications. For most violations, current maximum civil penalty is $1000 per violation. See midyear update for late 1988 changes to this list.

9. Failure to comply with special conditions of medical certificate—*90 day suspension to revocation.*

10. Operation with known physical deficiency—*90 day suspension to revocation.*

11. Failure to obtain preflight information—*30 to 90 day suspension.*

12. Deviation from ATC instruction or clearance—*30 to 90 day suspension.*

13. Taxiing, takeoff, or landing without a clearance where ATC tower is in operation—*30 to 90 day suspension.*

14. Failure to maintain radio communications in airport traffic area—*30 to 60 day suspension.*

15. Failure to comply with airport traffic pattern—*30 to 60 day suspension.*

16. Operation in TCA without or contrary to a clearance—*60 to 90 day suspension.*

17. Operation in ARSA without maintaining contact with ATC—*30 to 60 day suspension.*

18. Failure to maintain altitude in airport traffic area—*30 to 60 day suspension.*

19. Exceeding speed limitations in airport traffic area—*30 to 60 day suspension.*

20. Operation of unairworthy aircraft—*30 to 180 day suspension.*

21. Failure to comply with Airworthiness Directives—*30 to 180 day suspension.*

22. Operation without required instruments and/or equipment—*30 to 90 day suspension.*

23. Exceeding operating limitations—*30 to 90 day suspension.*

24. Operation within prohibited or restricted area, or within positive control area—*30 to 90 day suspension.*

25. Failure to adhere to right-of-way rules—*30 to 90 day suspension.*

26. Failure to comply with VFR cruising altitudes—*30 to 90 day suspension.*

27. Failure to maintain required minimum altitudes over structures, persons or vehicles:

    a. congested area—*60-180 day suspension.*

    b. sparsely populated areas—*30-120 day suspension.*

28. Failure to maintain radio watch while under IFR—*30 to 60 day suspension.*

29. Failure to report compulsory reporting points—*30 to 60 day suspension.*

30. Failure to display position lights—*30 to 60 day suspension.*

31. Failure to maintain proper altimeter settings—*30 to 60 day suspension.*

32. Weather operations:

    a. Failure to comply with visibility minimums in controlled airspace—*60 to 180 day suspension.*

    b. Failure to comply with visibility minimums outside controlled airspace—*30 to 120 day suspension.*

    c. Failure to comply with distance from clouds requirements in controlled airspace—*60 to 180 day suspension.*

    d. Failure to comply with distance from clouds requirements outside of controlled airspace—*30 to 120 day suspension.*

33. Failure to comply with IFR landing minimums—*45 to 180 day suspension.*

34. Failure to comply with instrument approach procedures—*45 to 180 day suspension.*

35. Careless or reckless operations:

    a. Fuel mismanagement/exhaustion—*30 to 150 day suspension.*

    b. Wheels-up landing—*30 to 60 day suspension.*

    c. Short or long landing—*30 to 90 day suspension.*

    d. Landing on or taking off from closed runway—*30 to 60 day suspension.*

    e. Landing or taking off from ramps or other improper areas—*30 to 120 day suspension.*

    f. Taxiing collision—*30 to 90 day suspension.*

    g. Leaving aircraft unattended with motor running—*30 to 90 day suspension.*

    h. Propping aircraft without a qualified person at controls—*30 to 90 day suspension.*

36. Passenger operations:

    a. Operation without approved seat belts—*30 to 60 day suspension.*

    b. Carrying passengers who are under the influence of drugs or liquor—*60 to 120 day suspension.*

    c. Performing acrobatics when all passengers are not equipped with approved parachutes—*60 to 90 day suspension.*

    d. Use of unapproved parachutes—*30 to 60 day suspension.*

    e. Permitting unauthorized parachute jumping—*30 to 90 day suspension.*

    f. Carrying passengers without recent flight experience—*30 to 120 day suspension.*

37. Operation while under the influence of drugs or alcohol, or consumption within 8 hours:

    a. Under the influence or .04 and above blood alcohol—*Revocation to emergency revocation.*

    b. Within 8 hours—*180 day suspension to revocation.*

38. Dropping of objects from an aircraft—*30 to 60 day suspension.*

39. Unauthorized towing—*30 to 60 day suspension.*

40. Acrobatic flight on airways, over congested areas, below minimum altitudes, etc.—*90 to 180 day suspension.*

41. Falsification of applications, certificates, records, etc.—*Revocation.*

42. Taking off with insufficient fuel—*30 to 150 day suspension.*

43. Operating so as to cause a collision hazard—*60 to 180 day suspension.*

44. Failure to produce pilot certificate, log, and records—*30 day suspension and thereafter until certificate, log, records are produced.*

45. Conviction for unlawful carriage of a controlled substance on an aircraft—*Revocation.*

46. Drug conviction when an aircraft is not involved—*180 day suspension to revocation.*

The Individual Flight Review

New FAA guidelines for conducting the BFR

Since November 1, 1974, Part 61 of the Federal Aviation Regulations has required a biennial flight review (BFR) for persons intending to act as pilot-in-command of any aircraft. This currency program was intended to be industry managed and FAA monitored, with no specific guidelines set on what the review should entail. An ad hoc industry task force was formed to recommend guidelines on conducting and determining the successful completion of a BFR.

In 1976 these ideas were put in pamphlet form and made available to the public. However, the content of the BFR was still left to the discretion of the reviewer (i.e., the certificated flight instructor).

Now, because of the growing complexity of the aviation environment, FAA has issued an advisory circular to supplement existing guidelines on the conduct of a review. AC 61–98, "Scope and Content of Biennial Flight Reviews," gives specific guidance to the flight instructor on how to structure a BFR to meet the needs of individual pilots. The circular takes into consideration the fact that each pilot has a different flying background, therefore needs individual attention. The review is separated into four sections as follows.

1. PRELIMINARY INTERVIEW

A personal interview between pilot and CFI is recommended before the BFR is initiated. This allows the CFI to become familiar with the pilot's flight experience and the customary scope of his or her flight activity. This will lead to the following considerations:

a. *Type of Equipment Flown.* The maneuvers and procedures reviewed will vary, depending on the make and model of aircraft used. For example, a review in a light twin-engine aircraft should be different from one conducted in a small two-seat tailwheel aircraft without radio or extra instrumentation. The instructor may wish to recommend that the pilot take the review in the aircraft usually flown, or in the most complex make and model if several aircraft are flown regularly.

Instructors should also consider their own experience and qualifications in a given make and model aircraft prior to giving a review in that model. For aircraft in which the CFI is not current or is not familiar, recent flight experience or sufficient knowledge of aircraft limitations, charac-

of flight operations is likely to change, or if other circumstances exist. For example, a pilot who normally conducts only local flight operations may be planning to begin flying to a location with a Terminal Control Area (TCA). Another pilot may operate only a two-seat aircraft without radio but in close proximity to a TCA. In both cases, the CFI should include TCA requirements and operating procedures as part of the biennial flight review.

c. *Amount and Recency of Flight Experience.* The CFI should review the pilot's logbook to determine total flight experience, plus type and recency of experience, in order to evaluate the need for particular maneuvers and procedures on the review.

For example, a pilot who has not flown in several years may require an extensive review of basic maneuvers from the practical test standards appropriate to that pilot's grade of certificate. This same pilot may also require a more extensive review of FAR Part 91, including recent changes in airspace and other requirements.

Another pilot who is upgrading to a newer or faster airplane should receive more emphasis on knowledge of aircraft systems and performance or in cross-country procedures appropriate to a faster airplane. Regardless of flight experience, the CFI should ensure that the plan includes all areas in which the pilot should receive a review in order to operate safely. In some cases the CFI may wish to recommend that the pilot undertake a complete refresher program, such as those described in the current issue of Advisory Circular 61–10, "Private and Commercial Pilots Refresher Courses."

d. *Scope of Review.* After completing the analysis above, the CFI should discuss it with the pilot and reach a mutually agreeable understanding regarding how the biennial review will be conducted. The CFI may wish to provide the pilot with reading or study materials or recommend such publications for study prior to actually undertaking the flight review. The CFI should also explain to the pilot the standards under which satisfactory completion of the review will be measured.

2. REVIEW OF FAR PART 91 OPERATING AND FLIGHT RULES

The CFI should tailor the review of general operating and flight rules to the needs of the pilot being reviewed. The objective

is to ensure that the pilot can comply with regulatory requirements and operate safely in various types of airspace and under various weather conditions.

As a result, the instructor should conduct a review that is broad enough to meet this objective, yet provide a more comprehensive and indepth review in those areas where the pilot's knowledge is not as extensive. In the later instance, the instructor may wish to employ a variety of reference sources, such as the *Airman's Information Manual*, to ensure that the pilot's knowledge is current.

The review of FAR Part 91 rules is critical due to the increasing complexity of airspace and the need to ensure that all pilots are familiar with TCA's, Airport Radar Service Areas (ARSA's), and other types of airspace. The biennial flight review may be the only regular proficiency and updating experience on those subjects that some pilots are exposed to. Accordingly, instructors should place appropriate emphasis on this part of the review.

The advisory circular will provide the instructor with a useful format for organizing the FAR Part 91 review and ensuring that essential areas are covered. The review should be expanded in those areas where the pilot's knowledge is less extensive.

3. REVIEW OF MANEUVERS AND PROCEDURES

The maneuvers and procedures covered during the review are those which, in the discretion of the CFI, are necessary for the pilot to demonstrate that he or she can safely exercise the privileges of their pilot certificate. Accordingly, the instructor should evaluate the pilot's skills and knowledge to the extent necessary to ensure the he/she can safely operate under a wide range of conditions and within regulatory requirements.

The instructor may wish to prepare a preliminary plan for the flight review based on a sequence of maneuvers which should be outlined in advance to the pilot take the review. For example, this may include a flight to the practice area or to another airport, with maneuvers accomplished while en route.

It could also include a period of simulated instrument flight time. The instructor should request the pilot to conduct whatever preflight preparation is necessary to conduct the planned flight. Examples of such activities may include checking weather, computing required runway lengths, calculating weight and balance, completing a flight log, filing a flight plan, and/or conducting the preflight inspection.

Prior to commencing the flight portion of the review, the instructor should discuss various operational subjects with the pilot. This oral review could include, but not necessarily be limited to, areas such as aircraft systems, speeds and performance; meteorological and other hazards, wind shear and wake turbulence; and operations in controlled airspace, such as TCA's and collision avoidance procedure. The emphasis during this discussion should be on practical knowledge of recommended procedures and regulatory requirements.

Regardless of the pilot's experience, the instructor may wish to review at least those maneuvers considered critical to safe flight, such as stalls, flight at minimum controllable airspeed, and takeoffs and landings. Based on his/her in-flight assessment of the pilot's skills, the instructor may wish to add other maneuvers from the practical test standards or flight test guide appropriate to the grade of pilot certificate.

The in-flight review need not be limited to evaluation purposes. The instructor may desire to provide additional instruction in weak areas or, based on mutual agreement with the pilot, defer this instruction for a follow-up flight.

4. POST-REVIEW

Upon completion of the review, the instructor should debrief the pilot and explain whether the review was satisfactory or unsatisfactory. Regardless of this determination, the instructor should provide the pilot with a comprehensive analysis of his/her performance, including any weak areas. The instructor should not endorse the pilot's logbook to reflect an unsatisfactory review, but should sign the logbook to record the instruction given.

A satisfactory BFR should be endorsed as follows: "*Ref: Far 61.57(a). Mr. /Ms. R. Jones, holder of Pilot Certificate (give number) has satisfactorily completed a Biennial Flight Review on* (date)."

This should be followed by the instructor's name, signature, CFI number, and expiration date.

NOTE: Following an unsuccessful review a pilot may continue to act as P.I.C. if a period of 24 months has not lapsed since the last successful BFR, or any of these other flight checks:

• Pilot certification or added rating.
• Any proficiency flight check required by the FAR.
• Pilot examiner flight check.
• Proficiency flight check given by a U.S. Armed Forces check pilot.

The AC is free from the DOT, M–443.2, Washington, DC 20590. ∎

Aviation Safety Commission
Executive Summary and General Aviation Recomendations
April 1988

After months of study, hearings, meetings around the country, and staff reports, the Aviation Safety Commission unanimously concludes that the nation's air transportation system is safe. However, safety is being maintained to an increasing extent through delays and other inconveniences.

Air transportation has changed during the past decade. Economic regulations that had shaped the industry since the 1920s were replaced by airline deregulation. The resulting increased competition has lowered fares, expanded service, and brought air travel to millions who had not previously been able to afford to fly. It has also made the FAA's job much more difficult.

The Aviation Safety Commission concludes that the present safety regulatory structure designed to ensure aviation safety is inadequate to deal with future growth and technological change. Now is the time to equip the regulatory system to accommodate changes in the numbers and kinds of aircraft, to take advantage of new technology in aircraft design and manufacture, to respond to heightened sensitivity on the part of the public to aviation safety, and to act on the backlog of potentially worthwhile safety improvements that have been languishing because of diffused authority and accountability. In short, *now* is the time for decisive action by Congress and the Executive Branch.

The Aviation Safety Commission believes that the Federal government must continue to play the central role in ensuring safe operation of the U.S. aviation system. We also share the common perception that, while the system is safe for now, the present governmental structure is not working effectively enough to ensure its safety in the future. Therefore, we agree unanimously that a major structural overhaul is essential. We believe that the regulatory process must remain governmental in character and should not be taken out of the Federal government or removed from public accountability.

The Commission's recommendations address in a constructive way all of the issues which have been raised in the current debate and reflect the input we have received from the Secretary of Transportation, the FAA, former FAA Administrators, Members of Congress, the NTSB, consumers, and industry experts.

Specifically, the Aviation Safety Commission recommends that FAA be transferred from the Department of Transportation and be established as a user-funded authority which is:

- overseen by a nine-member Board of Governors appointed by the President and confirmed by the Senate;
- managed by an Administrator who is appointed and confirmed for a term of seven years;
- subject to agency-wide regulatory oversight by a Director of Aviation Safety who is appointed and confirmed for a term of seven years;
- freed from the constraints of the federal civil service and procurement systems.

The Director of Aviation Safety has the authority to initiate rulemaking as well as disapprove regulations promulgated by the Administrator, and also has the authority to enforce compliance with the Administrator of existing rules and regulations. Decisions by the Administrator and the Director of Aviation Safety are appealable to a Safety Committee of the Board of Governors composed of the Administrator, the Director of Aviation Safety, the Secretary of Transportation, the Secretary of Defense and a public member, and hence are not subjected to OMB review.

The Aviation Safety Commission also recommends the following agenda for improving aviation safety:

Safety Inspection Programs

- national rather than regional certification programs for major and national jet carriers;
- establishment of a nationwide inspection program for all size carriers with a combination of regular, in-depth, and surprise inspections;

- separation of certification and surveillance functions in the new Authority;
- priority inspections for carriers undergoing major change;
- increasing the inspector workforce to accommodate these changes.

Regional Airline Safety

- reducing differences in equipment standards between regional and national carriers, with all aircraft providing scheduled service being required eventually to meet Air Transport Category Aircraft (Part 25) standards;
- reducing differences in operating practices between regional and national carriers, with all carriers eventually being required to meet Part 121 operations requirements.

General Aviation in the Air Traffic Control System

- requiring all aircraft to be equipped with a Mode C transponder in buffer zones around all large, medium, and small hubs;
- stronger enforcement against buffer zone violators with a separate radar position dedicated to tracking and notifying violators in each buffer zone.

FAA Rulemaking

- process must be streamlined and restructured to include clear and unambiguous responsibility and accountability.

Airport Safety and Capacity

- base airport certification on passenger volume rather than type of equipment;
- review of existing policies and requirements with particular emphasis on signage, directional indicators, and taxiway and intersection markings.

Use of Operations Research

- need to enhance operations research capabilities for better utilization in problem solving.

GENERAL AVIATION IN THE AIR TRAFFIC CONTROL SYSTEM

Aviation Safety Commission Recommendations:

The Aviation Safety Commission recommends that the FAA should designate around all large, medium, and small hubs, buffer zones whose horizontal boundaries are marked by clearly identifiable geographic or navigational landmarks and whose vertical boundaries begin at the ground and extend to the highest altitude used by civil aviation. Any aircraft that enters a buffer zone or flies above, below, or through a Terminal Control Area (TCA) must have an operating Mode C transponder.

The Aviation Safety Commission also recommends an increase in enforcement activities against aircraft who violate TCAs or buffer zones and a restructuring of enforcement procedures. A separate radar position dedicated to tracking violators should be established at each TCA.

Background and Discussion:

The public concern about general aviation aircraft in the air traffic control system stems mostly from the perceived risk of midair collision to commercial airliners posed by the presence of general aviation aircraft. Historically, such risks have been quite small but such accidents evoke considerable publicity. Between 1975 and 1986 there were 329 midair collisions in the U.S. of which about 60 percent involved fatalities. Of the fatal accidents during this period, two involved jet airliners. In 1978, a Pacific Southwest Airlines Boeing 727 collided with a four-seat Cessna 172 in San Diego. In 1986, an Aeromexico DC–9 collided with a four-seat Piper Archer over Cerritos, California.

Such accidents, coupled with the dramatic post-deregulation growth in air carrier traffic, have heightened concerns about midair collisions and those concerns have focused on reports of near midair collisions collected by the FAA and NASA-Ames. General aviation aircraft figure prominently in near midair collisions reported by pilots to the FAA. During the period 1985 through 1987, there were 1,064 near midair collisions reported to the FAA of which 758 (71 percent) involved a general aviation aircraft. Moreover, reports of near midair collisions between air carriers and general aviation aircraft more than doubled between 1985 and 1987.

The air traffic control system is designed to keep aircraft under its control from colliding with one another. Aircraft operating under instrument flight rules (IFR) are in constant radio contact with air traffic control and respond to controllers' instructions to change altitude and direction. Virtually all airline flights operate under instrument flight rules, even under conditions of clear weather and good visibility. Many general aviation flights, however, operate under visual flight rules (VFR) and are not in communication with air traffic controllers and cannot respond to their instructions. Because air traffic controllers do not communicate with VFR flights, they cannot prevent collisions between two VFR aircraft.

Controllers do, however, have the potential capability to prevent collisions between an airliner operating under IFR conditions and a general aviation aircraft operating under VFR conditions. If the controller is aware that the VFR aircraft poses a threat, the IFR aircraft can be instructed to change altitude or direction to avoid the threat. The difficulty is that not all general aviation aircraft are equipped to provide the controller with the information to determine easily if a threat exists. Aircraft without so-called Mode C transponders do not provide information to the controller about the altitude at which they are operating. Without Mode C, a controller cannot tell the difference between a general aviation aircraft that poses a collision threat to another aircraft and one that is several thousand feet above or below and thus poses no threat. When faced with a VFR aircraft without Mode C, the controller must make a subjective judgment about the severity of the potential threat. If that same aircraft had Mode C, the controller would know immediately if it was necessary to have the IFR aircraft change altitude or direction to avoid a collision.

The Aviation Safety Commission has concluded that safety could be improved while simultaneously reducing air traffic controller work load, especially during busy periods, if altitude information were available to the controller for all aircraft appearing on a controller's screen (primary visual display). While such information would be very useful at en route centers (ARTCCs), it is especially vital for safe operations in terminal areas (TCAs).

Therefore, the Aviation Safety Commission recommends that the FAA should designate around all large, medium, and small hubs, buffer zones whose horizontal boundaries are marked by clearly identifiable geographic or navigational landmarks and whose vertical boundaries begin at the ground and extend to the highest altitude used by civil aviation. The basic area for the buffer zones will begin at the point when normal descent into the airport by commercial jets reaches 12,500 feet or normal climb out of the airport reaches 12,500 feet, whichever is further from the airport. Any aircraft that enters a buffer zone or flies above, below, or through a TCA must have an operating Mode C transponder.

No aircraft or user will be exempt from the regulations regarding the buffer zones. Any aircraft equipped with a Mode C transponder will be required to have its Mode C equipment inspected annually. Currently, the FAA requires all aircraft that fly in IFR conditions get a biennial inspection of its transponder, encoding altimeter, and pilot-static system (which measures altitude). A mandatory inspection may be the only way to make the recommendation of a buffer zone effective. Because the controller will not receive verbal verification of aircraft's altitude if it is not under positive control, the controller must be confident that the information sent by the Mode C equipment is accurate.

This recommendation does not imply that current TCA boundaries need be changed nor that more aircraft need to be brought under positive air traffic control. Rather, the recommendation is intended to provide more information to the controllers about uncontrolled aircraft and to make it easier to avoid conflicts between controlled and uncontrolled aircraft.

The Aviation Safety Commission also recommends an increase in enforcement activities against aircraft who violate TCAs or buffer zones and a restructuring of enforcement procedures. The current system is too cumbersome to be effective and places too great of a burden on controllers whose primary responsibility is traffic separation. A separate radar position dedicated to tracking violators should be established at each TCA. The position would be staffed by a developmental controller only during periods with high potential for violations, typically during periods of high general aviation activity. A controller observing a violation would "hand off" the violator to the enforcement position whose responsibility would be to follow that aircraft until it could be identified.

The
Best of
✻ Flight**FORUM** ✻

FAA Explanations and Interpretations of the FARs.

a TAB/AERO exclusive

The Federal Aviation Administration publishes *FAA Aviation News* (formerly *FAA General Aviation News*) to promote flight safety by calling the attention of general aviation airmen to current technical, regulatory, and procedural matters affecting the safe operation of aircraft. One feature of this bimonthly magazine (available by subscription from the Superintendent of Documents, U.S. Government Printing Office, Washington, DC 20402) is the "Flight Forum" section, in which the FAA staff answers reader questions, often ones requiring an explanation or interpretation of Federal Aviation Regulations. Although "Flight Forum" is, in FAA's words, "advisory or informational in nature and should not be construed as having regulatory effect," the TAB/AERO staff believes all airmen will benefit from reading FAA's interpretations of its own regulations and publications.

On the following pages are edited excerpts from "Flight Forums" published between 1974 and 1988. The original month and year of publication are printed at the end of each answer. The questions and answers selected by TAB/AERO for reprint in this edition have undergone a general review by the TAB/AERO staff. However, regulations and procedures occasionally change, and it is the reader's responsibility to keep abreast of these changes. TAB/AERO disclaims any liability with respect to the use of this material. Please direct any substantive questions regarding this section to *FAA Aviation News,* AFS-810, Washington, DC 20591, and send a copy to Aviation Editor, TAB Books Inc., Blue Ridge Summit, PA 17294-0214. The TAB/AERO staff welcomes your comments but cannot answer substantive questions.

For your convenience, the FARs are keyed to "Flight Forum" question numbers where the ☞ symbol appears.

F1 • Tach Time or Hobbs Time?

In the . . . March/April issue, one paragraph states about student pilots, "He's got to have 40 hours, including 20 hours of solo, on the tach before he can apply for the private certificate." To my knowledge, nothing in the FAR's connects flight time to tach time.

"Flight Time" is defined in FAR Part 1 as "The time from the moment the aircraft first moves under its own power for the purpose of flight until the moment it comes to rest at the next point of landing (block-to-block time)." Flight time is sometimes loosely measured by the tachometer, which not only measures revolutions per minute but keeps a cumulative log of total revolutions on the engine and converts this data to hours of engine turnover at maximum power. This data is important in engine maintenance but does not always correspond to flight hours. Measuring flight time from the Hobbs Meter is a more accurate way of determining actual flight time.

The Hobbs Meter measures time in one of four ways: 1) electrically, by timing the period when the master switch is on; 2) dynamically, by air pressure from an air inlet when the aircraft is moving; 3) by oil pressure when it rises into the green area; or 4) by a tripping mechanism when the gear is lowered and retracted.
[see clarification below]   11-12/79

F2 • Measuring Flight Time

Regarding the letter on "Tach Time or Hobbs Time?" . . . the oil pressure actuated Hobbs meter on my airplane registers 20% less than an electronic timing device activated by an air switch—which always registers the same elapsed time as my wrist watch. Which should be used to measure "block to block" time?

Regulations refer to flight time as "block to block" time but do not specify how to measure it. Use whichever way you wish.   3-4/80

F3 • On Top

Is an instrument rating required for flying "VFR over the top" or "VFR on top"?

No and yes—in that order. The terms you mention refer to two entirely different types of operation. "VFR over the top" is flying above clouds, in VFR weather conditions and *not* on an IFR flight plan. If you are able to remain in VFR conditions at all times during your flight, including the climb and descent, you do not need an instrument rating.

On the other hand "VFR on top," sometimes called "VFR conditions on top," refers to an instrument operation requested by the pilot (in lieu of an assigned altitude) when he files his instrument flight plan. (If filing in person enter "OTP" in the altitude block of the flight plan.) Under certain conditions this type of flight plan provides a convenient way for an instrument-rated pilot to fly above the clouds, especially if the tops are known.

However, when you fly IFR/VFR on top, you must adhere to both the IFR and VFR flight rules. Flight must be conducted in VFR weather conditions; the altitude must be appropriate to the direction of flight and at least 1,000 feet above any cloud, smoke, haze or fog layer; and you are responsible for avoiding other aircraft, although ATC may advise of other known traffic. In addition you must follow applicable IFR flight rules, such as maintaining minimum IFR altitude; position reporting; maintaining radio communications; notifying ATC of flight plan changes; and adhering to ATC clearances. (Note: VFR on top operations are not authorized for positive control areas.)
                                          7-76

F4 • CFI Without a Medical?

I have been told that a certificated Flight Instructor who has lost his medical can still instruct in flight as long as he does not act as pilot-in-command. This seems contrary to FAR§ 61.19 which states that a flight instructor certificate is valid only while the holder has current pilot certificate and a medical certificate appropriate to the pilot privileges being exercised.

Would instruction from a CFI who has been denied a medical certificate be accepted for an additional rating?

That depends on the nature of the instruction. The need for a valid medical certificate and its relationship to the flight instructor certificate is discussed in Far Part 61 Certification Bulletin No. 61-2. The bulletin states that if a CFI is on board an aircraft for the purpose of providing flight instruction, but does not act as PIC or as a required crewmember, that CFI is NOT exercising pilot privileges which require possession of a current medical certificate under Far Part 61.

Accordingly, a CFI, who does not hold a current medical certificate, but is not acting as PIC as required flight crewmember, may give creditable flight instruction to a pilot who is fully qualified and currently rated to act as PIC for the aircraft. A non-medically current CFI is NOT permitted to give flight instruction to a student pilot; to a pilot whose BFR has lapsed or who is otherwise not qualified, rated, and current in the aircraft; to a non-instrument rated pilot in IMC or on an instrument flight plan; or to a pilot practicing instrument flight under a hood (which requires the presence of a safety pilot).
[see clarification below]   11-12/87

F4A • Class of Medical

What is not addressed [above] is the class of medical certificate required by a flight instructor in order to a) give creditable flight instruction or b) give creditable flight instruction for compensation. Can you clarify?

Whether flight instruction is given for compensation or not has no bearing on the class of airmen medical certificate required.   5-6/88

F5 • How To Change Your Name

I was recently married. When I wrote the FAA Record Center in Oklahoma City to have my name changed on my pilot certificate they replied that I had to send my marriage license and my old pilot certificate, and they would send me a corrected one. Is this a rule? If so, what do I do in the meantime? I would prefer not to part with my new marriage license, and even more importantly, won't I be illegal flying without my pilot certificate? I don't want to be grounded for an uncertain period of time while the paperwork is accomplished.

FAR 61.25 does make this requirement, but there is a simple way to accomplish the name-change. Take your two certificates (marriage and pilot) to your FAA District Office (GADO or FSDO) and fill out an application for a corrected certificate. They will take your airman certificate and give you a temporary one to use until you get your new one. They will examine your marriage license (a copy will do) and return it to you. 8/76

F6 • What is High Performance?

There is a difference of opinion at my airport on what constitutes a "high performance airplane". As I read the rules a high performance airplane is one that has more than 200 hp or that has retractable gear, flaps and a controllable propeller. I was checked out in a 180 hp Comanche (which has re-

tractable gear, flaps and controllable prop) and my instructor endorsed my logbook for "high performance airplanes". That instructor has transferred, and another instructor says my logbook is illegal because the Comanche did not have more than 200 hp.

I am also told that I cannot log pilot-in-command time if I am flying from the right seat, but I can find nothing in the rules to support this. Who is right?

You are right on both counts. FAR 61.31 (e) clearly states that a high performance airplane is one that has either more than 200 hp or retractable gear and flaps and a controllable propeller. Also, there is nothing in the regulations to prohibit logging pilot-in-command time if you are sitting in the right seat, as long as you are the "sole manipulator of the controls" of an aircraft for which you are rated.   10/77

F7 • High Performance or Complex

Another aviation magazine recently published the statement that a checkout in a 230 hp fixed-gear Cessna 182 made you legal to fly a 180 hp Cutlass RG and a 285 hp Bonanza. I interpret FAR 61.31 (e) to require a checkout and endorsement for (1) greater than 200 hp and (2) flaps, retractable gear, and controllable propeller as two independent issues. Who is correct?

The magazine is correct. FAR 61.31 (e) states that a pilot who is qualified for a high performance (greater than 200 hp) or complex (flaps, retractable gear, controllable prop) aircraft may act as PIC in either.   5-6/81
                                          7-8/81

F8 • Unacceptable Test Aids

I object to the FAA policy that prevents me from having an operating manual for my calculator with me while I am taking an airman written test. What's the statutory authority for this decision?

Federal Aviation Regulation §61.37(a)(5), states that no person may use any material or aid during the period when a test is being given except as authorized by the Administrator.

Operation booklets or manuals accompanying electronic or mechanical calculators sometimes contain information not directly related to the operation and use of the calculator. This information may pertain to subjects in the test items in FAA written tests. For this reason, such manuals may not be used during a written test.

In order that applicants for airman written tests may use scales, straight edges, protractors, plotters, navigation computers, and electronic or mechanical calculators while taking written tests, information not pertinent to the actual operation of the aids mentioned must be masked out by the applicant. Such information includes: FAR's, signals, cloud data, holding pattern diagrams, frequencies, weight and balance formulas, ATC procedures, etc.

Finally, while the use of electronic or mechanical aids is permissible as outlined above, it is the responsibility of the applicant to know how to use such aids.   7-8/86

F9 • Certification Requirements

FAR 61.71 (a) states that if a graduate of an FAA approved school presents his or her graduation certificate within 60 days after graduation, he or she may take the flight test for the certificate. FAR 61.39 (a) (5) requires a written statement from a certificated instructor within 60 days prior to taking the test. Does the graduation certificate meet the require-

---

Material on this page is advisory or informational in nature and should not be construed as having regulatory effect.

ment for the 60 day endorsement? If the applicant does not present the graduation certificate within the 60 days would he require only the instructor's statement, and would he still be considered a school graduate?

The graduation certificate and the instructor's written statement serve two separate purposes and are in no way interchangeable. Every applicant for a flight test must present a written statement from an instructor certifying that the applicant has been properly prepared for the test. The graduation certificate (from a pilot school that is certificated under FAR Part 141—either old or revised) allows the applicant to take the flight test with fewer hours of aeronautical experience than he would need if he did not attend an approved school (35 hours as compared to 40 for a private pilot). If he does not present his graduation certificate within 60 days he can not take advantage of the reduced flight hour requirement. In either case he needs the instructor's written statement for the flight test. **7/75**

F10 • From Sea to Land

An airman received his commercial pilot's certificate in a single-engine seaplane. Now he wants to add an airplane single-engine land rating to that commercial certificate, and this calls for a check ride in a complex aircraft (controllable prop, flaps, and rectractable gear). Must he take the check ride in a retractable gear airplane or does the fact that he met the complex portion in a seaplane suffice?

Under FAR 61.43 (Flight Tests), the applicant must be prepared to perform each of the required pilot operations listed in Part 61.127(a), Commercial Pilot Flight Proficiency. Therefore, the airplane furnished for the flight test must have retractable landing gear, flaps, and controllable propeller. Also, under Part 61.63(c)(2), the applicant must pass a flight test appropriate to the certificate and "applicable to the aircraft category and class rating sought." This does not allow any pilot operation required for the flight test to be ignored. **3-4/82**

F11 • Rental Aircraft

Can aircraft be rented by an FBO if they are not 100-hour inspected? I don't mean flying club aircraft, but those that are rented to pilots or solo students for pleasure or solo practice (not dual). Also, can an FAA flight check be taken in such an aircraft?

The 100-hour inspection is not required for rental aircraft, and the FAA flight check can be taken in such an aircraft. However, if paid instruction is given, or supervised solo work as a part of an overall instruction curriculum is flown, or if passengers are carried for hire, then the aircraft is required to have 100-hour inspections (FAR 91.169). **3/77**

F12 • Very Short Cross-Country

I have a commercial certificate, and I often ferry aircraft short distance, e.g., Leesburg, VA to Dulles Airport (10 nautical miles), for fuel or maintenance. Can I log such a short distance as cross-country time?

For a certificated pilot rated and current in the aircraft involved, any flight that ends with a landing at a point other than the point of departure—regardless of the distance—can be logged as cross-country time.

[see clarification below] **1-2/83**

F13 • Cross-Country Undefined

What is the regulatory basis of the definition of cross-country flight that [appears in "Very Short Cross-Country" above]?

Apart from regulations on cross-country requirements for private and commercial applicants (FAR Parts 61.109 and 61.129), there is no regulatory definition of what distance constitutes a cross-country flight. For airline transport pilot applicants, Flight Standards has determined that any flight with a landing at a point other than that of departure is a cross-country. By extension, this latter determination applies to general aviation flights logged other than for the purpose of certification. [see clarification below] **7-8/83**

F14 • Cross-Country Definitions

. . . you make reference to the distance constituting a cross-country flight as being undefined. Check FAR 61.93 and you will see it is any flight over 25 nautical miles.

FAR Part 61.93(a) refers to student pilots only, who can log trips of 25 NMi or less as cross-country but cannot apply it toward the private certificate. Part 61.109 specifies those distances precisely.

[see two preceding items] **9-10/83**

F15 • Log Mistakes

When making an entry in a pilot logbook, if an error is made, what is the proper procedure to follow in correcting it? Do the FARs address this?

The only rule on this subject, FAR Part 61.59(a)(2), says that no person may make an "intentionally false" entry in a logbook or alter an entry for fraudulent purposes. If an unintentional mistake is made, simply draw a line through the entire entry and move to the next available space. **5-6/82**

F16 • Right Seat PIC

Must a pilot have a logbook endorsement to fly as PIC in the right seat of an aircraft requiring only one pilot, such as a Piper 140, when flying VFR only?

Seat selection up front is at the pilot's option. **9-10/81**

F17 • Multi-Engine Instruction

Is there any way a non-multiengine rated pilot sitting in the right seat of a piston twin, less than 12,500 lbs., can log time toward a certificate or rating if the pilot occupying the left seat is a CFI single-engine only, but multi-engine rated?

Negative, but if your intention was to count the time toward a certificate or rating or to meet currency requirements. **11-12/81**

F18 • Instrument Time and Day Time

I have a question concerning the proper logging of flight time according to FAR 61.51(b)(3), "Conditions of Flight." If I make a two-hour IFR daytime flight, one hour only in clouds with the other in clear weather, I know I can log only one hour of actual instrument, but how is the daytime logged? One hour or two?

You flew two hours in daytime, so log two hours in the "Day" column of your logbook. Only

one of those hours was in instrument conditions, so log only one hour in the "Actual Instrument" column. **9-10/80**

F19 • Night Time

. . . when may a pilot legally log "night" time? I have heard several definitions and want to be sure I am doing it right.

In civil aviation the generally accepted definition of "night" time is that in FAR 61.57 (d), which refers to night experience as "the period beginning one hour after sunset and ending one hour before sunrise (as published in the American Air Almanac). Note that military logging practices may be different. **8/77**

F20 • Student-In-Command

Can a student pilot log his solo time as pilot-in-command time in the hopes of applying the PIC time to a later certificate, such as commercial?

Affirmative. FAA's General Counsel has interpreted FAR 61.51(c) to allow the logging of student pilot solo time as PIC time because the student is the sole occupant of the aircraft.

[see clarification below] **9-10/80**

F21 • Two Kinds of Students

I recently took an FAR test at my school and incorrectly answered a question that stated: "Can a student pilot log PIC time if he is the sole manipulator of the controls?"

I showed my instructor [the "Student-in-Command" item above] which said that a student can log PIC time. My instructor says I'm still wrong. Help! Who is right?

Both. A student (who does not have a private or better certificate) cannot log dual time as PIC, only as dual even if he were the sole manipulator of the controls. At times when the student is sole occupant, he may log the time as PIC and apply it to requirements for a subsequent certificate. [see clarification below] **7-8/81**

F22 • O Solo Mio!

. . . there have been a number of questions in the "Forum" regarding student logging of solo time, when he or she is the sole manipulator of the controls. My question is, can a student legally log solo time as PIC time?

Strictly speaking, a student can log sole occupant time only as solo time. However, after the student receives his private pilot certificate, he may apply his solo time as PIC time toward a higher rating.

[see clarification below] **1-2/82**

F23 • Student PIC?

Is a student pilot, when he or she is the sole occupant of an aircraft, the pilot-in-command? A flight instructor I know says an airplane with only a student pilot in it has no PIC. Is he right?

The definition of pilot in command, as given in FAR Part 1, is ". . . the pilot responsible for the operation and safety of an aircraft during flight time." Therefore the solo student would be PIC. [see 3 preceding items] **11-12/83**

F24 • IFR and Student PIC Time

FAR 61.51 says a pilot may log PIC time any time he is the sole manipulator of the controls of an aircraft for which he is rated. Suppose IFR conditions prevail when a pilot is taking instrument in-

---

Material on this page is advisory or informational in nature and should not be construed as having regulatory effect.

struction. *Don't the regulations expressly forbid anyone to fly as pilot-in-command under instrument conditions without an instrument rating?*

An instrument student who has a private or better certificate and is rated in the airplane may indeed log as PIC the time he is sole manipulator of the controls regardless of the meteorological conditions. However, if the flight is under actual instrument conditions the student must be accompanied by a pilot, who may be a flight instructor, holding an instrument rating and who is rated in the aircraft.

[see clarification below]   1-2/81

F25 • CFI as PIC

My question concerns . . . "IFR and Student PIC Time" [above]. Your answer says the student can log PIC time. Isn't it true that the instructor is the pilot-in-command?

Correct. As far as making any decisions necessary for safe flight, the flight instructor—when he is on board for the purpose of giving instruction—is always the pilot-in-command and may log the time as such. However, FAR Part 61.51(c)(2)(i) permits a pilot with a private or better certificate to *also* log as PIC time any period "during which he is sole manipulator of the controls of an aircraft for which he is rated."

[see clarification below]   3-4/81

F26 • Logbook Correction

In several issues of the "Forum" you have stated that a pilot with a private or better certification can log instruction as both dual and PIC time, provided he is rated in the airplane and current. I have been taking commercial instruction and logging it only as dual. I would like to make an entry in my log to correct this error. How do I go about it?

There is no error. Simply log the dual time that you were sole manipulator of the controls in the PIC column.   11-12/81

F27 • PIC in Name Only

In the September/October 1982 issue, you indicated . . . that no one can log pilot-in-command instrument time in instrument meteorological conditions unless they have an instrument rating. FAR's do not specifically say you can log this type of pilot time as PIC, but neither do they indicate you cannot. According to your interpretation, we now have to be rated for the condition of flight. Please advise if this is or is not correct.

Since publication of that answer . . . a new interpretation has been issued by the chief counsel's office: A non-instrument-rated pilot who is taking instrument instruction in IFR conditions may log that as PIC time under FAR 61.51, but may not actually serve as the PIC as defined in FAR 1.1. The other pilot must be the PIC and fully rated and current for that flight—not merely acting as a safety pilot.

[see clarification below]   7-8/83

F28 • PIC Under the Hood?

Your interpretation of 61.51(c)(2)(i) allows a private or better pilot to log time under the hood, when he/she is sole manipulator of the controls, as pilot in command time. Does this apply as well when the safety pilot is an instructor, on board for purposes of giving flight instruction?

. . . yes, a qualified pilot practicing IFR under a hood in VFR conditions, no flight plan, with an instructor on board may log as PIC that time when he is sole manipulator of the controls of an aircraft for which he is rated.   1-2/87

F29 • Hooded PIC

I have a question about two pilots, both fully qualified, flying an airplane in clear weather, no flight plan. One wears a hood and is practicing instrument procedures. The other is acting as the safety pilot.

My question is, who is the actual pilot in command? (Not who logs what.) Can a pilot wearing a hood act as PIC at the same time?

In the conditions you describe, either pilot may act as pilot in command. The decision as to who is to be in command should be decided between the two crew members in advance, in order to establish the responsibility. Wearing a hood in VFR weather, with no IFR flight plan and a qualified safety pilot in the cockpit, does not prevent you from acting as PIC.

[see clarification below]   3-4/87

F30 • More on Hoods, PIC's

Back to the subject of logging instrument time, how about these situations:

1. Pilots A and B are both current and rated for the airplane they are flying in VMC but on an instrument flight plan. Pilot A is instrument-rated and current, Pilot B is IFR rated but not current, Pilot B wears a hood and manipulates the controls. Who may be pilot in command of this aircraft and who can log what?

2. Both pilots are instrument-rated and current, as well as rated and current for the aircraft. Pilot B again wears a hood and manipulates the controls. Who is and/or logs what?

Finally, I understand that an instrument competency check may be given by a certificated instrument flight instructor in a ground trainer. When this takes place, may the instructor also sign off on the pilot's BFR renewal?

1. When an aircraft is on an instrument flight plan, the actual PIC must be a pilot who is instrument rated and current (as well as otherwise rated and current in the aircraft). In this case the PIC [would be Pilot A]. Pilot A would also be the safety pilot. However, Pilot B may *log* as PIC time that time during which he or she is the sole manipulator of the aircraft controls.

2. With both pilots fully rated and current for the aircraft and the flight conditions, either may be the actual pilot-in-command, in accordance with prior agreement. There is no reason why the actual PIC may not wear a hood or function as safety pilot (obviously he or she may only do one or the other, not both). The hooded pilot may, if so agreed upon, be the actual pilot-in-command and log the entire flight as PIC. The safety pilot may not log PIC time unless it is agreed that he is to be the actual PIC. In the latter case, the hooded pilot may only log as PIC the time he actually manipulates the controls.

Finally, a Biennial Flight Review may not be given in a flight simulator or trainer. Such a trainer may be used for the instrument competency check, but only in conjunction with a written letter of approval issued by the local Flight Standards District Office.   5-6/87

F31 • PIC Time Revisited

As a flight instructor I have tried to get a straight answer to this question, but only received contrary opinions from flight examiners and your publication. Perhaps you can set to rest this matter of logging dual time after a pilot is certificated for the flight involved.

For instance, say I am giving instruction to a private pilot seeking a commercial certificate. The private pilot is in the left seat as the sole manipula-

tor of the controls and is current and rated in the aircraft. If he logs the flight time as both dual and PIC time, which he is permitted to do, the sum of dual and PIC time on the certificate application could exceed the pilot's total time in his logbook. Would this not be misrepresentation and illegal?

Under the conditions described, a private pilot may log a period of flight time as both dual and PIC time. However, for purposes of certification or qualification, the total time logged for any given flight may not exceed the time shown in the "Duration of Flight" column in the pilot's logbook.   9-10/84

F32 • Logging Instrument Time

The situation is this: Pilot A is instrument rated and current. Pilot B is not instrument rated (but wants to practice for the rating). Both pilots are current for the category and class of aircraft, with current medicals and BFR's. Both hold private pilot licenses (only, no CFI for Pilot A). According to the FAR's, when Pilot B is in VFR conditions with a hood and on an IFR flight plan, with Pilot A in the right seat as check pilot, Pilot B may log this time as PIC and Pilot A may (only) log this time as "check pilot" or "second in command." (Right?)

My question is this: In the same situation, but in ACTUAL IFR conditions, who logs what? Can Pilot B still log PIC even though not rated for IMC? Or is Pilot A now the "official" PIC? Does the situation change if Pilot A is a CFI? (I would guess so since all "dual" time after the private license is also PIC.)

I have been unable to find a clear answer to this in the FAA publications (FAR's, AIM, etc.) and from various CFI's.

. . . P.S. One further question: Does any of this type of flying count toward the required time for Pilot B's instrument rating?

In answer to the first "situation" cited as we understand the conditions stated:

Pilot B *may* log the time during which he or she manipulates the controls of the aircraft flown as pilot-in-command in accordance with FAR § 61.51 (c)(2)(i).

Since Pilot B is not instrument rated, Pilot A *must* be pilot-in-command as defined under FAR § 1.1 (which identifies the pilot responsible for the operation and safety of an aircraft during flight time). It should be understood that although Pilot A is serving as "safety pilot" under FAR § 91.21(b), he or she must still be pilot-in-command under FAR § 1.1 and may log the time while serving as pilot-in-command in accordance with FAR § 61.51 (c)(2)(i).

In the second situation, the above circumstances are changed only by the flight conditions being changed to *actual* instrument meteorological conditions (IMC).

Pilot B may still log the time during which he or she manipulates the controls as pilot-in-command in accordance with FAR § 61.51 (c)(2)(i).

Pilot A *must* be pilot-in-command as defined in FAR § 1.1 and may, thus, log the time serving as pilot-in-command in accordance with FAR § 61.51 (c)(2)(i).

If Pilot "A" were to become a certificated flight instructor, he or she *must still* be pilot-in-command under FAR § 1.1 and would then be serving a dual role as both *safety pilot* and *flight instructor* and may log the time as pilot-in-command under either FAR § 61.51 (c)(2)(i) or § 61.51 (c)(2)(iii) as he or she chooses.

The flight experience gained under the above conditions is creditable toward an instrument rating and should be logged in accordance with the specific guidelines of FAR § 61.51 (a), (b), and (c), (2), (4), and (5).   9-10/86

Material on this page is advisory or informational in nature and should not be construed as having regulatory effect.

F33 • The Legal Operator

In the July Forum you state that a private pilot or better and his flight instructor may both log as pilot in command the time during which the pilot is the sole manipulator of the controls of an aircraft in which they are both qualified. Given this relationship, who is the legal operator of the aircraft?

That would depend upon numerous factors, such as what is said and done just prior to the flight, or prior to any incident that happened to occur. Sometimes after an accident the final determination is made in a court of law. If you are concerned about a specific situation, check with the FAA Regional Counsel for your area.
12/74-1/75

F34 • Dual Roles?

The FARs allow private pilots to log time as PIC when sole manipulator of the controls: does this apply during an instrument training flight with a certified flight instructor? Could you not also log the time as dual instruction? And does that mean that for a 2.0 hour flight you could log 2.0 hours as PIC and also 2.0 hours of dual? Would IFR conditions alter the situations?

A private pilot receiving instrument instruction during an instrument training flight in an aircraft for which he or she is rated may log the time spent actually manipulating the controls as PIC under FAR §61.51(c)(2)(i). It must be clearly understood, however, that flight experience acquired in this manner may be used solely to meet either the flight experience requirements for a certificate or rating, or the recent flight experience requirements under FAR Part 61. While the actual flight time may be logged under several columns in the pilot's log, i.e., pilot-in-command, instrument instruction received, cross-country, etc., no more than the total flight time may be logged. Thus a two-hour flight may not be logged for more than two hours total time.
9-10/87

F35 • Ground Time?

Can a flight instructor log as pilot-in-command time or instruction time any time he is on the ground but has supervised the student's flight?

Negative. The flight instructor logs as instructor or PIC time only that time he is physically in the airplane.
5-6/80

F36 • Second-in-Command

My question concerns the logging of second-in-command time by a safety pilot. FAR 91.21 (b) (1) requires that an appropriately rated safety pilot occupy the other front seat during simulated instrument flight. FAR 61.51 (c) (3) states that second-in-command time may be logged when the regulations under which the flight is conducted requires a second pilot. Does this mean that an appropriately rated safety pilot is allowed to log second-in-command time during simulated instrument flight in VFR conditions.

Affirmative. An appropriately rated pilot may log second-in-command time while acting as safety pilot during simulated instrument flight.
7-8/87

F37 • Logging VFR on Top

Can a pilot who is instrument rated, current and fully qualified, log as instrument time a portion of a flight which is conducted as VFR on top of clouds? Can this time be counted toward the currency requirement?

FAR Part 61.51(c)(4) states, in part, that "A

pilot may log as instrument flight time only that time during which he operates the aircraft solely by reference to instruments, *under actual or simulated conditions* [emphasis ours]." Of course, if you were accompanied by a pilot rated in the aircraft and acting as a safety pilot while you operated that aircraft solely by reference to the instruments, such time could be logged as instrument time. 5-6/83

F38 • IFR Instructor Currency

I am in need of an official interpretation of FAR 61.51(c)(4) (Instrument Flight Time) and 61.57(e)(1) (Recent IFR Experience). Can a flight instructor log approaches flown by his student while in instrument conditions, and may he use those approaches for his own currency under 61.57? When logging approaches under 61.57 for currency, does the approach have to be made to the minimum altitudes for that approach before leaving instrument conditions, or must the approach only be started to count toward currency?

An instrument instructor may log as instrument time the time he acts as an instructor in actual IFR conditions. This includes any approaches flown by his instrument student, and he may apply them toward his own instrument currency. In order to log approaches toward IFR currency the approaches must be carried at least through the so-called critical elements. This could include conducting the approach to a landing, to the minimum altitude and/or missed approach point, or through the approved missed approach procedure. 5-6/82

F39 • Clear Vision

I'm at the age where a glaucoma test is recommended as a part of the regular eye examination. I understand that some sort of desensitizing drops are put in the eye before the test, and I was wondering how long after the drops wore off before I could fly again.

If drops are used to "numb" the eyes for the glaucoma test, the effect should wear off within a few hours at most. Assuming an open cockpit is not being used, no flying restriction is necessary. Some practitioners now use a glaucoma screening device that does not touch the eye, and no drops are necessary. 7-8/80

F40 • Blood Donation and Flying

What is the regulation regarding pilots who recently donated blood, and where can I find it?

There is no regulation concerning how soon flying can be resumed after donating blood. The FAA Air Surgeon's office recommends, however, that pilots wait 24 hours after they have donated one unit (pint) of blood before flying again. 3-4/81

F41 • BFR and the Wings Program

I have a question regarding FAR §61.57. It states, in part, that a person who has completed a pilot proficiency check within the preceding 24 months need not accomplish the biennial flight review (BFR). Would getting one of the Phase Wings under the Pilot Proficiency Program count as a BFR? If so, does the CFI need to so annotate the logbook?

The flight time required—one hour of airwork, one hour of instrument flying, and one hour of takeoffs and landings per phase—to obtain either of the Phase I-V wings in Accident Prevention's Pilot Proficiency Program could, upon the discre-

tion of the flight instructor, be considered satisfactory for a biennial flight review. (Attendance at an FAA-sponsored safety meeting is also required to obtain a set of wings.) If the CFI determines that the qualifications of the BFR have been met, a specific BFR notation must be made in the airman's logbook. 1-2/86

F42 • Correspondence Course BFR?

I've heard rumors recently about a way to accomplish your BFR without flying—something about sending copies of your logbook to an examiner who will review its entries and sign you off if appropriate. Is this true? If so, I think it is seriously unsafe. Also, can a BFR be given by more than one instructor?

While the FAR's do not state that a biennial flight review must be given by a single instructor, the regulations do require that the person giving a review *certify* that the pilot has satisfactorily completed the review. This can only be done if a qualified instructor *observes* the pilot's actions and skills during an actual flight review. No one has been authorized to certify satisfactory completion of a BFR by mail. An actual flight demonstration is required. 9-10/85

F43 • Competency Check and BFR's

I'm writing in reference to FAR § 61.57 (Recent Flight Experience). I'm an instrument rated commercial pilot, and I've been meeting the recent IFR experience requirement by getting a competency check every six months. My question is does this IFR check count as a biennial flight review? If so, how is it noted in my logbook? If not, what do I have to do in addition to the IFR competency check?

It appears to me that if there is any way to combine the review and the competency check, I'd be better off legally, professionally, and financially.

The instrument competency check will satisfy the requirements for a biennial flight review. A flight review is not necessary if, within the preceding 24 months, you have had a pilot proficiency check (by the FAA, by an approved check airman, or by a U.S. armed forces pilot) for a pilot certificate, rating, or operating privilege. The instructor conducting the review notes successful or unsuccessful completion by an appropriate notation in your logbook. However, the decision to count the competency check as a BFR also is up to the flight instructor.
[see clarification below] 9-10/86

F44 • Unsuccessful BFRs?

I take issue with your answer to [the] question on "Competency Check and BFR's" . . . The latter two sentences in the paragraph indicate that a certificated flight instructor will determine in writing whether your biennial flight review or instrument competency check is successful. I see no such provision in FAR 61.57. Since the rule accepts the instrument check for a BFR, then I submit it is not up to the instructor to decide, as you say.

You are correct in that neither the instrument competency check or the BFR may be recorded as unsuccessful, or unsatisfactory. The instructor is free, however, to decline to endorse the pilot's logbook if in his opinion the applicant needs instruction or further practice or for any reason does not carry out the requested maneuvers.

Our statement regarding the acceptability of a competency check for a BFR was incomplete.

Material on this page is advisory or informational in nature and should not be construed as having regulatory effect.

FAR 61.57(b) provides that a BFR will require a review of certain rules, procedures, and maneuvers. If these items are covered during or after the competency check, the instructor may endorse the logbook for the BFR as well as the ICC. However, in that case, the endorsement must clearly encompass both. **1-2/87**

F45 • Blood Pressure Maximums

I would appreciate if you could advise me what guidelines exist concerning maximum allowable blood pressure in connection with private pilot licenses, and which FAA publication covering this issue would be available.

The maximum blood pressures permitted for airmen medical certification are contained in Part 67 of the Federal Aviation Regulations and in the Guide for Aviation Medical Examiners. After a normal cardiovascular examination, the maximum readings for first-class certificate applicants range from 140/88 to 170/100 (mm of mercury) depending on age. For second-class and third-class certificates, the applicant's blood pressure cannot exceed 170/100 mm mercury. If any antihypertensive medication is used, however, these values do not apply; eligibility for medical certification must be determined on an individual basis. **1-2/88**

F46 • Complex Instruction vs Practice

In FAR 61.129 (b)(1)(ii), it states that an applicant for a commercial pilot certificate must have "10 hours of flight instruction and practice given by an authorized flight instructor in an airplane having retractable landing gear, flaps, and a controllable pitch propeller." Further on in sections 61.129(b)(2)(i) and (ii) various references occur to "instruction" but without "and practice" being added.

So the question is, must all of the minimum 10-hours be in what are commonly known as "complex" airplanes be with a flight instructor, or may some of that time be solo?

The required 10 hours in a complex aircraft (FAR 61.129) may include as much solo time as your instructor finds acceptable in view of your experience and competence.

The terms "flight instruction and practice" in Section 61.129(b)(1)(ii) provide for both dual and solo flight time because it was determined that varying combinations of dual and solo time would be satisfactory for acquiring this experience. **9-10/88**

F47 • IFR Currency Rules

Has there been any discussion recently about revising the IFR currency rule? My instrument instructor used to tell me that while the six approaches required for currency represented a useful and valid measure of the IFR pilot's skill, the requirement of a minimum of six hours of instrument time logged for each currency period did not, since it is no real chore to fly straight and level on instruments once you've had the training. Approaches are much more demanding. These and actual time in instrument conditions are what really count.

So what can I do to prod the FAA into considering the merits of a change (I have a few ideas) in the regulations concerning IFR Currency?

At present, there are no plans to review the instrument currency requirements. However, any interested citizen may file a petition for rulemaking. Essentially the procedure consists of submit-

ting the proposed change of amendment in duplicate to the Rules Docket of FAA Office of the Chief Counsel. Set forth the substance of the rule, explain the petitioner's interest, submit any arguments supporting your position, and include a summary for publication in the *Federal Register*.

Complete details are contained in Part 11.25 of the Federal Aviation Regulations, which may be examined at your nearest flight standards field office, or purchased from Superintendent of Documents, Government Printing Office, Washington, DC 20402. Check price before ordering. **11-12/83**

F48 • Simulator Queries

I have a couple of questions about instrument time and simulators that I wish you could answer for me. First, does the FAA assume all simulator time is dual? FAR Parts 61.57, 61.129, and 61.155 legitimize the use of simulators in qualifying for certification but do not specify where to log that time—as dual or pilot in command. So, how do I log it?

Second, can an instrument rated pilot log for currency the time he spends in an approved simulator as sole operator (not receiving instruction from an appropriately rated flight or ground instructor)?

To answer your first question, time spent in a simulator or ground trainer may be logged as "instrument flight time" but not as dual or pilot in command. It may be counted in your total time.

Affirmative to your second question. An approved instrument simulator may be used for three of the six hours required under Part 61.57(e), as well as for the required six approaches for instrument currency. **1-2/84**

F49 • Instrument Competency Check

A pilot who has not flown for a year begins to fly again. Within a short period the pilot completes six hours of simulated instrument time of which at least three hours are in an airplane. This includes six instrument approaches with a safety pilot. At this point the pilot satisfies FAR § 61.57(e)(1)(i). Must the pilot still pass an instrument competency check under FAR § 71.57(e)(2)? It is assumed that the pilot has satisfied all other requirements such as biennial flight review, current medical, etc.

A pilot, who has not maintained instrument currency within the preceding 12 months, must take an instrument check under FAR§ 61.57(e)(2) even though he/she may have recently acquired the six hours and six approaches accompanied by a safety pilot. **11-12/86**

F50 • Instrument Currency

Due to having had conflicting input from a variety of sources, I seek your counsel and correct interpretation of FAR §61.57(e)(1) concerning instrument currency.

1. May all six instrument approaches be made in an "approved" simulator?

2. If so, must the simulator be in the same category as the instrument rating? That is, does someone with an "instrument helicopter rating" have to use a helicopter simulator or vice versa?

3. What is the FAR reference for question number 2?

FAR §61.57 states that an instrument pilot must have ". . . logged at least six hours of instrument time under actual or simulated IFR conditions, at least three of which were in flight in the category of aircraft involved, including at least six instrument approaches, or passed an instrument competency check in the category of aircraft involved."

All six approaches may be in an approved ground trainer, but all six approaches do not have to be in the same category as the instrument rating. There is no further reference in the FAR's, other than §61.57. The policy interpretation has been formulated by FAA's Airman Certification Branch and the Office of the General Counsel. **5-6/86**

F51 • Change of Address

I am a student pilot with a student certificate and 2nd Class Medical. I will be changing addresses shortly and would like to know if a student pilot must comply with FAR 61.60 requiring that Oklahoma City's Airman's Certification Branch be notified? If so, will I get a new certificate with whatever endorsements I already have?

Affirmative. Send your name, old address, new address, Social Security Number, and certificate number. No new certificate will be issued, but the address change will be in the computer. Unless this is done, the student certificate—or any pilot certificate, for that matter—is invalid after 30 days from the move. [see clarification below] **1-2/80**

F52 • "Invalid" Certificates?

In the letter "Change of Address" [above] . . . you reply that unless the Airman Certification Branch is notified of a change of address within 30 days of a move, the pilot certificate is "invalid." Is it truly invalid and thus illegal to fly until the notification is made?

FAR 61.60 provides that the holder of a pilot or flight instructor certificate who has made a change in his *permanent mailing address* may not after 30 days from the date of the change, exercise the privileges of his certificate unless he has notified the Airman Certification Branch in Oklahoma City of the new address. The word "invalid" was technically incorrect, however, the phrase "may not exercise the privileges of his certificate" is quite clear. Note that a pilot may move to a different residence without affecting his permanent mailing address. **5-6/80**

F53 • S/E, M/E and IFR

I presently hold an instrument rating for single airplanes, and I'd like to be similarly rated for multi-engine airplanes. But I've heard there's a new rule that says if I fail my multi-engine rating test I will lose my single-engine instrument rating as well. Is this actually true?

When the holder of a single-engine class rating with instrument privileges applies for a multi-engine class rating, with instrument privileges, and fails the instrument competency part of the test, he or she is issued a notice of disapproval indicating the area of failure. However, failure of any portion of the test requires a reapplication for the multi-engine class rating. In other words, the multi-engine class rating with VFR privileges is not automatically given to any pilot who passes the multi-engine part of the exam.

Furthermore an applicant who already holds an instrument rating in some other class or category of aircraft, and whose failure is due to lack of instrument competency (rather than specific multi-engine competency) may be required to demonstrate instrument competency prior to exercising instrument privileges in any aircraft. **11-12/86**

Material on this page is advisory or informational in nature and should not be construed as having regulatory effect.

F54 • Multi-Instrument Ratings

Supposing a pilot who is rated for single-engine land also holds an instrument rating. If he later acquires a multi-engine rating as well, does that mean he is qualified to fly multi-engine aircraft in instrument conditions?

Effective September 1, 1984, such a pilot will be required to demonstrate instrument competency as part of the practical test to exercise instrument privileges in multi-engine planes. However, should the pilot elect not to demonstrate instrument competency on the practical test, the pilot's multi-engine privileges will be limited to VFR only.
7-8/84

F55 • Instrument PIC Time

Suppose a pilot is applying for an instrument flight test and has the following qualifications: 15 hours instrument flight instruction from an instrument instructor (CFII); 10 hours instrument flight instruction from a flight instructor (CFI); and 15 hours PIC instrument hood time with an IFR-rated safety pilot.

Does he have enough time logged to apply for the rating? Since the application form asks for instrument PIC time, can the pilot log as instrument time the time flown with a safety pilot even though he doesn't have an instrument rating?

Although the airman rating application form provides a space for indicating instrument PIC time, such time is not required for an instrument rating. He may log the 15 hours with the safety pilot as PIC instrument time on the application form, provided he was the sole manipulator of the controls and was rated in the aircraft. In that case, he would appear to have the 40 hours of instrument flight time that is required to apply for the rating.
5-6/81

F56 • Instrument Written Tests

There is now an instrument rating for airplanes and another instrument rating for helicopters. FAR 61.65(f) requires a written test in each case. When a person already holds an instrument rating for either airplane or helicopter, and applies for the other, does he need to take another written test?

A second written test must be taken. (It will probably take less time to prepare for the second one since a certain portion of the material will have been also covered in the earlier test.) **10/75**

F57 • Military Flight Training

I was enrolled in Naval Aviation training and went through ground school plus 160 hours in complex aircraft, turbo-prop and jet. The requirements and standards were much higher than those for civilian pilots. After I disenrolled (for family hardship reasons), I've had trouble finding an instructor who will accept my background as sufficient for giving me a private pilot certificate—they indicate I don't have enough ground school. I thought the FAR's recognized military aviation training?

FAA issues special pilot certificates and/or ratings to those who have *completed* pilot training in the military. Partial military pilot training, however, may not cover all areas appropriate for civilian pilot training.

Your military flight time may be used in meeting aeronautical experience requirements, but you will have to meet all the other appropriate written and flight test requirements for at least a private pilot certificate before one can be issued to you.
1-2/85

F58 • Foreign Eligibility, Part I

I am a U.S. citizen, but I live in Canada and currently hold a Canadian commercial, multi-engine certificate and an IFR rating. What would be the process by which I could obtain equivalent U.S. qualifications so I can work as a pilot in the U.S.?

Under FAR Part 61.75 you can be issued a U.S. commercial certificate with appropriate ratings on the basis of your Canadian license without further testing. However, the U.S. certificate would bear the limitation: "Not valid for the carriage of persons or property for compensation or hire or for agricultural aircraft operations." To obtain an unrestricted U.S. commercial pilot certificate, which you must have to engage in commercial operations in the U.S., you would have to complete successfully the appropriate written and practical tests. Previous flight time experience may be applied toward the flight time requirements.
[see clarification below]
5-6/84

F59 • Foreign Eligibility, Part II

Your reply [in "Foreign Eligibility, Part I," above] included a statement that: "Previous flight experience may be applied toward the flight time requirements." However, I was wondering if the requirement of FAR 61.129(b)(2)(ii)—requiring 10 hours of instruction in preparation for the flight test—must be by an instructor holding a current U.S. Flight Instructor Certificate? How about instruction for the Instrument Rating?

There seems to be adequate justification to require this in order to indoctrinate the applicant into the U.S. airspace and ATC systems.

Affirmative to both questions. The instruction you refer to must have been given by the holder of an appropriate U.S. Instructor's Certificate.
11-12/84

F60 • The Doctor's Secret?

The names and addresses of pilot schools, airworthiness inspectors, designated pilot examiners, etc., are published as part of FAA's advisory circular system. However, the directory of Aviation Medical Examiners is a well-kept secret. Any reason why it could not be an advisory circular as well?

The demand for the entire list is not sufficient to warrant publishing it as an advisory circular. Names of AME's are often passed on by word of mouth or recommendation. Names of suitable doctors are available from any FAA regional air surgeon or field medical clinic or from the Office of the Federal Air Surgeon, FAA, AAM-1, 800 Independence Ave., S.W., Washington, DC 20591.
3-4/83

F61 • Pilot or Passenger?

Can a certificated and qualified pilot fly with a student pilot who is flying the aircraft? The qualified pilot does not have an instructor's rating but on the other hand he cannot be considered a passenger as he is qualified to legally fly the aircraft.

The holder of a private pilot certificate (or higher) who is qualified to act as pilot-in-command (PIC) in that aircraft and who is willing to assume the responsibility of PIC, can fly with a student pilot. In such a situation—even though the student sits in the left seat and manipulates the controls of the aircraft—the qualified certificate holder retains the responsibility as PIC. NOTE: If the PIC is not a certificated flight instructor the student pilot may not log the flight time toward the requirements for a pilot certificate.
1/76

F62 • Proper Endorsements

I would like clarification on the following: Student X soloed on February 16, 1985, continued to train, and received an endorsement to fly solo cross-country. Only one dual flight was taken between April 1 and June 16. On June 16 the student had a dual session, and the instructor noted in the logbook that ground reference maneuvers, minimum control airspeed, and landings were accomplished.

1. Does this endorsement constitute the 90-day review required in FAR Part 61?

2. When the student arrives to fly solo, how does the dispatcher or the chief flight instructor know that the 90-day review was completed satisfactorily?

Unless the flight instructor specified in the endorsement that he or she found the student competent for solo flight, the June 16 sign-off could not be considered a 90-day review endorsement.

A dispatcher or chief flight instructor would determine the student's status by examining the logbook for the proper endorsement. If any required endorsement for solo were not satisfactory, the student would have to fly with an appropriately rated instructor.

Guidance for proper endorsements, including necessary elements and suggested language, can be found in Advisory Circular 61-65B, "Certification: Pilots and Flight Instructors;" free from DOT's Subsequent Distribution Unit, M-443.2, Washington, DC 20590.
11-12/85

F63 • Overnight X-C

. . . there is apparently no requirement that you fly nonstop between . . . points on a crosscountry, so could a student pilot take an overnight crosscountry?

While not a general practice, an instructor could approve an overnight crosscountry flight . . .
1-2/80

F64 • Interrupted Training

Are student pilot training hours good indefinitely? I stopped flying in 1981 with 33 hours logged. Can those hours be applied to my total time after four years?

Affirmative. However, you might find that after four years, additional recurrent instruction will be needed to bring your skills up to par for certification as a private pilot.
11-12/85

F65 • A Question of Compensation

Is it considered an operation for compensation or hire if I, as a private pilot, take aerial photos from my experimental aircraft and sell them?

A private pilot may act as pilot in command of an aircraft in connection with any business or employment if the flight is only incidental to that business or employment and persons or property are not carried for compensation or hire. Therefore, your taking photos, which you subsequently sell, from your experimental aircraft would appear to be permissible *unless* you were carrying someone else who was taking the pictures and selling them [FAR § 91.42(a)(2)].

If there are further considerations which require discussion, consult your local flight standards district office.

Just a reminder: Since you are using an experimental aircraft, you may not operate over densely populated areas or in congested airways.
1-2/86

Material on this page is advisory or informational in nature and should not be construed as having regulatory effect.

F66 • Sharing the Fare

Regarding the FAR that allows the "sharing" of "actual expenses" of a flight that is piloted by a non-commercially-rated private pilot, and which prohibits the "charging" of the passengers for the flight, how does this FAR apply to flight in a rented aircraft? Specifically, if the pilot is flying at the sole request of the passengers (making a flight that otherwise would not have made), can the passengers be asked to absorb the entire cost of the rental, including fuel and prorated renter's insurance?

The type of operation you describe would be an air taxi flight, which legally could only be provided by the holder of an FAA air taxi certificate. 9-10/84

F67 • Spinning the Instructor

I have some questions regarding the flight instructor certification process. Is a CFI applicant required to demonstrate a spin on the flight test? Does he have to have a log-book entry endorsed by a CFI attesting to spin competency? The Flight Instructor Practical Test Guide (AC 61-58A) refers to the applicant being asked to demonstrate the recognition and recovery from a spin situation. What is meant by "spin situation?"

During the flight test, the CFI applicant may or may not be asked to demonstrate he is capable of recognizing and recovering from a spin. In place of a demonstration of spin competency, the examiner will rely on a log-book endorsement that the applicant is capable of recognizing and recovering from spins. A "spin situation" can be described as the examiner putting the aircraft into an attitude that could lead to a spin and expecting the applicant to recognize and recover. 11-12/81

F68 • Flight Test Roulette

What regulation by specific number requires a CFI applicant to demonstrate spins or bring an aircraft certificated for spins to the flight check?

FAR Part 61 does not state specifically that an applicant for a flight instructor certificate must demonstrate spins during the practical test. However, Section 61.187(a) requires the applicant's log-book to be endorsed to reflect that the prospective instructor is competent to pass a practical test on six subjects, one of which is recognizing and correcting common student pilot errors. A common student pilot error involves inadvertent spins resulting from failure to recognize and properly recover from stalls or slow flight.

Consequently, should the inspector or pilot examiner—acting as a student—place the aircraft in an attitude where a spin entry is imminent, the CFI applicant is expected to recover promptly and return the aircraft to normal flight.

Therefore, the aircraft provided by the applicant should be capable of performing all required maneuvers and procedures for the certificate being sought.

Advisory Circular 61-58A, "Practical Test Guide—Flight Instructor," provides this and additional information for the instructor applicant. 3-4/83

F69 • Simulator and Ground Instruction

I am a flight instructor, and I would like some help in clearing up an interpretation of FAR 61.189(a) concerning training of flight instructors. What is ground instruction meant to be, flight simulator or oral instruction?

Ground instruction, or "ground school," is usually interpreted to mean classroom or textbook training in aeronautical knowledge and aircraft systems. 5-6/81

F70 • Third Class CFI?

Can a CFI with only a third class medical certificate give non-profit flight instruction to friends and family?

Affirmative. 7-8/81

F71 • Instrument Instruction

Instrument instructors holding only a flight instructor, single-engine certificate have been giving instrument instruction in multi-engine aircraft for a long time. The only requirement was that they hold a multi-engine rating and meet the currency requirements. There is some question . . . whether according to new Part 61 this can still be done. I would appreciate a correct interpretation.

A certificated flight instructor need not hold a multi-engine class rating on his *instructor's* certificate in order to give instrument flight instruction in a multi-engine airplane, provided the pilot receiving the instruction holds a multi-engine rating on his *pilot* certificate. The CFI must, however, hold a multi-engine rating on his pilot certificate in order to serve as a safety pilot when giving instrument instruction under simulated instrument conditions. 9-10/80
3-4/81

F72 • Who Pays for AD Repairs?

The title for this letter should be "the great AD Note rip-off." In November I received AD 74-24-13 relating to altimeter defects. My altimeter was included. I contacted my A & P, the instrument was removed and sent to the shop for repairs. Several days later it was reinstalled—at a cost of $50 to me, the owner.

I realize the directive was issued in the interest of safety, I understood the urgency of the repair and gave it swift attention. The directive . . . made it clear that the failure was a manufacturing defect. I think the Government should have taken steps to insure that the company incurs the loss. As it stands, hundreds of aircraft owners, including myself, had to pay for the mistakes of a manufacturing company.

FAA has the responsibility for assuring that aircraft are airworthy, but lacks the statutory authority to require the manufacturer to incur the expense of correcting. Whether or not the manufacturer can be compelled to replace defective equipment is a matter for civil courts to decide. 6/75

F73 • Aircraft/Engine Logs

Please clarify the procedures for signing off "annual" inspections on aircraft and engines operated under Part 91. Some mechanics say that only the aircraft—not the engine—must have the logbook entry for an "annual"; others say the aircraft and each logbook for an engine must have the "annual" entry.

Annual inspections should be recorded in both aircraft and engine logs. Requirement for such entries are found in FAR 91.173(a)(1) and in FAR 43.11.
[see clarifications below] 2/75

F74 • Signing Off the Annual

When we have an annual inspection done, should the holder of an Inspection Authorization (IA) sign both the airframe and engine logbooks, or should he sign only the airframe book, naming all the items he has covered in the inspection?

If separate logs are maintained, each should be signed. Individual items need not be indicated for inspection.
[see clarification below] 9/76

F75 • Separate Logs?

I would like to know if a mechanic would be in violation (in case of a subsequent accident or incident) for not recording an annual inspection in the engine logbook, even though he did so in the aircraft log?

The mechanic is not in violation of the rules if he enters the annual inspection only in the aircraft maintenance logbook, since this inspection is required by regulation to include the engine (thus a single signoff indicates the engine has been inspected).

However, the *aircraft owner/operator* is required by FAR 91.173 to keep maintenance records for "each aircraft (including the airframe) and each engine, propeller, rotor and appliance of an aircraft." Furthermore, he is required to pass on the pertinent maintenance record when any of these items, such as the engine, is disposed of. For those reasons we suggest that aircraft owners keep separate records for these various items, and that they have their mechanic make annual inspection entries in each logbook that is kept. And we recommend that mechanics, in the best interest of their customers, make the entries in this manner. 2/77

F76 • Visual Standards

There has been a revision in the *Guide for Aviation Medical Examiners* regarding applicants for first or second class medical certificates who fail to meet FAR Part 67 vision standards. If the applicant's uncorrected distant visual acuity is worse than the current requirement of 20/100 but no worse than 20/200, a special ophthalmologic evaluation report will no longer be required, and the AME will be authorized to issue a certificate if the applicant is otherwise qualified.

This change applies only if in the course of the clinical examination required for certification the AME finds no evidence of significant underlying pathology. A Statement of Demonstrated Ability will be issued if the Aeromedical Certification Branch concurs.

For those whose uncorrected distant visual acuity is worse than 20/200 or whose vision does not correct to 20/20, FAA is retaining the current procedure which requires non-issuance by the AME, completion of a special ophthalmologic evaluation (FAA form 8500-7), and submission of all documentation to the FAA Aeromedical Certification Branch in Oklahoma City, OK. Procedures for granting special issuances to third-class applicants, where necessary, remain unchanged. 5-6/87

F77 • Contacts for Pilots

. . . FAA now allows pilots to use contact lenses instead of glasses, without a waiver. I think FAA should reconsider this rule. If a pilot should get something under a lens—and I've seen this happen countless times—while he is in the air, I can foresee many problems. Removal takes two hands: who flies the plane? The head must be bent for removal; this could induce vertigo. With one lens out, depth perception is impaired. Finally, there is no guarantee the lens will be reinserted properly on the first try!

FAA has no evidence that the use of contact lenses, as permitted, compromises aviation safety, and we know of no accidents or incidents where their use was a factor. Although it is painful to get something under a lens, the chance of this happening during normal flight is remote. If it should occur, most users would be able to quickly and easily remove the lens. In normal, stabilized flight the head movements required for removal should not induce vertigo. Also, loss of depth perception with one lens removed should not present a significant problem.

FAA is interested in hearing from airmen with regard to any problems known to arise in connection with this rule change.　　**6/77**

F78 • Flare-ups

Six months ago I was diagnosed as having a duodenal ulcer, but with the proper medication and diet, I am fully recovered. My aviation medical is due soon and I was wondering if I needed to do anything special for it. I did not fly while on medication, and I'm hoping this will clear me fully to resume.

The Federal Air Surgeon's Office requires that if you have had an active ulcer within the last three months, or a bleeding ulcer in the last six, you must prove that it is healed in order to be medically certified. Evidence required includes a statement from your physician with X-ray confirmation. A list of the medications you require—if any—to control symptoms is also necessary. Simple anti-acids are generally acceptable for pilots.　　**5-6/84**

F79 • Vision Standards for Pilots

To be an instructor you need at least a second class medical. I have only a third class medical, the reason being that my eyes are poorer than the minimum 20/100 uncorrected required for the second class medical. I do not understand the rationale behind this rule. With glasses on I can see as well as any other pilot who has a second class medical certificate. Why are you so concerned about unaided vision when I always fly with glasses on?

In flight operations, there is always a possibility of displacement of spectacles or a contact lens during a critical phase of flight. This is one important reason why higher vision standards are required for holders of second class medical certificates, since their duties and responsibilities may be greater or more complex than those associated with third class certification. It is also a fact of concern that distortion of peripheral vision is greater in lenses of high refractive power.

However, while FAA believes that a higher medical standard must be applied to persons involved in commercial operations (including flight instructors) we do recognize the need to evaluate each case independently. Thus, FAR 67.19 provides for the special issuance of medical certificates to applicants who do not meet certain of the established medical standards. To find out if you qualify for such special issuance, contact DOT/FAA, Civil Aeromedical Institute, AAM-100, P.O. Box 25082, Oklahoma City, Oklahoma 73125.　　**1/77**

F80 • Bypassed Pilots

I would like information on how a pilot who has had bypass surgery can return to flying. Does FAA recognize modern day surgical techniques in this area?

Indeed it does. In order for a pilot who has coronary bypass surgery (where a blocked or damaged cardiac blood vessel is replaced with a healthy vessel from another part of the body) to resume flying, he must apply for an exemption from aeromedical regulations.

No sooner than one year after his surgery, the pilot's cardiologist must send the results of a cardiovascular evaluation with treadmill stress test and a post-operative angiogram to the Federal Air Surgeon's office. If the stress test results are satisfactory and if the angiogram shows the bypasses are still functioning and there is no new or unbypassed heart disease, the pilot can receive a Class III medical and resume flying as a private pilot. Mandatory semiannual reevaluations (stress tests) are also required indefinitely.　　**1-2/81**

F81 • Nix on Pacemakers

Considering the advances made in modern heart disease treatment with pacemakers, can a pilot who has a demand type pacemaker return to flying under the same conditions as one who has had successful bypass surgery?

Negative, regretfully. Pilots who must rely on cardiac pacemakers continue to have the underlying condition that required insertion of this mechanical device. There is always the possibility that a pacemaker will fail without notice and result in sudden incapacitation. Bypass surgery, on the other hand, is designed to restore normal blood flow to the heart muscle. Although the underlying disease may continue, special diagnostic tests are used to identify serious progression and minimize the risk of sudden incapacitation.　　**5-6/81**

F82 • Medical Reinstatement

[A year ago] I suffered an anterior myocardial infarction for which I was hospitalized about 10 days. Since that time I have been taken off all medication for my heart and have returned to work. My doctor says I am getting along fine, but informed me I have to wait two years to have my private pilot's certificate reinstated. This seems to be an excessive amount of time. Could you give me more specific information?

The Federal Air Surgeon's office normally requires a two-year period following an infarction before considering certificating an airman. This is the usual amount of time for full recovery and stabilization of the heart disease. The interval has been reduced to a year for some patients who have undergone cardiac catheterization and coronary angiography and these tests did not reveal evidence of further problems. Consult your local aviation medical examiner or FAA office for details on how to apply for an exemption.　　**1-2/82**

F83 • Flight Into Ice

. . . your article on "Light Plane Anti-icing and De-icing Procedures" referred to FAA Regulations which specify that before an aircraft is flown into known icing conditions it must be properly equipped and approved for such flight.

This regulation, however, appears in Part 91 under "Subpart D" which applies to "Large and Turbine-Powered Multi-Engine Airplanes" and I believe it also appears in Part 135. Does that mean that I, as a Part 91 pilot (personal flying) can fly a light single-engine airplane legally into known icing conditions without anti-ice or de-ice equipment?

Not exactly. The rules are spelled out more specifically for Large Aircraft and Commercial Operations, since the safety of more persons is apt to be at stake. However, pilots who deliberately fly small aircraft into known icing conditions without protective de-icing or anti-ice equipment, thus endangering "life or property of another" could be subject to violation of FAR 91.9 (Careless or Reckless Operation).　　**6/78**

F84 • Forecast Remarks

1. In determining alternate requirements under FAR 91.23, does one have to use the "Remarks" section of the forecast?

2. If the forecast is obviously wrong (judging from several sequence reports and other data), must it still be used if it is restrictive, e.g., if it makes an alternate necessary?

FAR §91.5 (Preflight Action) requires that all available information on weather reports and forecasts must be used by a pilot before flight. Since the "Remarks" portion of the forecast is part of that available information, it should be considered.

Aviation forecasts of the National Weather Service are developed from a broad base of current inputs. When a cold front or some other weather phenomenon moves faster or slower than originally predicted, terminal forecasts may, indeed, be incorrect. You can ask for updated forecasts, but if the only available forecast indicates an alternate airport is required, you must file and plan your fuel load accordingly. Better safe than sorry.　　**3-4/85**

F85 • Airport Vicinity

Part 91.5(a)—Preflight Action—states that for ". . . a flight not in the vicinity of an airport . . ." certain information must be obtained by the pilot. What distance is implied by "not in the vicinity of an airport?"

As a matter of general practice anything less than 25 nautical miles is considered in the vicinity of an airport.　　**11-12/82**

F86 • Coffee, Tea or ?

In a corporation airplane, is the serving of meals to passengers by the co-pilot a violation of FAR 91.7, "Flight Crewmembers at Stations?" As far as company definition of duties is concerned, the serving of meals is considered to be a part of the operation of the airplane.

If the type certification of the aircraft requires a co-pilot, he must be at his station during take-off, landing and enroute, unless his absence is necessary in the performance of his duties in connection with the operation of the aircraft or in connection with his physiological needs. Serving meals to passengers is not recognized as being in connection with the operation of the aircraft as far as application of the regulation is concerned.　　**2/75**

F87 • Uncontrolled IFR?

. . . I have been tempted to fly IFR in uncontrolled airspace. Please clarify how best to do this safely, since controlled airspace begins here on the east coast at either 700 or 1,000' AGL. Also, what is the best procedure for low-level air navigation using electronic NAVAIDs? Finally, please clarify any regulations in Part 91 that I should know about so I can fly legally.

IFR flight in uncontrolled airspace is not prohibited by regulation. (see Part 91.) As long as the pilot is IFR qualified, and the aircraft is appropriately equipped, he may fly in uncontrolled airspace in instrument meteorological conditions. However, because of terrain hazards, unreliability of NAVAIDs at low altitudes, and absence of traffic separation, it may not be possible to do so except in a careless or reckless manner—which is a violation of Part 91.9 of the Federal Aviation Regulations.　　**1-2/82**

Material on this page is advisory or informational in nature and should not be construed as having regulatory effect.

F88 • Arthritis and Flying

I have just begun taking indomethacin for arthritis pain. What are the possible side effects of this medication on my flying? I haven't noticed any side effects as yet.

FAR 91.11(a)(3) states that no person may act as a crewmember of a civil aircraft "while using any drug that affects his faculties in any way contrary to safety." Indomethacin is an effective anti-inflammatory drug used commonly for arthritis sufferers. Even though you do not yet feel any affects, you could suffer any of the following during flight: dizziness, blurred vision, impaired hearing, and, rarely, hallucinations.

The Federal Air Surgeon recommends that you abstain from flying while taking any medication—if even you apparently experience no side effects. You should discuss your condition with your aviation medical examiner.　　**5-6/84**

F89 • Room to Spare

Behind the back seat of my four-place Cherokee there is a roomy baggage area. The space is completely open to the cabin and has windows and safety belts (installed to restrain the luggage). When the seats are filled, can I let two small children (ages five and six) sit back there during a flight, perhaps on cushions, if their weight does not exceed the placarded limit of 200 pounds and the weight and balance checks okay?

You may not. FAR 91.14 requires that, during takeoff and landing, "each person on board must occupy a seat or berth with a safety belt [and shoulder harnesses, if installed,] properly secured about him." (Exceptions are children under age two, being held by an adult; or persons on board for the purpose of sport parachuting, who may sit on the floor, but must use safety belts.)　　**6/76**

F90 • Doubling Up Children

With regard to FAR 91.14, can two children share one seat and seatbelt if the weight limitations are not exceeded, or must each person have their own seat and seatbelt. I have received conflicting answers on this and need it clarified. Our plane has four seats, and we have four children.

Separate seats and belts are required for persons two years old and older only in air carrier operations (Part 121) and in passenger operations for certain large, privately owned aircraft (Part 125). In general aviation (Part 91) and commuter operations (Part 135) a safety belt may be used for more than one person in a single seat (such as a bench seat) as long as the strength limits of the seat and safety belt are not exceeded and the aircraft remains within allowable weight and balance limits.
　　1-2/81
　　5-6/81
　　9-10/81

F91 • Not So Easy Rider?

I recently overheard a conversation about a check ride given at my airport that sent me to my rule book to see what regulations had been violated. To my surprise I could not find one to cover this situation, so could you help out? I'm sure this must be illegal.

A pilot, while taking a check ride in a high performance single engine aircraft, was carrying a passenger in the rear seat. The pilot was accompanied by a properly certificated flight instructor. The check ride included unusual attitudes.

*So, is it legal to carry the passenger? And, given the unusual attitudes, were the pilot and flight in-*structor in violation of FAR 91.15(c) (Parachutes and Parachuting)?

There are no provisions in the FARs against carrying passengers during check rides. However, arrangements should be made among all parties beforehand to make sure it is agreeable.

Part 91.15(c) is more for aerobatic flight than for unusual attitudes used on check rides, i.e., ". . . unless each occupant of the aircraft is wearing an approved parachute, no pilot of a civil aircraft, carrying any person (other than a crewmember) may execute any intentional maneuver that exceeds (1) a bank of 60° relative to the horizon; or (2) a nose-up or nose-down attitude of 30° relative to the horizon."

Unusual attitudes are not *necessarily* aerobatic maneuvers, providing they do not exceed the parameters in FAR 91.15(c) and 91.71 (Acrobatic Flight). However, 91.15(d) does not require parachutes if the above mentioned maneuvers are included in a flight test for a pilot certificate or rating.　　**1-2/84**

F92 • Appropriately Rated

For simulated instrument flight FAR 91.21(b)(1) requires an "appropriately rated" safety pilot. Please explain the term "appropriately". Does it mean any pilot with an airplane rating, or does he have to be qualified in the particular airplane used? Must he be instrument rated? Can a student pilot act as a safety pilot?

The safety pilot is there with his dual controls so he can take over control of the aircraft if necessary to avoid other traffic, or recover from hazardous attitudes. Thus he must be qualified to operate the aircraft under conditions as they exist during the flight. If the flight is conducted in a light twin-engine land airplane the safety pilot needs at least a private pilot certificate with airplane multi-engine land ratings. A single-engine pilot would not qualify. When simulated instrument flight is conducted under VFR, neither the pilot under the hood nor the safety pilot is required to have an instrument rating. A student pilot is not acceptable as a safety pilot; for purposes of this rule the student is not considered "rated". Furthermore a student pilot may not have the judgment, experience and skill needed to be a safety pilot.
[see clarification below]　　**3/77**

F93 • Safety Pilot Qualifications

We would like a clarification of FAR 91.21(b)(1) as regards the following situation. Assume the airplane is high performance, is being flown in VFR conditions with no flight plan, with the instrument conditions simulated by a hood, worn by the PIC, who is fully qualified and current in the airplane and has a current BFR and medical. Our question is, does the safety pilot also have to be fully qualified and current?

Under a Part 91 operation, if the pilot wearing the hood and flying the airplane is qualified and current as you describe, as is acting as pilot in command, the safety pilot must have a current medical certificate. Should the safety pilot elect to assume PIC, he must also be current in his BFR and qualified for the high performance airplane. If passengers are carried, the PIC must also meet the general experience requirements of FAR 61.57(c) and (d).
[see clarification below]　　**1-2/87**

F94 • Qualified But Not Current

*Your response [in "Safety Pilot Qualifications" above] stated simply that "the safety pilot must have*a current medical certificate." Does that mean that a fully qualified pilot, conducting practice instrument approaches under a hood under VFR in a multi-engine aircraft, could legitimately utilize a single-engine pilot with a current medical certificate as a safety pilot, if at no time that single-engine pilot were to act as pilot-in-command?

A safety pilot is a required crewmember for a certain type of flight, and as such must be appropriately rated and qualified for the aircraft, as well as in possession of a valid medical certificate. Thus, in the example you cited, a single-engine pilot may not act as safety pilot since he/she does not hold a multi-engine class rating.
[see two preceding items]　　**7-8/87**

F95 • Alternate Airports Reviewed

If the terminal forecast for a given airport calls for above minimum conditions, but also contains phrases such as "occasionally" or "variable" or "chance of," with regard to below minimum conditions, is it permissible to designate that airport as an alternate when filing an IFR flight plan?

. . . the Office of Chief Counsel, FAA, has . . . reviewed this subject and determined that an airport covered by such a terminal forecast would not qualify as a legal alternate under FAR 91.83(c). The forecast must indicate that the weather conditions will be at or above appropriate minimums at the anticipated ETA. Accordingly, any indications in the forecast as to the possibility of below minimum conditions prevailing at the ETA would disqualify the airport.　　**9-77**

F96 • Mode C Is Not On With "ON"

I wish to direct your attention to a somewhat misleading statement . . . The news article entitled, "Transponders Must Be 'On' in Controlled Airspace," implies that transponders must be turned to the "on" position while in controlled airspace. Most general aviation transponders that I have used display "off", "standby", "on" and "alt" (altitude) positions. The "on" position does not provide Mode C altitude information. The "alt" position transmits that information to air traffic control . . .

. . . FAR 91.24 specifies that "each person operating an aircraft equipped with an operable ATC transponder . . . shall operate the transponder, including Mode C equipment if installed . . ." Turning a transponder to "on" is sufficient for non-Mode C equipped aircraft; however, I submit that most transponders in use today have Mode C capability.

Your point is very well taken. A Mode C equipped transponder should be on "alt" in controlled airspace, and wherever else you need to transmit your altitude.　　**3-4/86**

F97 • VOR Receiver Error

Neither FAR §91.25 (VOR equipment check for IFR operations) nor the Airman's Information Manual addresses the question of what, if anything, should be done with the bearing error found in a VOR receiver check when the error is within the tolerances allowed by regulation. Should the error be applied as a correction to the setting used in flying a prescribed radial or disregarded?

If the bearing error obtained during the VOR check is within the tolerances allowed (plus or minus 4° error for ground checks, plus or minus 6° for airborne checks), the error may be disregarded, since it holds true only for that particular point on

---

Material on this page is advisory or informational in nature and should not be construed as having regulatory effect.

the ground or in the air where the test was performed. Federal airways on VOR radials are configured with these tolerances in mind. If you, the pilot, made the VOR check, and it fell within the error allowed, you must enter the date of the test, place, and error in the aircraft logbook (or other record) and sign it. Of course, if the error is greater than the tolerances, no further IFR flight should be conducted, and the system should be checked as soon as possible by an avionics technician. 5-6/85

F98 • VOR Test via Radar?

Is it permissible under FAR 91.25 to check VOR equipment by ATC confirmation of position along the centerline of a Victor airway or other usable VOR radial in a radar environment? If so what is the maximum permissible error? It seems as though this would be a more accurate method than that described in FAR 91.25(b)(4)—prominent point on an airway.

Negative. The many variables which affect radar accuracy would prohibit its use as an alignment reference. For example, unless the radar and VOR sites were co-located, it would be difficult to apply an error tolerance. 5-6/81

F99 • Safely Over-Gross?

If I fly my Bonanza five pounds over gross on a non-commercial flight and think I can do so safely, am I flying illegally? If so, would most insurance policies be void during that period?

Any over gross flight operation, even one pound, would be illegal according to FAR 91.31 which provides in part that "no person may operate a civil aircraft without compliance with the operating limitations for that aircraft." Many insurance policies contain a waiver which disallows payment of a claim if noncompliance with the FAR's can be proved in connection with an accident. 5-6/80

F100 • Automatic Shutdown

I have been informed that while operating my single engine aircraft under FAR Part 91 VFR operations, should one of my original equipment systems, such as a radio, gyro compass, or carburetor temperature gauge, etc., become inoperative, my aircraft is grounded until that item is fixed.

Is this a true statement?

Under current regulations that is correct. The agency is drafting a rule change that may provide operators of single engine aircraft with an approved minimum equipment list that should provide some relief. 5-6/87

F101 • Right-of-way Responsibility?

I seem to recall being told that when a plane has the right-of-way (is to the right in a converging situation, or is being overtaken) it must maintain speed and heading and possibly altitude. However, FAA regs (Part 91.67) do not state the responsibilities of the plane with the right-of-way, as the maritime rules do for a ship with the right-of-way. Please comment.

The FARs do not clearly spell out the action required of the pilot who legally has the right of way under circumstances such as you describe. However, each person operating an aircraft has the responsibility to see and avoid the other aircraft, by maneuvering if necessary, which could involve changes in either heading, altitude or airspeed. 1/78

F102 • Acrobatic Practice Areas

Aerobatic flight seems to be gaining in popularity around the country. Would you please state in no uncertain terms where, when, in what aircraft, and who may practice such maneuvers. I am especially interested in areas surrounding uncontrolled (unicom) airports.

Section 91.71 of the Federal Aviation Regulations, which establishes rules pertaining to acrobatic flight, prohibits acrobatic flight "over any congested . . . area of a city, town or settlement; over an open air assembly of persons, within a control zone or federal airway; below an altitude of 1,500 feet above the surface, or when flight visibility is less than three miles."

(Acrobatic flight is defined as "an intentional maneuver involving an abrupt change in an aircraft's attitude, an abnormal attitude, or abnormal acceleration, not necessary for normal flight.)

Any pilot can perform acrobatic flying with any aircraft that is not prohibited from acrobatic flying, and can do so just about any time provided the conditions of this regulation are complied with. Common sense suggests that you have dual instruction in acrobatics before you venture out alone. 11/75

F103 • Acro vs. Aero

Can you tell me if "aerobatic," and "acrobatic" are interchangeable terms? A while back I was called to task for speaking of "acrobatic maneuvers". My critic informed me that the correct term was "aerobatic," since such maneuvers were not part of a circus.

The FARs speak only of "acrobatic flight" which is defined in FAR 91.71 as "an intentional maneuver involving an abrupt change in an aircraft's attitude, an abnormal attitude or abnormal acceleration, not necessary for flight." As to "aerobatics", these are defined in Webster's College Dictionary as "stunts performed in an airplane or glider." 3/77

F105 • Left Turn OK?

My question is very basic—to wit, where do I find the authorization to depart a controlled airport without specifically requesting a direction of flight? Everyone in my local area—including the FAA—assumes that a 45 degree left turn out is standard, and no request for it need be made. That's fine, if it is universally accepted—but it isn't written down anywhere that I can find. Pilots say they make left turnouts automatically and always have. They seem to assume it is somewhere in an Air Traffic Controllers Directive. I for one won't "assume" anything where Air Traffic Control is concerned.

This is probably a nit-picking point, but if you can't tell me why it is standard to depart to the left without a request at a controlled field, why can't I make a right turn out and not request it either?

This point is covered by FAR 91.87(f)(1) which states that pilots taking off from an airport with an operating control tower "shall comply with any departure procedures established for that airport by the FAA." There are no "standard" departure procedures common to all controlled airports. Such procedures are established locally, taking into consideration such conditions as available runways, terrain, traffic, other airports nearby, etc., as well as the overriding concerns of air traffic control: safety and expediency. Tower personnel would be able to tell you of procedures for your airport.

Obviously, in the interest of reducing radio congestion unnecessary transmissions should be eliminated. However, one unfailing rule for pilots at tower controlled airports—or anywhere under air traffic control—is: When in doubt, ask. [see clarification below] 8/75

F106 • Departure Procedures

It is my understanding that the only FAA-established airport traffic patterns are those published in Part 93 and they cover exactly eleven airports or seaplane bases. No other airports have FAA-established departure procedures. Therefore, although not recommended, a pilot departing from any airport (except those eleven and unless noise abatement rules apply) can make any turn he desires (including one of 180 degrees) or none at all, unless local airport management has established procedures—right?

There are many airports with FAA-established procedures, in addition to those named in Part 93. Some are for noise-abatement purposes; others exist because of terrain, traffic off parallel runways or satellite airports, etc. The pilot departing a controlled airport must comply with these procedures or instructions received from the tower. If there are no established procedures for the runway in use, and the tower has not specified a direction of turn, then the pilot can make any turn he desires after takeoff. However, the controller may be able to provide you with traffic information if he knows the direction you intend to fly. 11/75

F107 • ARSA Questions

Must I comply with no-transponder identification vectors and radar sequencing, prior to entering the ARSA, or may I decline such "service" and ask the tower for a direct route to the airport or other destination? Secondly, is there no provision for a no-radio aircraft in an ARSA as there was in an airport traffic area with light gun signals?

In practicality, to me the Baltimore ARSA appears to be operated as a mini-TCA.

In response to your first question, the answer is *NO*. FAR 91.88 does not require you to participate in ARSA services outside of ARSA airspace. However, if your destination airport lies within the ARSA, or if you wish to transit the ARSA airspace, you must participate; you cannot simply proceed toward the airport and contact the tower for landing instructions.

If you wish to avoid vectoring for radar identification, which is part of the service, you can make your initial contact at one of the VFR reporting points for the approach control jurisdiction. A list of such reporting points will be provided by the facility.

[Editor's Note: Mode C transponders will be required within and above ARSAs beginning December 30, 1990.]

Regarding your second question, procedures for operating a no-radio aircraft into an ARSA are the same as for an Airport Traffic Area. Prior authorization is required in either case. [see clarification below] 7-8/85

F108 • Free to Choose

Your response to the "ARSA Questions" [above] fails to address the real substance of the question from the non-transponder equipped pilot who has been given vectors to establish identification and position before entering an ARSA and

---

Material on this page is advisory or informational in nature and should not be construed as having regulatory effect.

wants to know if he must comply. Isn't the answer really a "yes"? Doesn't that make the ARSA a mini-TCA, like the writer says?

If he wishes to enter ARSA airspace, yes, he must comply with ATC instructions. However, he may have the option of proceeding to his destination without penetrating the busy airspace of the ARSA.

The only requirement for operation in an ARSA is two-way radio communication with the air traffic facility in charge. [Editor's Note: Mode C transponders will be required within and above ARSAs beginning December 30, 1990.] The requirements for operation in a Terminal Control Area (TCA) are 1) an appropriate authorization from ATC; 2) a private pilot certificate [for a Group I TCA]; 3) an operable VOR or TACAN; 4) two-way radios; and 5) an altitude encoding transponder.

1-2/86

F109 • The Right Approach

It seems that many pilots completely ignore traffic patterns at uncontrolled airports. Please clarify the following: At an uncontrolled airport, where right or left turns are indicated by a segmented circle (or light at night) in VFR conditions, is a pilot in violation of FAR 91.89 or any FAR if he makes a straight-in approach, or if he flies a wrong-way pattern?

Although it is not always considered a "good operating practice" it is not necessarily a violation of any FAR to make a straight-in approach to an uncontrolled airport under VFR conditions, where airport displays indicate left or right turns. It would, on the other hand, be a violation for a pilot to disregard the segmented circle and make left turns where right turns are indicated, or vice versa. FAR 91.89 does not require that turns be made in approaching to land at a non-tower airport, but if turns are made they must be in the appropriate direction, i.e., to the left, unless airport markings or light signals indicate they should be made to the right.

3-4/74

F110 • Decision Height Penalties

In a recent discussion concerning inoperative components and remote altimeter settings, the question came up as to which penalty should be used to determine the appropriate approach minimums for an ILS in a situation where the middle marker was inoperative and the local altimeter setting was not available: the increase in the decision height for an inoperative component or the increase in decision height for a remote altimeter setting?

The approach minimums adjustment for an inoperative component on an ILS is found in the "Inoperative Component and Visual Aids Table" of the instrument approach plate booklets. For a remote altimeter setting, the decision height is increased on the approach plate at least five feet for every mile greater than five miles that the reporting station is located from the airport of landing. In a case where you have both situations, combine the two penalties to arrive at an adjusted decision height.

5-6/85

F111 • Below Minimums Approach

Is it legal for Part 91 aircraft to commence an instrument approach even if the report ceiling is below DH or MDA and the visibility is below published minimums? If the approach can be commenced, is it legal to land if the flight visibility, as determined by the pilot, allows identification of visual references

as per §91.116? Also, when "flight visibility" is mentioned in §91.116, does it refer to visibility from the cockpit?

A pilot operating under FAR Part 91 may commence an instrument approach regardless of the reported ceiling and visibility. Descent below the MDA/DH and landing can be completed as long as the following three conditions prevail:
(1) "The aircraft is continuously in a position from which a descent to a landing on the intended runway can be made at a normal rate of descent, using normal maneuvers . . ."
(2) "The flight visibility is not less than the visibility described in the standard instrument procedure being used."
(3) Except for a Cat II or III approach . . . at least one of the following visible references for the intended runway is distinctly visible and identifiable to the pilot: threshold, threshold markings, threshold lights, runway end identifier lights, visual approach slope indicator, touchdown zone or touchdown zone markings or lights, runway or runway markings, or the runway lights.

Flight visibility, as used in FAR §91.116 and defined in FAR Part 1, ". . . means the average forward horizontal distance, from the cockpit of an aircraft in flight, at which prominent unlighted objects may be seen and identified by day and prominent lighted objects may be seen and identified by night."

3-4/86

F112 • ILS FAF and MAP

I have a question concerning FAR §135.225. Specifically, paragraph (c)(1) states that an ILS approach has a final approach fix (FAF), which seems to me to be the wrong wording. I thought than an ILS approach has no FAF per se. If, in fact, this is correct, where exactly would one abandon the approach if the weather goes below the authorized landing minimums once the approach has begun?

Your notion that an ILS approach has no FAF is a common one that has recently been clarified. Traditionally, this fix was often considered to be either the outer marker (wrong), the Maltese Cross (wrong), or the published Glide Slope Intercept Altitude (close). The FAF for an ILS approach is actually the intersection of the Glide Slope Intercept Altitude (GSIA) with the glide slope, and is designated on the profile section of the approach chart by a lightning bolt symbol . . .

As to your question about when to abandon the approach, for a Part 91 operation a missed approach must be executed when the pilot reaches the decision height and cannot see and identify at least one of the visual references indicated in FAR §91.116(c)(3) for the designated landing runway. Commercial operators must discontinue an approach if the reported weather goes below landing minimums *prior* to their reaching the FAF, as required by FAR §135.225(b) and §121.651(b).

9-10/85

F113 • ADF but No VOR

Assume the following: (1) A published NDB approach to an airport that has surveillance radar; (2) a missed approach procedure that uses a VOR holding fix, and (3) an aircraft with only an ADF and two-way radio (no VOR receiver). My question: May a legal instrument approach be made, provided ATC is notified of the nav/com equipment status before the approach is initiated? I would appreciate an answer and also a reference to the regulations that apply.

Affirmative. The approach can be made under the conditions you describe. Applicable rule is

FAR 91.116 which states in part: "Operation on unpublished routes and use of radar in instrument approach procedures. When radar is approved at certain locations for ATC purposes it may be used not only for surveillance and precision radar approaches, as applicable, but also may be used in conjunction with instrument approach procedures predicated on other types of radio navigational aids."

6/77

F114 • Riding with a Student

Please advise if I, an FAA licensed private pilot, can ride the right seat with a student pilot (just to keep him company and bolster his confidence).

The holder of a student pilot certificate may not carry passengers. A student may be accompanied by a certificated pilot (private pilot or better) who is fully qualified for and current in the aircraft, but the latter must be the pilot-in-command. The student may not credit the flight time toward certification requirements.
[see next question]

5-6/88

F115 • Instruction from Non-CFI's

In the letter "Riding with a Student" in the May/June 1988 issue, you responded to the question of whether a private pilot could fly in the right seat with a student pilot ". . . to keep him company and bolster his confidence..." by saying that the pilot (not yet a flight instructor) must be fully qualified and current in the aircraft and that pilot would be pilot-in-command. I don't think your answer went far enough.

Over the years I have tried to discourage the idea of non-flight instructors giving instruction. First and foremost they are not certificated to do so. Second, most have not received any training in the educational process necessary to provide constructive and purposeful training. Third, many well-intentioned pilots have no experience in anticipating possible dangerous situations.

When my students ask if they can practice with a licensed pilot friend, I recommend that the student just ride as an observer. I explain that unless the person with them is a certificated flight instructor their time may be wasted and might even be detrimental to the learning process.

From a professional standpoint, what is the purpose of being trained for and certificated as a flight instructor, if any FAA licensed pilot could do it just as well?

We might add that there are some liability aspects for a non-CFI pilot who flies with a student pilot that bear consideration.

11-12/88

F116 • Two Stall Speeds

. . . you defined . . . V_{so} . . . as stalling speed. In my training as a student pilot, I was taught two different stall speeds. V_{so} (stuff out), was stall speed with landing gear down and full flap extension. This was located at the bottom of the white arc on the airspeed indicator. The other stall speed, V_s, was at the bottom of the green arc. This speed was considered stall speed for the airplane in a "clean" configuration, flaps and landing gear retracted. Is this correct?

"Stalling Speed" or V_s refers to the minimum steady flight speed at which the airplane is controllable.

Material on this page is advisory or informational in nature and should not be construed as having regulatory effect.

V_{so} is the stalling speed, or the minimum steady flight speed, in the landing configuration.

V_{sl} is the stalling speed, or minimum steady flight speed, obtained in a specific configuration. These configurations are defined in the airplane flight manual. They are not confined to the "clean" configuration. **3-4/88**

F117 • Below MEA/MOCA Altitudes

While on an IFR flight plan on a Victor airway, I have been assigned an altitude as low as 2,000 feet below the MEA by a radar approach control facility. The segment of the airway that I was on had no MOCA listed and my position was approximately 10 miles from the VOR I was enroute to. I asked the controller if this altitude assignment would be for radar vectors and he replied, "Negative, fly the airway." I was not given any lost communications instructions and reluctantly I descended to the assigned altitude.

I believe that if I am to "fly the airway" I am responsible for both terrain clearance and for radio navigation reception. The only way I can be assured of meeting this requirement is to fly at or above the MEA or MOCA when within 25 statute miles of the VOR. I also believe that FAR 91.119(a)(1) requires me to fly at or above the appropriate minimum altitude for that segment of the airway. I also reference that Airman's Information Manual paragraph 401(b)(2) and FAA Handbook 7110.65D paragraph 4-44(a) which indicate that this type of operation is not a proper operating practice. If I am incorrect in my belief, please provide me with a specific reference which indicates this procedure is authorized in a radar environment.

Without knowing the identity of the airway in question, it is difficult to provide you with a specific answer as to the propriety of the controller's direction. FAR § 91.119 does allow for IFR operation below an MEA, but only where a MOCA is also published for this segment of the airway. A controller may also vector an aircraft at an altitude below the MEA or descend an aircraft to an initial approach altitude which is below the MEA.

If you as pilot in command feel at any time that a controller advisory would create unsafe circumstances for you, you should so inform the controller immediately and, if necessary, maneuver the aircraft in the manner you deem necessary for safety. In this event you should keep ATC informed of what you are doing, and be prepared to justify your actions fully.

Any questions you have over such an incident should be taken up with the facility manager as soon as possible, in order to have access to voice tapes of the actual transmission. These tapes are routinely erased every 15 days, unless otherwise requested. **7-8/86**

F118 • Commercial Venture

I came across an advertisement which promised that I could start a business in aerial photography without holding a commercial pilot certificate.

Is there any validity to this promotion? I have a private pilot certificate with about 150 hours total time. As a side business, I would plan to rent an aircraft to shoot aerial pictures with a photographer. We would then sell the finished photo and frame for a profit. As pilot in command, using the plane for a side business, can I fly without the commerical certificate?

The operation you describe would appear to be a commercial enterprise, and therefore you would be required to have a commercial pilot certificate. **5-6/88**

F119 • Below DH or MDA

I would like a better understanding of FAR 91.116(c)(3)(i) as regards the sighting of approach lights when descending below the Descent Height or Minimum Descent Altitude during an instrument approach. Some pilots I know interpret it to mean that during any category of approach you can descend to 100 feet above the touchdown zone elevation if any approach lights are visible, but you cannot go below 100 feet above the TDZE unless you can see the red terminating or siderow bars of lights.

That is essentially correct, unless you are making a category II or III approach where the visual references specified by the Administrator do not include the approach lights.

Bear in mind that the pilot must also be in compliance with the requirements of paragraphs (1) and (2) of FAR §91.116(c).

(1) requires that the aircraft be continuously in a position from which a descent to a landing on the intended runway can be made, using normal maneuvers. For Part 121 and 135 operations the aircraft must touchdown within the TDZ of the runway of landing.

(2) requires that you maintain the minimum flight visibility for the instrument approach procedures being used. **3-4/88**

F120 • Night Flight

During what time period is night flight logged? Do you log night flight between sunset and sunrise or the period beginning one hour after sunset and ending one hour before sunrise, as suggested by FAR 61.57(d).

Night flight, for purpose of satisfying the PIC recency of experience requirements for carrying passengers under Part 61.57(d), is defined by that rule: The period beginning one hour after sunset and ending one hour before sunrise.

Night flight for other purposes means the time between the end of evening civil twilight and the beginning of morning civil twilight, as published in the *American Air Almanac*, converted to local time. **9-10/88** **11-12/88**

F121 • Chapped Lips and Oxygen

My question concerns the possible fire or other hazards involved with using supplemental oxygen in aircraft when the pilot/user has a case of chapped lips, and has applied a petroleum based anti-chapping product. The oxygen sets that I have seen all are labeled "use no oil" on the controls.

Using supplemental oxygen when your lips are coated with an anti-chapping medication, petroleum based or otherwise, should not be a cause of concern. FAA research on this subject found that even in the presence of 100% oxygen it was not possible to ignite such materials with a static spark. The burning point of these medications is in the order of 250° C. The agency has no confirmed reports of an inflight fire due to using an anti-chapping medication with an oxygen mask. The warning about "use no oil" may reflect a concern about oil seepage impeding normal oxygen flow. **5-6/88**

F122 • ATP Crosscountry

When applying for an ATP doesn't any flight between two airports, no matter what the distance between them, get counted as crosscountry time?

Affirmative. According to FAR 61.155, an applicant for an ATP certificate must have at least 500 hours of crosscountry time, of which 100 hours is as PIC. The leg distance is not specified as it is for private or commercial certificates. **11-12/81**

F123 • Other Than Student Pilot

You have pointed out that FAR § 61.65(e)(1) requires an applicant for an instrument rating to have ". . . a total of 125 hours of pilot flight time, of which 50 hours are as pilot in command in cross-country flight [in a powered aircraft] with other than a student [pilot] certificate." Does the phrase, "other than a student pilot," refer to the 50 hours of cross-country time only, or also to the 125 total hours of flight time?

The phrase, "other than a student pilot" in the flight time requirements for an instrument rating applies only to the 50 hours of cross-country time. (Student pilots are not allowed to log PIC time for cross-country flight—only solo, or dual, as appropriate.) The 125 hour total may include some student pilot flight time. **1-2/87**

F124 • Multi-Engine PIC

May a private pilot, with Airplane Single-Engine Land rating, log PIC time while receiving dual instruction in a multi-engine land airplane from a multi-engine flight instructor?

Can a pilot with the same qualifications (ASEL) log PIC time while flying a multi-engine land airplane after a flight instructor has endorsed his or her logbook for multi-engine flight?

A private pilot may log as PIC only that time during which he is sole manipulator of the controls of an *aircraft for which he is rated* [Part 61.51(c)(2)]. Thus a pilot who is not yet multi-engine rated (i.e., taking instruction) may not log dual multi-engine instruction as PIC time.

For your second question: yes—with proper endorsement, that time may be logged as PIC time. [see clarification below] **5-6/83**

F125 • Multi-Engine Endorsement

Please review your answer to the [preceding] question on multi-engine PIC. I do not believe a flight instructor can endorse the logbook of a single-engine pilot to fly multi-engine PIC. Doesn't this require an endorsement on the pilot certificate by an authorized examiner?

With an appropriate endorsement from a flight instructor, a single-engine rated pilot may fly a multi-engine aircraft solo (with no other occupants). Successful completion of a multi-engine flight test and endorsement by an FAA inspector or pilot examiner is required before the pilot can carry passengers. **9-10/83**

F126 • Estimated Time En Route

I am concerned about the estimated time en route figure that goes into block 10 of the instrument flight plan. The Airman's Information Manual provides no

---

help. The only rule I can find that speaks of this issue is FAR 91.83, which says you must give ". . . the point of first intended landing and the estimated lapsed times until over that point."

Some pilots interpret this "point" to be the initial or the final approach fix. Some estimate from takeoff to landing. Which is correct?

The actual approach fixes you will use may not be known in advance. You should interpret the rule literally and estimate your ETA based on the actual distance to the airport. Air Traffic Control will protect your airspace for 30 minutes after that point in time, in the event of radio failure.
[see clarification below]   **5-6/88**

F127 • Late IFR Arrivals

I have some difficulty with your response to the [preceding] question . . . Your response indicated that you should use the distance from takeoff point to the airport, since you won't know your approach fixes in advance; and that ATC would protect approach airspace for 30 minutes beyond your ETA.

Don't you know your approach fix when you receive your clearance? Doesn't ATC protect airspace longer than 30 minutes in the event of radio failure? It seems to me that unexpected weather en route could change your ETA, and how could you advise ATC if your radios have failed?

As indicated in FAR 91.83(a)(6) and in FAA's Instrument Flying Handbook your flight plan indicates ETA to an airport and city, not to a fix. The actual approach will be determined by ATC in accordance with traffic conditions when you report in.

If you are unable to contact Approach Control by the time given as your ETA, Approach Control will protect the approaches at your destination for an additional 30 minutes. ATC is not required to restrict the airspace any longer than a half hour, but might do so, at the discretion of the responsible controller, under some circumstances.

If you experience radio failure during an IFR flight you would be expected to continue the flight VFR if possible, land as soon as practicable and contact ATC. If you cannot go VFR, continue to your destination in accordance with FAR 91.127, which covers radio communications failures procedures for IFR aircraft.

Note: It is virtually impossible to provide regulations and procedures for every possible situation involving communications radio failure during an IFR flight—including weather induced delays. In situations not covered by regulation the pilot would be expected to use his own best judgment.   **11-12/88**

F128 • SVFR for Practicing?

I recently flew a large multi-engine, turbine-powered airplane into the Salem (OR) airport during daylight hours. The airport lies within a 24 hour a day control zone. The current weather observation was ceiling 1,100 broken, visibility seven miles. Upon completion of a practice IFR approach I wanted to remain in the traffic pattern and conduct several VFR practice landings. However there was a problem. The pattern altitude for my large aircraft was 1,500 AGL, which would have put me in the clouds. To fly VFR

in the zone I would have had to be 500 feet below the clouds, which would have put me at 600 feet AGL or less.

In order to fly at a safer, higher, and quieter altitude I requested a Special VFR clearance. The tower advised me that SVFR could not be granted because the field was above VFR minimums. I was told that if I was requesting a takeoff clearance that was okay.

Were they correct in denying me SVFR, and if so, why? It seems to me that if ATC can deny SVFR because the weather is above VFR, then they could also refuse to give out an IFR clearance on the same grounds.

The authority for the controller's decision is vested in FAA Order 7110.65. This order only authorizes granting a SVFR clearance when the controlling airport has IFR weather conditions.
[see clarification below]   **3-4/88**

F129 • SVFR Clearances

With reference to the . . . request for a special VFR clearance to practice touch and go's within a control zone—which was denied—I don't feel you adequately answered the question.

I understand that SVFR is available to help pilots exit an airport in a control zone where the ceiling is below one thousand feet and visibility is less than three miles. When a pilot can stay clear of clouds and maintain one mile visibility he may be issued a SVFR clearance.

To my knowledge SVFR is not intended for practice in the control zone or to enable VFR pilots to enter clouds, but just to exit the control zone.

As indicated in the Airman's Information Manual (C4-S4-2) a VFR pilot may request and be granted a clearance to "enter, leave, or operate within most control zones in special VFR conditions, traffic permitting and providing such a flight will not delay IFR operations."

"Special VFR conditions" means weather that is less than the VFR minimums but affords at least one mile ground visibility if taking off or landing, and one mile flight visibility elsewhere in the control zone. Helicopters are excepted from the visibility minimums; they are only required to remain clear of clouds. All SVFR flights must remain clear of clouds.   **9-10/88**

F130 • When Is a CZ Not a CZ?

Some of my flight instructor colleagues teach that a control zone does not "exist" except when weather is below VFR minimums for the CZ? I maintain that a CZ always "exists" where depicted on the charts. What is your opinion?

A control zone is not contingent upon weather minimums for its existence. The zone and its hours of operation are established by regulation in Part 71 of the Federal Aviation Regulations.
[see clarification below]   **11-12/83**

F131 • CZ Cautions & Questions

Isn't the answer to the [preceding letter] . . . a bit misleading?

While a control zone does exist even in good weather, it has no practical meaning as no restrictions are imposed on traffic in the CZ unless the weather is below basic VFR (1,000' ceiling and/or three miles visibility). Are CZ's dependent upon weather reporting capability, and if so, when the weather observer goes home and the CZ hours of operation

end for the day, can the VFR pilot penetrate the existing non-operational CZ? Or must he be a certificated weather observer so he won't be in violation of some FAR?

Even in VFR weather, the fact that a control zone exists should alert you to the possible presence of aircraft making IFR approaches or departures in the area. Also, control zone weather minimums are always those of controlled airspace, which are more stringent than those of uncontrolled airspace.

As for the absence of a weather observer, or the closing of a ground facility, the CZ still exists as per its regulatory established times, as indicated on sectional charts and facility directories. In the absence of local information, weather for IFR flights may be obtained from nearby reporting sites through air traffic, but the instrument approach is penalized somewhat, e.g., the DH or MDA is increased by so many feet per mile from the altimeter reporting site. For VFR flights once the facility is closed for the day, it is up to the pilot to maintain VFR conditions in the CZ.   **3-4/84**

F132 • Traffic Separation in CZs

I have heard pilot instructors make the statement that VFR and IFR traffic are separated by air traffic control in a control zone. I keep telling them this is nonsense—isn't it?

That could be a dangerous and misleading statement. In VFR weather the pilots of both VFR and IFR aircraft must scan continuously for other traffic. If the ceiling is less than 1,000' or the visibility is less than three miles in a control zone, VFR aircraft are not allowed in the control zone without a special clearance from ATC. In a control zone IFR traffic is given separation from other IFR traffic and any SVFR traffic.
[see clarification below]   **1-2/86**

F133 • CZs Revisited

In your reply [above], you probably didn't write all that you were thinking. This resulted in a slightly misleading answer. A person doesn't have to "stay out" of a control zone if the ceiling is less than 1,000', he simply may not fly beneath a ceiling of less than 1,000', nor may he enter the airport traffic area, as you explained, but he can fly in the control zone above the airport traffic area if he maintains a cloud separation per FAR 91.105 and has 3 miles, or 5 miles, as appropriate, visibility. Even in visual meteorological conditions ATC can refuse entry into a lot of airspace, but the control zone isn't one of them.

You are correct. A VFR pilot may transit a control zone above a cloud layer, if he has appropriate cloud clearance and flight visibility, regardless of the ceiling height over the airport.   **3/4-86**

F134 • Control Zone Weather

Your "Instrument Corner" is a most useful addition to a very good publication. Please clarify: The official weather observation at the primary airport determines the visibility at all other airports within the control zone. Yes or no?

Not so. With regard to **visibility**, FAR 91.105(d) prohibits operating VFR (except Special VFR) to or from an airport within a control zone unless ground visibility at that airport is at least three statute miles (regardless of existing visibility at the primary airport). If ground visibility is not reported at that

---

Material on this page is advisory or informational in nature and should not be construed as having regulatory effect.

airport, flight visibility during landing, taking off, or while operating in the traffic pattern is used.

IMPORTANT: The rule differs regarding ceiling. FAR 91.105(c) prohibits VFR operation (except Special VFR) "within a control zone beneath the ceiling when the ceiling is less than 1,000 feet," as reported by the primary airport.   **10-77**

F135 • Special VFR in Transit?

If I am transitting a control zone while operating under VFR, and I determine that the flight visibility in the area where I intend on operating is less than that required for VFR operations, can I obtain a Special VFR clearance even if the primary airport is reporting VFR conditions?

If you determine that the flight visibility in the control zone ahead will be marginal or less than required for VFR operations, you may request a Special VFR clearance, even when the primary airport is reporting VMC. However, if this would interfere with the existing flow of IFR traffic in the zone, you can expect a delay, or possibly a denial.

Maintain VFR conditions until a clearance is received. If a clearance is denied, select another route of flight.
                                                  3-4/86

F136 • Control Zones—"Weather" or Not

Isn't it true that when weather reporting is not available, the control zone is not in effect? Or is my interpretation of control zones incorrect?

The hours of operation of control zones are established or rescinded only by regulation or by NOTAM if provided for in the rule that establishes the specific control zones.
[see next question]   **5-6/82**

F137 • CZ Criteria

The local hangar flight crew were discussing control zones recently when the correlation between control zones and airports with instrument approaches

was reviewed. We discovered, to my surprise, that at least eight airports in Washington State have instrument approaches but no control zones! Some of the Minimum Descent Altitudes extend below the transition area into uncontrolled airspace where the instrument pilot can legally meet the VFR pilot in a one-mile and clear-of-clouds environment.

Will you please tell us the criteria used for establishment of a control zone?

Control zones are established at airports to provide controlled airspace fore instrument procedures. There are two criteria which the airport must meet: (1) a controlling ATC facility (on or off field) with two-way radio communications extending to the runway surface, and (2) Weather observing and reporting by a Federally certificated observer. (Note that once a control zone has been established by regulation, if the airport temporarily loses either required criteria, it is NOTAMed to that effect but retains its control zone.)

The condition you describe—arriving at your MDA in uncontrolled airspace—certainly demands extra pilot vigilance but is not an uncommon occurrence.   **5-6/85**

MODE C TRANSPONDER REQUIREMENTS

EN ROUTE FLIGHT

When — As of July 1, 1989

Who is affected — All aircraft flying in the adjoining 48 states and the District of Columbia.

Where — At and above 10,000 feet mean sea level (MSL) except in that airspace which is below 2,500 feet above ground level (AGL).

Exceptions — Gliders, balloons, and other aircraft without an electrical system are excepted below the Positive Control Area (commonly at 18,000 feet MSL).

Note: — See following definitions of airspace which may be encountered en route and in which Mode C equipment is required.

AIRPORT RADAR SERVICE AREAS

When — As of December 30, 1990

Who is affected — All aircraft operating within and above an ARSA, up to and including 10,000 feet MSL.

Exceptions — None

10,000 Feet MSL

DESIGNATED AIRPORT AREAS

When — As of December 30, 1990

Who is affected — All aircraft flying within a 10 mile radius of specially designated airports (currently only Billings, MT and Fargo, ND). Designated airspace extends from the surface to 10,000 feet MSL, excluding the airspace which is outside of the Airport Traffic Area and below 1,200 feet AGL.

Exceptions — Gliders, balloons, and other aircraft without an electrical system are excepted.

10,000 Feet MSL

TERMINAL CONTROL AREAS

When — As of July 1, 1989

Who is affected — All aircraft operating within a 30 NM radius of a TCA primary airport, from the surface to 10,000 feet MSL.

Exceptions — Within the airspace outside of the lateral boundaries and/or below the floors of a TCA, gliders, balloons, and other aircraft without an electrical system are excepted.

☐ Aircraft Without Electrical Systems Are Exceptions to the Requirement   ▨ No Exceptions

10,000 Feet MSL

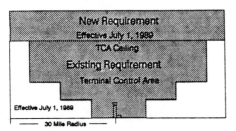

GENERAL NOTE: — Deviations from all of the above provisions may be authorized by the controlling ATC facility on a case-by-case basis. These may include agricultural aircraft operations, medical emergencies, aerial photography, approaches and departures from airports underlying controlled airspace, etc. There is nothing in this rule that would change the operational rules for ultralights.

Quick-Reference Index
to
Frequently Consulted Regulations*

---

*This index is abridged to allow easy access to frequently consulted regulations. Complete indexes are provided at the beginning of each Part.

FAR Index

FAR Index